More Praise for

THE SKY IS FOR EVERYONE

"I thought I had written the definitive book about women in astronomy, but I bow to *The Sky Is for Everyone*. Its authors are the observatory directors, the university professors, the leaders of the international research teams that launch telescopes into space and draw down the secrets of the cosmos, each one telling her own story of a life in science."
—DAVA SOBEL, author of *The Glass Universe*

"A rich and captivating array of personal stories that provide a welcome addition to the history of astronomy. Here are the women who revealed the stars' chemical compositions, designed cutting-edge telescopes, mapped the universe's large-scale structure, measured cosmic expansion, and hunted for extrasolar planets. They are the pathfinders who helped define our modern universe."
—MARCIA BARTUSIAK, author of *The Day We Found the Universe* and *Black Hole*

"This thought-provoking collection of stories by an absolutely amazing set of women astronomers worldwide is motivating, at times heartbreaking, and truly inspirational."
—ANNE-CHRISTINE DAVIS, University of Cambridge

"What makes this book priceless is that each astronomer has written her own words. Every story is different, and all the details matter, yet the commonalities are impossible to miss. I can't imagine a better resource for learning what it's like to be a successful woman in science."
—SEAN CARROLL, author of *Something Deeply Hidden: Quantum Worlds and the Emergence of Spacetime*

"Through the individual voices of extraordinary women, Trimble and Weintraub tell a humbling and inspiring story of the evolution of astronomy and the struggle of women to enter the field."
—JENNY GREENE, Princeton University

"These eloquent memoirs by prominent women astronomers span sixty years of social and scientific progress. They are pointed, poignant, and wise. Read them, ponder their common themes, and let us all learn from them."
—ROGER BLANDFORD, KIPAC, Stanford University

"This is a book whose protagonists have traversed with joy, perseverance, and pluck the entire golden age of modern astronomy and astrophysics. Together, their thrilling and moving essays combine into an eloquent, detailed history of scientific ideas, education, and aspirations, the magnificent instruments and places of discovery they have inhabited, and the doors they have opened to others and onto the universe."

—DIANA KORMOS BUCHWALD, general editor of the Collected Papers of Albert Einstein

"These stories will absolutely inspire our next generation of female scientists. . . . [*The Sky Is for Everyone*] is an absolute must-read for any young lady who is thinking of starting a career in astronomy. The same applies to anybody who is interested in the history of women in astronomy."

—MARY MCINTYRE, *Journal of the British Astronomical Association*

"The essays in *The Sky Is for Everyone* illustrate the progress women have made in the field, although that push to equality is not complete."

—JEFF FOUST, Space Review

"This anthology of 37 short autobiographies covers more than six decades of astronomy and shows the varied paths of female astronomers and the roadblocks that can slow or sideline their success."

—ASHLEY YEAGER, *Science News*

"The particulars of each woman's experience are unique, which is part of what makes their stories compelling. . . . I was affected (sometimes deeply) by each woman's story in a different way, and I often felt compelled to take note of some phrase or piece of sage advice."

—NICOLLE ZELLNER, *American Scientist*

The Sky Is for Everyone

The Sky Is for Everyone

Women Astronomers in Their Own Words

Edited by VIRGINIA TRIMBLE
and DAVID A. WEINTRAUB

PRINCETON UNIVERSITY PRESS
PRINCETON AND OXFORD

Published by Princeton University Press
41 William Street, Princeton, New Jersey 08540
99 Banbury Road, Oxford OX2 6JX

press.princeton.edu

First paperback printing, 2023
Paper ISBN 978-0-691-25391-6
Cloth ISBN 978-0-691-20710-0
ISBN (e-book) 978-0-691-23736-7

British Library Cataloging-in-Publication Data is available

Editorial: Ingrid Gnerlich & Whitney Rauenhorst
Production Editorial: Ali Parrington
Text Design: Karl Spurzem
Jacket/Cover Design: Pamela L. Schnitter
Production: Jacqueline Poirier
Publicity: Sara Henning-Stout & Kate Farquhar-Thomson

Jacket/Cover image: Nobel-laureate Andrea Ghez (UCLA) guides Keck Observatory's twin infrared lasers into the observable universe during a setting Full Moon in the early morning at 13,000 ft. elevation on Maunakea. Photo by Andrew Richard Hara.

This book has been composed in Arno

To the memory of Maud Worcester Makemson (1891–1977),
the first woman astronomer I ever met, and one of the most remarkable
—Virginia Trimble

For Kelly Lee Weintraub and Lennon Leigh Weintraub
—David A. Weintraub

*Women hold up half the sky**
and some day it will be so in astronomy!

* MAO ZEDONG, 1949

Contents

Illustrations

Abbreviations

Acronyms and Other Codes Decoded

AAPT	American Association of Physics Teachers
AAS	American Astronomical Society
AAUW	American Association of University Women
AGN	active galactic nuclei
AIP	American Institute of Physics
ALMA	Atacama Large Millimeter/submillimeter Array (Atacama Desert, Chile)
ApJ	*Astrophysical Journal*
APS	American Physical Society
ARA&A	*Annual Review of Astronomy & Astrophysics*
ASP	Astronomical Society of the Pacific
AURA	Associated Universities for Research in Astronomy
A&A	*Astronomy & Astrophysics*
BLR	Broad Line Region (of quasars)
CAMK	Copernicus Astronomical Center (Poland)
CARMA	Combined Array for Research in Millimeter-wave Astronomy (Cedar Flat, CA)
CCAT	Cerro Chajnantor Atacama Telescope (Atacama Desert, Chile)

CCD	charge-coupled device
CEA	Commissariat à l'énergie atomique (Atomic Energy and Alternative Energies Commission) (France)
CERN	Conseil Européen pour la Recherche Nucléaire (European Organization for Nuclear Research) (Geneva)
CfA	Center for Astrophysics (Cambridge, MA)
CFHT	Canada-France-Hawaii Telescope (Maunakea, HI)
CNPq	Agency Conselho Nacional de Pesquisas (National Council for Scientific and Technological Development) (Brazil)
CNRS	Centre national de la recherche scientifique (National Centre for Scientific Research) (France)
COSPAR	Committee on Space Research
CSST	China Space Survey Telescope (China, to be launched into Earth orbit)
CSWA	Committee on the Status of Women in Astronomy (of AAS)
CTIO	Cerro Tololo Inter-American Observatory (Headquarters: La Serena, Chile)
DAO	Dominion Astrophysical Observatory (British Columbia, Canada)
DGS	director of graduate studies
DOE	Department of Energy (U.S.)
EIRO	European Intergovernmental Research Organizations
ELT	Extremely Large Telescope (Cerro Amazones, Chile)
ESA	European Space Agency (Headquarters: Paris)
ESO	European Southern Observatory (Headquarters: Garching, Germany)
FYST	Fred Young Submillimeter Telescope (Cerro Chajnantor, Chile)

Gemini	Gemini Observatory North (Maunakea, Hawai'i), South (Cerro Pachón, Chile)
GMT	Giant Magellan Telescope (Las Campanas, Chile)
HEAO-1	High Energy Astrophysics Observatory 1 (NASA, in Earth orbit)
Herschel	Herschel Space Observatory (ESA, in solar orbit)
HST	Hubble Space Telescope (NASA, in Earth orbit)
IAG-USP	Instituto Astronômico e Geofísico University of São Paulo
IAS	Institute for Advanced Study (Princeton, NJ)
IAU	International Astronomical Union (Headquarters: Paris)
IfA	Institute for Astronomy (Honolulu, HI)
IoA	Institute of Astronomy (Cambridge, UK)
IR	infrared
IRTF	Infrared Telescope Facility (Maunakea, HI)
ISO	Infrared Space Observatory (ESA, in Earth orbit)
ISOCAM	Infrared Space Observatory Camera
JCMT	James Clerk Maxwell Telescope (Maunakea, HI)
JWST	James Webb Space Telescope (NASA, launched into solar orbit, December 2021)
KAO	Kuiper Airborne Observatory (NASA, high-altitude airplane)
KPNO	Kitt Peak National Observatory (Tucson, AZ)
LAMOST	Large Sky Area Multi-Object Fiber Spectroscopic Telescope (China)
LIGO	Laser Interferometer Gravitational-Wave Observatory (Hanford, WA and Livingston, LA)
LISA	Laser Interferometric Space Antenna (ESA, to be in solar orbit)

LLO	LIGO Livingston Observatory (Livingston, LA)
LMC	Large Magellanic Cloud
LRIS	Low Resolution Imaging Spectrograph (instrument built for Keck Observatory)
LSC	LIGO Scientific Collaboration
LSST	Large Synoptic Survey Telescope (now the Vera C. Rubin Observatory; Cerro Pachón, Chile)
MIRI	mid-infrared instrument (instrument built for JWST)
MtW	Mount Wilson Observatory (Los Angeles, CA)
NAS	National Academy of Sciences (U.S.)
NASA	National Aeronautics and Space Administration
NOAO	National Optical Astronomy Observatory (Headquarters: Tucson, AZ)
NRAO	National Radio Astronomy Observatory (Headquarters: Charlottesville, VA)
NRC	National Research Council (U.S.)
NSB	National Science Board (U.S.)
NSBP	National Society of Black Physicists
NSF	National Science Foundation (U.S.)
NSO	National Solar Observatory (Headquarters: Boulder, CO)
NWO	Nederlandse Organisatie voor Wetenschappelijk Onderzoek (Dutch Research Council)
OVRO	Owens Valley Radio Observatory (Big Pine, CA)
Palomar	Palomar Observatory (Palomar Mountain, CA)
PASP	*Publications of the Astronomical Society of the Pacific*
PI	principal investigator
POSS	Palomar Observatory Sky Survey
RAS	Royal Astronomical Society

SAAO	South African Astronomical Observatory (Cape Town, South Africa)
SALT	Southern African Large Telescope (Sutherland, South Africa)
SDSS	Sloan Digital Sky Survey (Apache Point, NM)
SKA	Square Kilometre Array (Headquarters: UK; telescopes in South Africa and Australia)
SOAR	Southern Astrophysical Research Telescope (Cerro Pachón, Chile)
SST	Spitzer Space Telescope (NASA, in solar orbit)
STEM	Science, Technology, Engineering, Mathematics
STScI	Space Telescope Science Institute (Headquarters: Baltimore, MD)
TAO	Tokyo Astronomical Observatory
TWAS	The World Academy of Sciences (Headquarters: Trieste, Italy)
UKIRT	United Kingdom Infrared Telescope (Maunakea, HI)
URM	underrepresented minorities
USP	University of São Paulo (Brazil)
UTC	Coordinated Universal Time
UV	ultra-violet
VLT	Very Large Telescope (Cerro Paranal, Chile)
WIYN	Wisconsin-Indiana-Yale-NOAO Telescope (Kitt Peak, AZ)
WMAP	Wilkinson Microwave Anisotropy Probe (NASA, in solar orbit)
XMM	X-ray Multi-Mirror Mission (ESA, in Earth orbit)

Acknowledgments

Astronomy has been our world for most of our adult lives, and the world of women in astronomy for one of us. We are grateful to the many teachers and mentors who opened the doors for us and for the opportunities in astronomy we have had. We hope you enjoy and, perhaps, learn from following us into these worlds on a path made of words written by thirty-seven very interesting women (present company excepted, of course). And we thank all of our contributors for their generosity of spirit in looking back at their careers, sharing their stories with us, and allowing us to share them with you. Thanks, especially, to Shazrene Mohamed for permitting us to borrow the title of her chapter, which she also uses as part of her work and outreach, as the title of this book.

Prelude

Welcome to the worlds of women astronomers! Your editors have had the very good fortune to be able to collect here thirty-seven autobiographies. This means that some very distinguished scientists, now deceased, are missing (see chapters 1 and 39). Also missing are a dozen or more colleagues whom we asked for their stories but who were just too busy, otherwise committed, or, sometimes, just not sufficiently interested to contribute. Also missing are at least as many important other watchers of the skies, because the publisher limited the number of pages you now hold. For these many reasons, this volume is not all-inclusive and encyclopedic. Since there may someday be a volume 2 or an online version (with fewer grams per chapter), please let us know about anyone, including yourself, who really should be here.

As it is, we hope these chapters cover a wide enough range of generations (PhDs from 1963 to 2010), nationalities, and subdisciplines to give some feel for the territory. Every chapter writer tells of some triumphs and some failures, barriers overcome, some mild and some monstrous, and a life that has brought, on balance, fulfillment. Yes, there have been other women who started out to be astronomers, but for whom the barriers were simply too high. They necessarily are not here, though we know at least a few.

To paraphrase the late Supreme Court justice Thurgood Marshall, we cannot know the names of all the people who fought to open doors for us, though many are in the timelines of chapter 1. Nor can we know the names of the people who will walk through the doors we ourselves had to kick and shove to get through—sometimes while getting our hands dirty. But they will surely include students of many of our chapter writers. Our job is to keep working to open those doors anyway, not for the thanks or the glory but for the sake of doing what is right.

What was clear before we started was that things do change. Time was when a woman who wanted to study the stars had to have a father, brother, or husband to provide entry. (This road is still not completely closed.) Then came a time when the best available was to be a "computer" at a large observatory, examining positional measurements, photographs, or spectrograms obtained by male observers. None of these human computers is here either, though the history of astronomy includes some spectacular achievements by women whose job descriptions did not actually include "research" or "discovery."

Both the world wars, the space race, and the Cold War contributed to changes, partly because men were drawn away from observatories, laboratories, and calculating machines and partly because these events brought about a need for more education, more research, and more knowledge in fields that at least overlapped astronomy.

Opportunities for women to attend graduate school, to receive fellowships and prizes, and to lead projects and teams have definitely increased more or less monotonically from the 1950s onward. Does this mean that the authors of the later chapters found a more welcoming environment than the writers of the first few chapters? That you must judge for yourself and decide if not, why not, and what still needs to be done and by whom.

Virginia Trimble and David A. Weintraub

The Sky Is for Everyone

Chapter 1

Beginnings

In France, Dorothea Klumpke (1861–1942) earned her Docteur-ès-Sciences at the University of Paris in mathematical astronomy in 1893, after completing her thesis, "L'etude des Anneaux de Saturne" (A study of the rings of Saturn), thereby becoming the first woman to achieve the academic distinction of earning an advanced degree for work done in astronomy. She then began her distinguished career as the director of the Bureau des Measures at Observatoire de Paris, leading the effort there to produce a section of the great photographic star chart known as the *Carte du Ciel*. In 1901, she married astronomer Isaac Roberts and moved with him to England. After his death in 1904, Klumpke Roberts returned to Observatoïre de Paris, where she continued to carry out astronomical research, and in 1929 published *The Isaac Roberts Atlas of 52 Regions, a Guide to William Herschel's Fields of Nebulosity*. In 1934, she was elected Chevalier de la Légion d'Honneur.

Professional successes like hers, built on her own credentials, accomplishments, and intellectual merits, were rare for women before 1900. Instead, most of the few women who in previous centuries had any role in astronomy typically gained access through their husbands, brothers, or fathers.

More than two centuries earlier, in Danzig (now Gdańsk), Poland, in 1663, sixteen-year-old Catherina Elisabetha Koopman (1647–93) married fifty-two-year-old astronomer Johannes Hevelius, which gave her access to his observatory, Stellaburgum, where she became his assistant. In 1687, as Elisabetha Hevelius, she would publish the catalogue she and

Johannes had worked on for decades, the *Catalogus Stellarum Fixarum*, which included the positions of 1,564 stars, including 600 newly identified stars and a dozen newly named constellations.

Beginning in 1774, in Bath, England, Caroline Lucretia Herschel (1750–1848) assisted her older brother William with his observational work, cataloging thousands of previously unknown star clusters and nebulae, and learning to make her own observations and astronomical calculations. On her own, she discovered eight comets and the galaxy NGC 205. But her most important work was done in helping her brother put together the *Catalogue of One Thousand New Nebulae and Clusters of Stars*, which was first published in 1786 with William as sole author. They added another 1,000 objects to this catalogue in a 1789 paper, and 500 more in a third version published in 1802, each time with William as the only author credited. Later, she helped her nephew John Herschel expand the inventory to more than 5,000 objects in his 1825 *Catalogue of Nebulae and Clusters of Stars*. For her efforts on this project, in 1828 the Astronomical Society of London (which in 1831 would become the Royal Astronomical Society) honored her with their Gold Medal and then, in 1835, elected her as an honorary member (according to the organization's by-laws at that time, women were ineligible for election as regular members). Her career, important and limited as it was, was only possible because her brother and nephew needed her help.

Mary Fairfax Somerville (1780–1872), known in her time as the "Queen of Science," was also elected as an honorary member of the Royal Astronomical Society in 1835. Somerville was an exception to the rule that women who contributed in a significant way to astronomy before 1900 did so because they were the wife, sister, or daughter of an astronomer. Her second husband, William (a surgeon), encouraged her, but she forged her professional career independent of his. In 1826, Somerville's "The Magnetic Properties of the Violet Rays of the Solar Spectrum" was the first research article written by a woman published in the *Philosophical Transactions of the Royal Society*. She would later cement her reputation as a popularizer of science, writing several books, including *The Mechanism of the Heavens* in 1831 and *On the Connexion of the Physical Sciences* in 1834.

A half century later, Margaret Lindsay Huggins (1848–1915) was instrumental in furthering the successful career of her husband, William. Beginning in the late 1870s, she was a participant with him in photographic and spectroscopic research, including his studies of the Orion Nebula and the planets, and conducted some of her own research at their private observatory in London, England. With William, she wrote *Atlas of Representative Stellar Spectra*, published in 1899, and was elected (an honorary member) to the Royal Astronomical Society in 1903.

Also in the late nineteenth century, Mary Proctor (1862–1957) learned to help her father, the famous English astronomer and popular science writer Richard Proctor, with his manuscripts and in founding and publishing the journal *Knowledge*, which led to her own career lecturing about astronomy and writing about astronomy for children. She was elected as a member of the British Astronomical Association (BAA) in 1897 and to the Royal Astronomical Society (a society with, relatively, more professionals than the BAA) in 1916 as one of its first female Fellows.

Another British woman of this era, Mary Evershed (1867–1949), worked closely with her husband, John, who directed the Observatory at Kodaikanal in India. Her observations of solar prominences were published in a single-author paper in the *Monthly Notices of the Royal Astronomical Society* in 1913.

The only other women named honorary members of the Royal Astronomical Society prior to the change in the rules in 1915 that permitted women to be elected as ordinary members were Anne Sheepshanks (1789–1876), whose claim to fame was donating significant sums of money to the University of Cambridge to promote astronomical research, Agnes Mary Clerke (1842–1907), who was considered Somerville's successor as a well-received popularizer and historian of astronomy, and two human computers at the Harvard College Observatory, Williamina Fleming (1857–1911) and Annie Jump Cannon (1863–1941), who made measurements and computed the positions, brightnesses, and spectral types of stars whose images were found on photographic glass plates obtained by (male) Observatory astronomers.

Fleming began her astronomical work because she was in the right place at the right time—in 1881, she was the housekeeper for Harvard

College Observatory director Edward C. Pickering when he invited her to start working at the Observatory as a computer. She assigned spectral types to, and calculated positions and brightnesses for, most of the nearly 10,000 stars included in the first *Draper Catalogue of Stellar Spectra*, published in 1890. She continued her work as a computer until her death in 1911.

Cannon, who worked at the Observatory from 1896 until 1940, is perhaps the most famous of the Harvard College Observatory computers. She put her stamp on modern astronomy through her work classifying the spectra of more than 225,000 stars and reinventing the stellar classification system, which was formally adopted for use by the international astronomy community in 1922. Her system of stellar spectral classes, used in the nine-volume expansion of the *Henry Draper Catalogue* published in the years from 1918 to 1924, is still in use today. Cannon, who in 1938 was appointed as the William C. Bond Astronomer and Curator of Astronomical Photographs at the Observatory, was the first woman to receive an honorary doctorate from the University of Oxford.

In 1888, Antonia Maury (1866–1952) also became one of Pickering's famous computers (she continued to work at the Harvard College Observatory until 1933), in large part because she had a family connection: her aunt Anna Palmer Draper (1839–1914), the widow of Henry Draper, had donated the money to the Observatory that supported the work of producing the *Draper Catalogue* and the *Henry Draper Catalogue*. Maury discovered that some stars had narrow and others had broad spectral lines. And though her boss was not interested in this way of classifying stars on the basis of the widths of their spectral lines, others were. Maury's discovery has been used by astronomers for more than a century to help distinguish evolved stars with extended envelopes (giants, which have narrow lines) from stars still living on hydrogen fusion in their cores (dwarfs, or main-sequence stars, which have broad lines).

Henrietta Swan Leavitt (1868–1921), who worked as another of Pickering's computers from 1902 until 1921, made perhaps the most important astronomical discovery of her time when in the 1900s she identified a few handfuls of stars known as Cepheid variable stars—stars that cycle from bright to faint and then to bright again in a regular but peculiar and identifiable way—and showed that their periods of variation in brightness were positively correlated with their maximum brightnesses. She published a paper, "1777 Variable Stars in the Magellanic Clouds," with these results in the *Annals of the Harvard College Observatory* in 1908 as the sole author. It included the understated conclusion that "it is worthy of notice that . . . the brighter stars have the longer periods." By 1912, Pickering recognized the importance of her work, and so her follow-up paper, "Periods of 25 Variable Stars in the Small Magellanic Cloud," was published under his name, only, though he was kind enough to add the note "prepared by Miss Leavitt" to the publication. This relationship, now known as the Leavitt Law, was the critical discovery that allowed Harlow Shapley to discover the structure of the Milky Way, Edwin Hubble to discover that the Milky Way Galaxy is not the entire Universe and that most of the galaxies that make up the Universe are moving apart, and modern astronomers to use the Hubble Space Telescope to improve measurements of the distance scale of the Universe.

Cannon, Fleming, Maury, and Leavitt are just four of the more than forty women who were hired to make measurements of stars found in the Harvard College Observatory photographic plate collection in the years prior to World War I, starting in 1875 when the Observatory changed a policy such that it could hire women onto the staff. The first woman hired was Director William Rogers's wife, Mrs. R. T. Rogers, who would work at the Observatory for twenty-three years. That same year, Anna Winlock (1857–1904), the eldest daughter of the deceased former Observatory director Joseph Winlock, was put on staff, where she would remain a computer for twenty-eight years. Anna's sister Louisa (1860–1916) started her twenty-nine-year career as a computer in 1886. And in 1879, Selina Cranch Bond (1831–1920), the daughter of the first Observatory director, William Cranch Bond, joined the staff, where she would continue working for twenty-seven years.

Harvard computer Adelaide Ames (1900–1932) was Observatory director Harlow Shapley's first graduate student (starting in 1921) and in 1924 became the first woman to earn a MA in astronomy at Radcliffe. Together, they published "A Survey of the External Galaxies Brighter than the Thirteenth Magnitude" (generally known as the *Shapley-Ames Catalogue of Bright Galaxies*) in 1932.

Another woman who started as a Harvard computer (1907–12) after receiving her AB degree from Radcliffe in 1907 was Margaret Harwood (1885–1979). She, however, spent most of her career elsewhere. In 1912, after receiving a fellowship from the Maria Mitchell Association on Nantucket Island, she stayed, becoming the director of the Maria Mitchell Observatory in 1916 after completing her MA degree at the University of California, thus becoming the first woman astronomer to direct an independent observatory in the United States. She remained as director until 1956 and spent her career measuring the variability of light received from stars and asteroids and mentoring several generations of up-and-coming women astronomers.

Margaret Walton Mayall (1902–95) spent the first thirty years of her career as a Harvard computer, but she is best known for the subsequent twenty years, which she spent as Director of the American Association of Variable Star Observers (AAVSO). She started working at the Observatory in 1924 and continued until 1955. After earning her MA in astronomy from Radcliffe in 1927, she worked in partnership with Annie Jump Cannon until 1941 and then completed Cannon's unfinished work, which she published in 1949 as *The Henry Draper Extension: The Annie J. Cannon Memorial*. Finally, beginning in 1954 and continuing until 1973, she served as Director of the AAVSO and oversaw the transition of the organization from an activity headquartered and run by the Harvard College Observatory into an independent, nonprofit, scientific organization.

Harvard was not the only observatory that used talented women as inexpensive labor. Annie Scott Dill Maunder (1868–1947) was hired as a human computer in 1890 at the Royal Observatory in Greenwich, thereby becoming one of the first women in England to be paid for her work in astronomy. Maunder is best known for the work she did with her husband, Walter, discovering the pattern in which over a period of eleven years sunspots appear at successively lower solar latitudes and then repeat this cycle. This so-called butterfly pattern is a manifestation of the twenty-two-year-long solar sunspot cycle. Along with Mary

Proctor, she was elected with the first cohort of women to become Fellows of the Royal Astronomical Society in 1916.

After completing a master's degree in astronomy at Mount Holyoke College in 1906, Jennie Belle Lasby (1882–1959) was immediately hired as the first woman computer at Mount Wilson Observatory (MtW), in California, where she would work until 1913, publishing an article on the rotation of the Sun and writing several articles on spectroscopy with Walter Adams. She then worked in 1915 with J. C. Kapteyn at Potsdam Observatory, taught at Carleton College for the duration of World War I as a war-replacement instructor, and then taught astronomy at Santa Ana College in California from 1919 until 1946.

Just months after Lasby started at MtW she was joined by Cora Gertrude Burwell (1883–1982), who, after graduating from Mount Holyoke in 1906, became the second of the female computers at MtW, where she would work until 1949. She would publish nearly forty papers in her career, including two catalogues of hot stars with bright hydrogen emission lines.

Many of these women had extraordinary careers as computers at the end of the nineteenth century and in the first decades of the twentieth. But they rose to distinction almost entirely by carrying out tasks assigned to them by male astronomers. Their educations and opportunities for leadership and advancement were limited. The glass ceiling above them was thick and impermeable.

Maria Mitchell (1818–89) carved a more independent path as America's first important female astronomer. Mitchell became famous after discovering a comet in 1847, for which she was awarded a gold medal by King Christian VIII of Denmark. She then worked for the U.S. Coast Survey, beginning in 1849, tracking the positions of planets, before being appointed to the faculty at Vassar College in 1865 and thereby becoming the first woman to hold a position as a professor of astronomy.

In the years immediately after Mitchell joined the faculty at Vassar, a few other women would receive faculty appointments as astronomers in the United States at small colleges, mostly in the Northeast. Immediately after graduating from Mount Holyoke College in 1866, Elisabeth

Bardwell (1831–99) became an instructor there and would remain at Mount Holyoke for the next thirty-three years. She published her observations of meteor showers in the magazines *Sidereal Messenger* and *Popular Astronomy*.

Susan Jane Cunningham (1842–1921), who studied at the observatories of Vassar, Harvard, Princeton, and Cambridge, helped found both the mathematics and the astronomy departments at Swarthmore College in 1869, becoming Swarthmore's first professor of astronomy that year and later also a professor of mathematics.

A decade later, in 1876, Sarah Frances Whiting (1847–1927) began her career on the faculty at Wellesley College as their first professor of physics, where she started their physics department and would teach Annie Jump Cannon (Wellesley, Class of 1884). Later, after learning astronomy under the tutelage of Edward Pickering at the Harvard College Observatory, she established and became the founding director of Wellesley's Whitin Observatory, built with funds donated by Mrs. John C. Whitin. During her career, she wrote popular articles about astronomy and a textbook for teaching astronomy.

In 1883, after studying astronomy for a year with Edward Pickering, Mary Emma Byrd (1849–1934) took up a position as first assistant at the Goodsell Observatory at Carleton College in Minnesota. She would eventually earn a PhD in astronomy at Carleton in 1904. In 1887, Smith College lured her away to become both Professor of Astronomy and Director of the Smith College Observatory, where she worked until 1906. Byrd wrote two textbooks on astronomy and contributed articles to *Popular Astronomy*.

Winifred Edgerton Merrill (1862–1951) was the first American woman to earn a doctoral degree in mathematics, doing so at Columbia in 1886. Her dissertation work, which focused on both mathematics and astronomy, included the calculation of the orbit of a comet. In 1887, she was offered a professorship at Smith College, but her marriage that year led her to give up on a career in mathematical astronomy; however, she did teach mathematics at the Oaksmere School for Girls in New York until 1926. The inscription on her portrait, which hangs in Columbia's Philosophy Hall, reads "She opened the door."

In the United Kingdom, one woman who found her own path to scientific accomplishment, despite her lack of formal education, was Mary Adela Blagg (1858–1944). Beginning in 1907, working first with Samuel Saunder and then with Karl Müller, she established the modern nomenclature used for lunar features, cleaning up the mess left behind by a handful of nineteenth-century astronomical mapmakers. Blagg was one of only four women (the others were Cannon, Leavitt, and Klumpke Roberts) who were among the 207 Individual Members of the International Astronomical Union at the end of its first General Assembly, held in Rome in 1922, she as part of Commission 17 on Lunar Nomenclature.

One place where, more than a century ago, women managed to have some presence in professional astronomy was in the founding in 1899 of the Astronomical and Astrophysical Society of America (AASA; renamed the American Astronomical Society [AAS] in 1914), with 11 women[1] among its 113 charter members. By 1910, women numbered 20 out of the 180 additional members,[2] and in 1912 Annie Jump Cannon was elected as treasurer (she would be re-elected six times), becoming the first woman astronomer to hold office in the nascent AAS.

Emily Elisabeth Dobbin (1874–1949) was one of the original 113 members of AASA and, with Caroline Furness, one of two women of the 39 members who were actually at Yerkes Observatory for that first 1899 pre-meeting. After earning her master's degree in astronomy at the University of Chicago in 1903 and publishing two papers in 1904, she briefly taught mathematics, first at the University of Missouri and then at a high school in St. Paul, Minnesota. While in Minnesota, she became deeply involved in the suffrage movement and published a newspaper, the *Minnesota Socialist*.

During the two decades that preceded the entrance of the United States into World War I, only ten women received PhDs in astronomy in the United States. Many of them followed their degrees with long, productive professional careers. In addition, at least two women earned PhDs in mathematics doing astronomy research and then spent long careers working in astronomy.

Just a year after Dorothea Klumpke completed her PhD in Paris, Margaretta Palmer (1862–1924) earned her PhD as an astronomer (the degree was issued by the mathematics department) at Yale in 1894, calculating the orbit of Comet 1847 VI, the comet whose discovery had made Maria Mitchell, her undergraduate mentor at Vassar, famous. She later would compute orbits for three comets discovered by Caroline Herschel, those of 1786, 1788, and 1797, and, with Frank Schlesinger and Alice Pond, write the first *General Catalogue of Trigonometric Stellar Parallaxes,* published in 1924. Palmer spent her entire career at the Yale Observatory, holding titles of computer (1894–1912) and research assistant (1912–24) before being promoted to the rank of instructor in 1923.

Anna Delia Lewis (b. 1870) was likely the first woman to earn a PhD in astronomy in the United States, completing her dissertation "Variable Stars" at Carleton in 1896 using data collected at Goodsell Observatory. She published a single paper, in *Popular Astronomy* in 1915, on sunspot observations made in 1914. She taught astronomy for four decades at Carleton, Mount Holyoke, Albert Lea, and Lake Erie colleges before retiring in 1936.

Flora Ellen Harpham earned her PhD in astronomy at Carleton College in 1908 and was later appointed as professor of astronomy and mathematics at the College for Women in Columbia, South Carolina. She published one research article in 1892, on Comet Swift, and three later articles in *Popular Astronomy,* including one on some of the earliest astronomical photographs ever taken, the Rutherfurd photographs taken in 1864 and 1865. She later translated the book *Historical Sketch of the Foundation of the Metric System* from the original French edition.

Caroline Ellen Furness (1869–1936) earned her doctorate in astronomy at Columbia in 1900. Furness joined the faculty at Vassar College in 1903, where she remained until 1933. Furness studied variable stars, comets, and minor planets, wrote the textbook *Introduction to the Study of Variable Stars,* and was elected a Fellow of the Royal Astronomical Society in 1922.

Harriet Williams Bigelow (1870–1934) earned her PhD in astronomy at the University of Michigan in 1904—the same year Mary Byrd earned her doctorate at Carleton—and published eight papers, most on comets, over a twenty-four-year span. She was appointed Director of the Smith College Observatory in 1906, was promoted to professor and department chair in 1911, and served as both Vice President and President of the AAVSO.

Anne Sewell Young (1871–1961) earned her doctorate in astronomy at Columbia in 1906. She began her professional career in 1899 as Director of Williston Observatory at Mount Holyoke. Later, with her PhD in hand, she transitioned from instructor to professor and remained at Mount Holyoke until her retirement in 1936. She was one of the eight astronomers, and the only woman of the eight, who in 1911 formed the AAVSO. She would serve as its vice president in 1922–24.

Edith Dabele Kast (1880–1967) was awarded her PhD in astronomy in 1906 at the University of Pennsylvania and continued work there, at Flower Observatory, as a University Fellow for Research in Astronomy for one year. Her dissertation, "The Mean Right Ascensions and Proper Motions of 130 Stars," was her only professional publication. She was elected to membership in AASA in 1904, among the second dozen women to join. Following her marriage in 1907 and move to Utah, her membership lapsed before 1910.

Ida M. Barney (1886–1982) spent her career in astronomy after earning a PhD in mathematics at Yale in 1911. After teaching at several women's colleges for a decade, she was hired in 1922 as the chief assistant to Frank Schlesinger, Director of Yale Observatory, and with him produced the twenty-three-volume *Yale Zone Catalogues* of stellar proper motions, containing data for the movements of more than 147,000 stars. Barney completed the last thirteen volumes over a period of fifteen years after Schlesinger's death in 1943.

Emma Phoebe Waterman (1882–1967; later Haas) and Anna Estelle Glancy (1883–1975) both earned their doctorates in astronomy at the University of California, Berkeley in 1913. Waterman's career ended almost as soon as it started—she published two papers on bright Class A stars and one titled "The Present Status of the Problem of Stellar Evolution" in 1913 but none afterward—marrying in 1914. She did, however, make variable star observations for many more years, and beginning in 1953 she assisted Director Margaret Mayall in the work of the AAVSO. Glancy published her observations of comets and on the distance to the star Polaris, working from the Observatorio Nacional in Córdoba,

Argentina, through the 1910s and then left astronomy to spend the rest of her professional career working for the American Optical Company, applying her knowledge of optics to commercial research.

Mary Murray Hopkins (1878–1921) earned her PhD in astronomy at Columbia in 1915, after which she published a short series of papers on the star 61 Cygni. Hopkins, who started teaching at Smith College in 1906, was appointed associate professor in 1916.

In the 1920s through 1930, only five more women earned PhDs in astronomy in North America.

In 1921, Priscilla Fairfield Bok (1896–1975) completed her PhD in astronomy at Berkeley. Immediately afterward, she took up a position on the faculty at Smith College while working, unpaid, as a researcher at Harvard. According to her astronomer-husband Bart, "My wife Priscilla did a lot of work at Harvard, and didn't even want a job. We felt we were both better off if she was a free agent and could do what she wanted to do." Of course, Shapley, the director at Harvard College Observatory, was happy to have her work for free, and Harvard University was not about to offer a faculty position to a woman. In later years, she took up teaching positions at Wellesley College and then at Connecticut College before giving up her paid professional career and accompanying her husband when he accepted positions first at Mount Stromlo Observatory in Australia and then at the University of Arizona. In work done over three decades in partnership with her husband, she became an expert on the Milky Way (writing *The Milky Way* with Bart in 1941) and showing that the process of star formation is active in the modern Universe.

In 1925, Cecilia Helena Payne (1900–1979; later Payne-Gaposchkin) completed one of the most important doctoral dissertations in astronomy of the twentieth century, working with Harlow Shapley at the Harvard College Observatory. In her dissertation, she showed that most stars have the same chemical composition and are composed primarily of hydrogen and helium.[3] Shapley had recruited her to be his graduate student in order to claim that Harvard had a graduate program in astronomy, not just an observatory; her degree, however, was awarded by Radcliffe College because at that time Harvard did not grant degrees to women.

Only a year later, in 1926, Allie Vibert Douglas (1894–1988), who previously had studied under and written a paper with Arthur Eddington, at Cambridge in 1923, earned her PhD in astronomy at McGill University, becoming the first Canadian woman to do so. She also was the first woman to serve as President of the Royal Astronomical Society of Canada. She later would write the first biography of Eddington, *The Life of Arthur Eddington*, and produced papers about cyanogen bands in stellar spectra and the CH^+ molecule in interstellar space. She continued at McGill as a lecturer for thirteen years before moving to Queen's University, Ontario, as a professor of physics and retiring in 1964.

Emma Williams Vyssotsky (1894–1975), working with Payne, earned her PhD in astronomy at Radcliffe College in 1930. Immediately after completing her degree, Williams Vyssotsky relocated to McCormick Observatory at the University of Virginia, where her husband, Alexander, had recently been promoted to assistant professor. She was hired as an instructor and would publish twenty-seven papers, most on the motions of stars and all of them jointly with her husband, over the years from 1933 through 1951.

Maud Worcester Makemson (1891–1977) received her PhD in astronomy at Berkeley in 1930. In 1932, she joined Furness on the faculty at Vassar as an assistant professor of astronomy and navigation and became director of the Vassar Observatory in 1936 (where she was the first astronomy instructor of Vera Florence Cooper [Rubin] [1928–2016]). Makemson was a pioneer in archaeoastronomy, in particular Polynesian and Mayan astronomy, and astrodynamics. Her monographs include *The Morning Star Rises: An Account of Polynesian Astronomy* (1941), *The Maya Correlation Problems* (1946), and *The Book of the Jaguar Priest* (1951). After her retirement from Vassar in 1957 she began appointments as a lecturer of astronomy and astrodynamics at the University of California, Los Angeles and a consultant to Consolidated Lockheed-California, at which time she cowrote the textbook *Introduction to Astrodynamics*. When she took up her appointment at UCLA, she was the only female faculty member there in astronomy, mathematics, or physics (and Virginia Trimble's [chapter 6] first astronomy instructor).

In this same era, Louise Freeland Jenkins (1888–1970) found a path to success in astronomy despite completing her education in 1917 with only a master's degree, earned under the tutelage of Anne Young at Mount Holyoke. Hired at Yale Observatory in 1932, she was assistant editor of the *Astronomical Journal* for sixteen years and wrote with Frank Schlesinger the second edition of the *Yale Catalogue of Bright Stars* in 1940 and the second edition of the *General Catalogue of Stellar Parallaxes* in 1935. She compiled the third edition of *Stellar Parallaxes* in 1952, as well as a 1963 supplement, on her own.

Merrill, Klumpke Roberts, Palmer, Lewis, Harpham, Furness, Byrd, Bigelow, Young, Kast, Barney, Waterman, Glancy, Hopkins, Bok, Payne, Douglas, Vyssotsky, and Makemson were all working to crack the glass ceiling that kept women in low-level helper roles in astronomy. In completing doctoral dissertations, they were asking their own scientific questions and using their considerable intellectual skills to make progress toward answering them. Nevertheless, for most of them their achievements combined with their professional knowledge and skills were not enough to allow them to join the old boys' club.

World War II triggered major employment opportunities that led to new careers for women, not just in industry but also in academia. After more than half a century during which women astronomers were limited almost entirely to research assistant positions at a small handful of observatories or faculty positions at women's colleges, a few women found themselves in the right places at the right times to do important work. Sometimes, they were able to take a step up the academic ladder; other times, the solidly entrenched rules that limited career opportunities for women won.

One woman who made enormous contributions, first to radar work for the Australian government's Council for Scientific and Industrial Research during World War II and then as a pioneer in the new field of radio astronomy, was Ruby Payne-Scott (1912–81). In 1933, Payne-Scott was the third woman to earn a BSc in physics at the University of Sydney; she would then earn a MSc in physics three years later. After discovering several kinds of radio bursts from the Sun and making important engineering contributions to radio frequency detectors, she was forced into early retirement from astronomy in 1951 simply because she was married. A decade later she did return to work, but not to astronomy, teaching at an Anglican school for girls.

Henrietta Hill Swope (1902–80) earned a master's degree in astronomy at Radcliffe in 1928 and then worked for fourteen years as a Harvard Observatory computer making measurements of variable stars. During and after World War II, Swope worked for the U.S. Navy Hydrographic Office, computing long range navigation (LORAN) tables. After the war, Swope taught at Barnard College for five years and then teamed up with Walter Baade, at the Carnegie Observatories, as his research assistant. With Baade, she determined the distance to the Andromeda Galaxy to be 2.2 million light years,[4] discovered the Draco dwarf galaxy, and showed that the Cepheids in Andromeda are more like those in the Milky Way than like those in the Small Magellanic Cloud. Notably, several of the most important "Baade and Swope" papers were published in the years from 1961 through 1965, well after Baade had retired in 1958 and died in 1960.

As presaged by Annie Maunder, Margaret Huggins, and a few others, our more recent era also includes some women whose involvement in astronomy was furthered or even initiated by working with their better-known husbands, including some who also got their starts contributing to the war effort in the 1940s. These distaff astronomers included Martha Betz Shapley (1890–1981), Barbara Cherry Schwarzschild (1914–2008), Katherine C. Gordon Kron (1917–2011), and Antoinette Piétra de Vaucouleurs (1921–87).

After marrying Harlow Shapley in 1914, Betz Shapley abandoned her PhD work in German literature at Bryn Mawr College to follow Harlow Shapley in his career, moving first to Mount Wilson and then to Harvard. Self-taught in astronomy, she specialized in the study of eclipsing binary stars and managed to publish more than two dozen professional papers over a dozen years before abandoning astronomy in 1932 due to the pressure of family life. She returned to astronomy and wrote four more papers after 1946. She also calculated munitions trajectories during World War II and served on IAU Commissions 27 and 42.

Barbara Cherry was well on her way toward a career in astronomy, completing her coursework at Harvard in astronomy while a graduate student at Radcliffe and working at the MIT Radiation Lab, when she married Martin Schwarzschild in 1945. Although she wrote two research papers and a *Scientific American* article in the 1950s, all with Martin, her marriage limited and effectively ended her career.

Katherine C. Gordon studied astronomy at Vassar, Lund University, and again at Harvard College Observatory, before working at Lick Observatory from 1941 into 1943, at the California Institute of Technology in 1943–44, and at the Naval Ordnance Test Station in China Lake, California, in 1944–45. After marrying Gerald Kron in 1946, she followed him as he pursued his career but also managed to publish twenty-four papers about variable stars and edit the *Publications of the Astronomical Society of the Pacific* (1961–68).

Antoinette de Vaucouleurs published nearly eighty papers across half a century, despite having only an undergraduate degree from the Sorbonne. After marrying Gérard de Vaucouleurs in 1944, she worked as the assistant director at Mount Stromlo Observatory and then helped her husband and his collaborators at Lowell Observatory and the University of Texas, where she held a position as research scientist associate in the astronomy department for twenty-five years.

Helen Dodson-Prince (1905–2002), who completed her PhD in astronomy at the University of Michigan in 1934, was one of the first women for whom her advanced degree opened up a path for joining the faculty of a major research university. She held a faculty position at Wellesley College (1933–45) before being appointed to the faculty of Goucher College in 1945. Dodson-Prince relocated again in 1947, taking up a professorship at the University of Michigan. She specialized in the study of solar flares, with much of her early work done at the solar observatory at Meudon, near Paris.

After completing her PhD in astronomy at Radcliffe in 1931, Helen Sawyer Hogg (1905–93) followed her husband to the University of Toronto and joined the staff at David Dunlap Observatory, where she established her reputation by discovering variable stars in globular clusters. After serving as acting chair of the astronomy department at Mount

Holyoke in 1940–41, she returned to the University of Toronto and began teaching, as a substitute for the male faculty who had been called to military duty during World War II. She continued in this position after the war and in 1951 was belatedly promoted to a permanent faculty position.

Also in 1931, Charlotte Emma Moore-Sitterly (1898–1990) earned her PhD in astronomy at Berkeley. Moore had begun her career in 1920 at Princeton as a computational assistant to Henry Norris Russell and would continue to work with him into the 1950s. Moore established herself as an expert in atomic spectroscopy, working at Princeton until 1945 before relocating to the National Bureau of Standards and then the Naval Research Laboratory. Moore's *A Multiplet Table of Astrophysical Interest* and *An Ultraviolet Multiplet* became basic references for all astrophysical line identifications and also of great value to atomic physicists and chemists. Without these references the analyses of stellar atmospheres would not be possible.

E. Dorrit Hoffleit (1907–2007) completed her PhD at Radcliffe in 1938 working under the supervision of Bart Bok. She continued working at the Observatory until she was recruited to Aberdeen Proving Ground in 1943 to compute army artillery firing tables during the war and to White Sands Missile Range to work on Doppler tracking of captured V-2 rockets after the war. In 1948 she took a large pay cut in order to return to a computing position at Harvard Observatory, because that allowed her to return to her work with the Harvard plate collection. Then, in 1956, she accepted Yale Observatory director Dirk Brouwer's offer of a position that had her on the Yale Observatory research staff for six months and as director of the Maria Mitchell Observatory for the other half of each year. At Yale, she took charge of updating and editing what became the third (1964) and fourth editions (1982) and supplement (1983) of the *Yale Bright Star Catalogue* and wrote (with W. F. van Altena and J. T. Lee) the fourth edition (1995) of *The General Catalogue of Trigonometric Stellar Parallaxes*. She continued as director of Maria Mitchell Observatory until 1978, and during those years she left her imprint on modern astronomy by mentoring 102 women and three men in summer undergraduate research programs, including about 25 who became professional astronomers,[5] among them Janet Akyüz Mattei (1943–2004; PhD in astronomy from Ege University in Turkey in 1982), who would go on to become the director of the AAVSO for thirty years.

Payne-Gaposchkin, meanwhile, persevered at the Harvard College Observatory for three decades, continuing to work in low-paid, low-prestige research positions because Harvard University did not allow women to be appointed to the faculty. The male astronomers who controlled the hiring of faculty at most American colleges and universities early in the twentieth century had little interest in diversifying their faculties by hiring women, no matter how talented. Nevertheless, Shapley had Payne-Gaposchkin giving lectures, though without an official appointment, and working with graduate students in their research beginning almost immediately after completing her degree. In 1945, after nearly two decades of teaching, her name was printed in the university course catalog for the first time. Another decade later, in 1956, she was at last appointed to a professorship. A few months later, she was tapped as Chair of the Department of Astronomy.

At the University of Chicago, a combination of World War II opportunities and the presence of Subrahmanyan Chandrasekhar led to a quintet of women earning PhDs at that institution in the 1940s.

Margaret Kiess Krogdahl (1920–2013) was the first woman to earn a PhD in astronomy at the University of Chicago, completing her degree in 1944 under the supervision of Chandrasekhar. After a three-year research fellowship at Yerkes Observatory, she moved with her husband, Wasley Krogdahl, who had also been a Chandra student, to Lexington, Kentucky, where he had been appointed to the faculty at the University of Kentucky. She published nine of her own papers, the last in 1960, and then edited his publications, including *The Astronomical Universe: An Introductory Text in College Astronomy*.

Merle Eleanor Tuberg Gold (1921–2017), who studied the Sun, followed quickly in Krogdahl's footsteps. She earned her PhD in astronomy in 1946 at Chicago, also working with Chandrasekhar, publishing her only two research papers along the way. She then took up a postdoctoral position in Cambridge, UK, where she met and married astrophysicist Thomas Gold, with whom she had three daughters. Her career ended at about the time her family started, and it was said in her obituary that she had chosen family over science.

Marjorie Hall Harrison (1918–86) also completed a PhD in astronomy under Chandrasekhar, in 1947. After publishing a half-dozen papers in the 1940s on models of stars with hydrogen-depleted cores, she taught astronomy at Sam Houston State University in Huntsville, Texas.

The last two women who would complete PhDs in astronomy at the University of Chicago in the 1940s went on to very distinguished careers. Anne Barbara Underhill (1920–2003), who also worked with Chandrasekhar, completed her degree in 1948 and Nancy Grace Roman (1925–2018), who worked with W. W. Morgan, did so in 1949.

Underhill, born in Vancouver, Canada, worked as a postdoctoral scholar at Copenhagen Observatory for a year and then did research at Dominion Astrophysical Observatory in Canada until 1962, when she was recruited for and accepted a full professorship at the University of Utrecht in the Netherlands, finally joining NASA in 1970 for the remainder of her career. She specialized in studying hot (early spectral type) stars and computing numerical models of the atmospheres of stars.

Roman worked at Yerkes and McDonald Observatories until 1954 before moving to the Naval Research Laboratory, where she became head of the microwave spectroscopy group in the newly formed radio astronomy program. She then joined NASA in 1959.

Martha Elizabeth Stahr (Patty) Carpenter (1920–2013) earned her PhD in astronomy at Berkeley in 1945, taught at Wellesley from 1945 until 1947, and then was hired onto the faculty at Cornell University, where she was the first female faculty member there in the College of Arts & Sciences. She relocated from Cornell to the University of Virginia in 1965, where she remained until her retirement in 1985. As a radio astronomer, she studied the Sun and 21-centimeter line emission from the Milky Way. An additional, important part of her legacy may have been mentoring Vera Rubin's master's thesis on the kinematics of galaxies and thereby encouraging Rubin to pursue a career in astronomy.

Vera Florence Cooper Rubin completed her PhD in astronomy at Georgetown University in 1954, taught at Montgomery College for a year, and then did research again at Georgetown. Beginning in 1958, she also assumed some lecturing duties there and was appointed as an assistant professor in 1962. After relocating to the Carnegie Institute of Washington in 1965, Rubin carved out one of the most distinguished careers in astronomy in the twentieth century. She set out to study the kinematics of spiral galaxies by measuring their rotation speeds, expecting to find that their rotation curves in their outer parts would be correlated with morphology; to her disappointment, they nearly all turned out to have similar rotation curves. Instead, her measurements demonstrated that a tremendous amount of unseen matter, dark matter, must exist in these galaxies in order for them to hold together.

One unusual PhD recipient was Frances Woodworth Wright (1897–1989), who earned her doctoral degree in astronomy at Radcliffe College in 1958, after earning her bachelor's degree from Brown nearly forty years earlier, in 1920. Hired as a computer at Harvard College Observatory in 1928, she continued working there as a research assistant until 1961. She made her reputation teaching celestial navigation to military personnel during World War II, beginning in 1942, and was later hired as a lecturer at Harvard in 1958 to teach celestial navigation to Harvard students and sailors, where the course continues to be taught by the Frances W. Wright Lecturer on Celestial Navigation. Wright wrote four books, three on celestial navigation, the last one published in 1980, and a fourth, *The Large Magellanic Cloud* (1967), written with Paul Hodge.

Martha Locke Hazen Liller (1931–2006), who also received her PhD in 1958, hers from the University of Michigan, specialized in observations of variable stars and planetary nebulae. In 1969, after working as a research fellow at the Harvard College Observatory, she was appointed Curator of Astronomical Photographs and in that position took charge of the Harvard Plate Archives for the next several decades.

Thus, at about the time when the space age began, in October 1957, a total of as few as eight women astronomers (Sawyer Hogg, Dodson-Prince, Payne-Gaposchkin, Douglas, Worcester Makemson, Carpenter, Rubin, and Wright)[6] held faculty appointments at any rank at major research universities in North America.

Only four more similar appointments would follow in the 1960s: Beverly Turner Lynds (b. 1929; PhD, UC Berkeley, 1955) was hired by the University of Arizona in 1961; E. Margaret Peachey Burbidge (1919–2020; PhD, University College London, 1943) began as a member of the faculty at the University of California, San Diego in 1962; Elske van Panhuys Smith (b. 1929; PhD, Radcliffe, 1956) joined the University of Maryland faculty in 1963; and Ann Merchant Boesgaard (chapter 3) got started at the University of Hawai'i in 1967.

Thousands of dark nebulae are named for Lynds (e.g., Lynds 1551, or just L1551), as are the thousands of stars discovered within them (e.g., L1551 IRS 5), as a result of her publication in 1962 of *Catalogue of Dark Nebulae*. Lynds would serve in 1976–77 as the first female assistant director of Kitt Peak National Observatory.

Burbidge was the lead author on a landmark paper in 1957 on how heavy elements are made through nuclear fusion processes in stars. Generations of astrophysicists learned, only half jokingly, that the Big Bang made hydrogen and helium; then Burbidge, Burbidge, Fowler, and Hoyle made all the rest. Her list of firsts, which is nearly endless, includes first woman President of the International Astronomical Union's commission on galaxies, first female director of the Royal Observatory Greenwich, first woman President of the American Astronomical Society, and Inaugural Fellow of the American Astronomical Society. She also served as the founding director of the Center for Astrophysics and Space Science at UC San Diego, where she helped develop the Faint Object Spectrograph, one of the first instruments launched on the Hubble Space Telescope and which she used to find evidence for a supermassive black hole at the center of the galaxy M82.

Smith studied solar flares and the solar chromosphere and wrote a widely used undergraduate textbook, *Introductory Astronomy and Astrophysics*. After moving into university administration, she served as assistant provost and assistant vice chancellor at the University of Maryland and then as a dean at Virginia Commonwealth University.

In this same time period, Susan Kayser (b. ~1939) became the first woman to earn a PhD in astronomy at Caltech, in 1966. She was informed in the letter that initially rejected her from the program, "Women aren't really suited for observing on the long, cold, lonely nights." Upon completing her degree work, the Associated Press described her accomplishment with the words "Astrophysicists getting prettier." Nevertheless, Kayser would carve out a distinguished career as a program manager at NASA and the National Science Foundation, working on the Helios and ISEE-3 (later called ICE) spacecraft and radio astronomy experiments at NASA and the Gemini Telescopes at NSF, before finishing her career at the Fermi National Accelerator Laboratory.

Also in the mid-1960s, Sachiko Tsuruta, born in Yokohama in about 1936, earned her PhD in astrophysics from Columbia in 1964, working with A.G.W. Cameron on the cooling of neutron stars, objects which, at that time, were not yet known to exist. Since then, she has become an expert on neutron stars, black holes, and supermassive black holes. After holding positions at the Harvard-Smithsonian Center for Astrophysics and the Max Planck Institute in Munich, she assumed a full-time faculty position at Montana State University in 1989.

In Europe, Julie Marie Vinter Hansen (1890–1960) received an appointment as observer at the University of Copenhagen Observatory, where she spent her entire career, making her the first woman to hold an appointment at the University of Copenhagen. Specializing in calculating orbits of comets and asteroids, she was first appointed as a computer in 1915 and began her university appointment after she finished her education in 1919. Beginning in 1920, she became the Danish editor of *Nordisk Astronomisk Tidsskrift* (Nordic Astronomy Review), and in 1922 she stepped into the role of assisting Elis Strömgren, the editor for the IAU's International News Service, in disseminating both the IAU Circulars (postcard announcements giving information about astronomical phenomena requiring prompt dissemination) and the Central Bureau of Astronomical Telegrams (for announcing new astronomical discoveries). In 1947, she replaced Strömgren as director of the International News Service and as editor of the Circulars and the Telegrams. In 1921, Vinter Hansen and Helene Marie Emilie Kempf (the wife of German astronomer Paul Kempf, who died in 1920) were the first two women elected as members of the Astronomische Gesellschaft (the German Astronomical Society).

In 1939, Frida Elisabeth Palmér (1905–66), having been trained by Knut Lundmark at Lund University, became the first Swedish woman to earn a PhD in astronomy, publishing her measurements on the positions and proper motions of semi-regular variable stars. During World War II, she worked for the signals intelligence agency Försvarets Radioanstalt decoding Soviet navy messages; after the war, no longer active in astronomy, she taught mathematics and physics at the Halmstad grammar school.

In 1945, Wilhelmina Iwanowska (1905–99), who earned her habilitation degree from Stefan Batory University (now Vilnius University) in Poland in 1937, was a founding faculty member of the Nicholas Copernicus University in Toruń, Poland,[7] served as the first director of the Institute of Astronomy in Toruń, and was the first woman to serve as a vice president of the International Astronomical Union (1973–79).

In 1946, shortly after completing a doctoral degree in physical sciences at the University of Rome, Giusa Cayrel de Strobel (1920–2012) became the first woman appointed as Astronomer in Italy, taking up a position at Asiago Astrophysical Observatory and the University of Padua; after completing a second PhD in 1966, this time in astronomy, she would work for the rest of her career at the Observatoire de Paris and would serve as President of the IAU Commission on Stellar Spectra from 1982 to 1985.

Alla Genrikhovna Massevich (1918–2008) received her doctorate from Moscow University in 1952. She was the first Soviet astrophysicist to undertake theoretical calculations of the structure and evolution of stars and mass loss from stars. After completing her degree, she became Deputy Chairman of the Astronomical Council of the Soviet Academy of Sciences, a position she held for thirty-five years. Among her responsibilities was, beginning in 1957, the optical tracking of Sputnik satellites. In 1972, she would add to her duties a faculty appointment at the Moscow Institute of Geodesy and Cartography.

In 1962, Edith Alice Müller (1918–95), who earned her PhD in mathematics at the University of Zurich in 1943, joined Underhill as a female astrophysicist and faculty member at a European university when she was appointed to the faculty at the University of Neuchâtel, Switzerland. In 1965, Müller joined the professoriate at the University of Geneva. In a major work published in 1960 jointly with Leo Goldberg and

Lawrence Aller, Müller determined the abundances of the elements in the Sun.

In 1964 Margherita Hack (1922–2013), who completed a thesis on Cepheid variable stars at the University of Florence in 1945, was appointed as full professor at the University of Trieste, Italy, where she remained, as the only female astronomy professor in Italy through the time of her retirement in 1992. She was the first Italian woman to direct the Astronomical Observatory of Trieste.

In the 1970s, many astronomy departments at research universities in North America began to hire women. A more correct understanding, however, would be that many of them decided the time had arrived to hire one woman each.

In 1971, both Virginia Trimble (chapter 6) and Erika Böhm-Vitense (1923–2017; PhD, University of Kiel, 1951) landed faculty positions, Trimble at the University of California, Irvine and Böhm-Vitense at the University of Washington. Trimble may well have been the first woman to be the subject of a hiring competition for astronomy faculty positions, entertaining multiple offers both in 1968 and again in 1971.

Then Roberta Humphreys (chapter 7) broke through at the University of Minnesota in 1972, the same year Sandra Moore Faber (b. 1944; PhD, Harvard, 1972) joined the faculty at the University of California, Santa Cruz and Judith Pipher (chapter 12) entered the professorial ranks at the University of Rochester.

Silvia Torres-Peimbert (chapter 8) joined the faculty at Universidad Nacional Autónoma de México in 1973.

Two years later, in 1975, Yale University hired Beatrice Muriel Hill Tinsley (1941–81; PhD, University of Texas, 1966) and Michigan State University offered an appointment to Susan M. Simkin (1940–2021; PhD, University of Wisconsin-Madison, 1966). Tinsley, who likely was the first woman astronomer an Ivy League school had to compete for in a faculty search, chose Yale over an offer from the University of Chicago.

In 1979, Caltech hired Judy Cohen (chapter 11) and Arizona State University added Susan Wyckoff (b. 1941; PhD, Case Western Reserve University, 1967) onto their respective faculties.

Thus, one needed all the fingers on two hands, but no more, to count the number of female astronomers hired into professorships at major research universities in North America in the 1970s.

In 1980, Smith College renewed its century-long commitment to women and astronomy by hiring Suzan Edwards (b. 1951; PhD, University of Hawai'i, 1980).

In 1983, after a several-year hiatus in the hiring of women astronomers onto research university faculties in North America, Cornell hired Martha Haynes (chapter 16) and the University of California, Berkeley added Imke de Pater (b. 1952; PhD, Leiden University, 1980).

In 1984, Princeton University added Jill Knapp (chapter 13), the University of Massachusetts hired both Susan G. Kleinmann (b. 1945; PhD, Rice University, 1972) and Judith Young (1952–2014; PhD, University of Minnesota, 1979, and the daughter of Vera Rubin), Indiana University added Phyllis Lugger (b. 1954; PhD, Harvard, 1982), and Dartmouth College added Mary K. Hudson (b. 1949; PhD, UCLA, 1974).

In 1985, the University of Texas at Austin hired Harriet Dinerstein (b. 1954; PhD, UCSC, 1980); and in 1986, the University of California, Los Angeles added Jean L. Turner (b. 1954; PhD, UC Berkeley, 1984).

Also in 1986, the University of Delaware hired Barbara A. Williams, making her the first Black woman astronomer to hold a professorship in the United States. Williams, who made radio telescope observations of compact clusters of galaxies, was also the first Black woman to earn a PhD in astronomy in the United States, doing so in 1981 at the University of Maryland.[8]

Then, in 1988, Johns Hopkins University added Rosemary Wyse (chapter 20), in 1989 Penn State University hired France Córdova (chapter 17) and Princeton University added Neta Bahcall (chapter 9), and in 1990 Ohio State University hired Kristen Sellgren (b. 1955; PhD, Caltech, 1983). That made a total of fourteen more women hired into professorial appointments at major research universities in the United States through 1990.

Amazingly, through 1990 identifying and counting the women hired onto research university faculties in astronomy was fairly easy because the number remained so small. A few major astronomy programs didn't diversify until the late 1990s or even later, but after 1990 tracking the many women both earning PhDs and receiving faculty appointments in astronomy in the United States becomes much harder because so many more women were earning doctoral degrees, completing postdoctoral appointments, and entering the professoriate.

Among the many scientific accomplishments during this period by these and other women, a few stand out, among them the discovery of pulsars (neutron stars) by S. Jocelyn Bell (later Burnell) (chapter 5) in 1967; the discovery in the late 1960s by Beatrice Tinsley that the luminosities and chemical make-ups of galaxies evolve over cosmic time; Vera Rubin's measurements from the early 1970s onward of the rotation curves of spiral galaxies, which later made clear that nearly all galaxies contain more dark matter than stars and gas, especially in their outskirts; the discovery in 1976 by Sandra Faber and Robert Jackson that the luminosities and orbital speeds of the stars near the centers of galaxies are correlated (the Faber-Jackson relation); and the discovery in 1989 by a team led by Margaret Geller (b. 1947; PhD, Princeton, 1974) and John Huchra of the Great Wall of galaxies that is part of the filamentary structure of the Universe.

Other firsts occurred over these years on the management and leadership side of the profession.

In 1937, Maude Verona Bennot (1892–1982; MS, Northwestern University, 1927) became the acting director of Chicago's Adler Planetarium, a position she would hold until 1945, making her the first woman in the United States, and likely the world, to head a planetarium facility; Bennot was forced out of her job and of a career in astronomy in 1945 when the president of the Chicago Park District board engineered her removal.

In 1959, Nancy Grace Roman was appointed as the first chief of astronomy and relativity in the Office of Space Science at the newly created organization called NASA. She is credited with nurturing the development of several space-based telescopes, including the International Ultraviolet Explorer (IUE), the Cosmic Background Explorer (COBE), and the Hubble Space Telescope (HST), and is widely known as the "mother" of the Hubble. In her honor an infrared telescope due to be launched by NASA in the mid-2020s has been renamed the Nancy Grace Roman Space Telescope.

In 1976, Edith Müller became the first woman to be elected as general secretary of the International Astronomical Union, serving in that role until 1979.

In 1976 Sidney Wolff (chapter 4) was appointed as associate director of the Institute for Astronomy at the University of Hawai'i; she served as acting director in 1983–84, became director of Kitt Peak National Observatory from 1984 until 1987, and then was appointed director of and led the National Optical Astronomy Observatory until 2001.

In 1980, Andrea Kundsin Dupree (b. 1939; PhD, Harvard University, 1968) became the first woman and youngest person to serve as associate director of the Harvard-Smithsonian Center for Astrophysics.

In 1983, Neta Bahcall (chapter 9) was appointed as chief of the General Observer Support Branch for the newly formed Space Telescope Science Institute.

And in 1985 Catherine Cesarsky (chapter 10) became the first woman selected as a principal investigator on a European Space Agency project.

As with the progress over the last several decades toward gender equity in management and leadership opportunities, the awarding of important national and international honors is also evolving. Some examples follow.

1959: E. Margaret Burbidge and her husband, Geoffrey, shared the Helen B. Warner Prize for Astronomy, given annually since 1954 by the American Astronomical Society "for a significant contribution to observational or theoretical astronomy during the five years preceding the award." No other woman had previously received this honor, nor would any other woman receive this honor again for nearly fifty years, until Sara Seager (chapter 33) did so in 2007.

1976: Cecilia Payne-Gaposchkin was recognized with the Henry Norris Russell Lectureship, given annually since 1946 by the American Astronomical Society "on the basis of a lifetime of eminence in astronomical research." She was the first woman to present this lecture. The others who have been recognized as Russell Lecturers are E. Margaret

Burbidge (1984), Vera Rubin (1994), Margaret Geller (2010), Sandra Faber (2011), and Ann Boesgaard (2019, chapter 3).

1989: Jocelyn Bell Burnell (chapter 5) received the Herschel Medal, awarded by the Royal Astronomical Society annually since 1974 for "investigations of outstanding merit in observational astrophysics." As of 2021, no other woman has received this award.

1989: Martha P. Haynes (chapter 16) became only the second woman to receive the Henry Draper Medal, awarded every four years since 1886 by the U.S. National Academy of Sciences "for investigations in astronomical physics." She shared this prize with Riccardo Giovanelli for their work in constructing "the first three-dimensional view of some of the remarkable large-scale filamentary structures of our visible Universe." Annie Jump Cannon was the first woman to receive the Draper Medal, nearly sixty years before, in 1931. No other women have won this award.

1994: Edith Müller became the first woman to receive the Prix Jules-Janssen, for her work in solar spectroscopy. The Prix Jules-Janssen has been awarded annually since 1897 by the Société Astronomique de France for international distinction. Elizabeth Nesme-Ribes (1942–96) shared the award (posthumously) in 1997, as did Thérèse Encrenaz (b. 1946) in 2007. In 2009, Catherine Cesarsky (chapter 10) was the fourth woman so recognized; Suzanne V. Débarbat (b. 1928) received the award in 2013, Suzy Collin-Zahn (b. 1938) did so in 2015, as did Ewine van Dishoeck (chapter 22) in 2020.

1996: Vera Rubin was awarded the Gold Medal by the Royal Astronomical Society, this society's highest honor, for her work in studying galaxy rotation curves. She was the first female awardee since Caroline Herschel received her Gold Medal 168 years earlier. In 2005, E. Margaret Burbidge and Geoffrey Burbidge shared the Gold Medal. In 2020 Sandra Faber and in 2021 Jocelyn Bell Burnell (chapter 5) received the Gold Medal.

2003: Virginia Trimble (chapter 6) became the first person designated as President of a second International Astronomical Union Division (XII, Union-Wide Activities), after serving as President of Division VIII (Galaxies and Cosmology) from 2000 to 2003.

2004: Vera Rubin was the first woman to receive the James Craig Watson Medal, awarded every two years since 1887 by the U.S. National Academy of Sciences "for outstanding contributions to the science of astronomy," for "her seminal observations of dark matter in galaxies, large-scale relative motions of galaxies, and for generous mentoring of young astronomers, men and women." Margaret Geller was honored with this award in 2010 ("for her role in critical discoveries concerning the large-scale structure of the Universe, for her insightful analyses of galaxies in groups and clusters, and for her being a model in mentoring young scientists"), as were Ewine van Dishoeck in 2018 (chapter 22, "for her many important contributions to the field of molecular astrophysics and astrochemistry") and Lisa Kewley (b. 1974) in 2020 ("for her fundamental contributions to our understanding of galaxy collisions, cosmic chemical abundances, galactic energetics, and the star-formation history of galaxies").

2006: Catherine Cesarsky (chapter 10) was elected President of the International Astronomical Union (IAU) for the 2006–9 term. She was the first woman to lead the IAU since it was founded in 1919. Silvia Torres-Peimbert (2015–18; chapter 8) was the second, Ewine van Dishoeck (2018–21; chapter 22) the third, and Debra Elmegreen (b. 1952; 2021–24) the fourth.

2012: Jane Luu was the first woman to win or share the Shaw Prize in Astronomy, which was established in 2002. She shared this prize with David Jewitt for their discovery and characterization of trans-Neptunian bodies. In 2021, Victoria M. Kaspi and Chryssa Kouveliotou became the second and third women so honored when they shared the Shaw Prize in Astronomy for their contributions toward our understanding of magnetars.

2017: Cathie Clarke (chapter 25) was the first woman recipient of the Eddington Medal in Theoretical Astrophysics, awarded annually since 1953 by the Royal Astronomical Society for "investigations of outstanding merit in theoretical astrophysics." In 2018, Claudia Maraston (b. 1966) was the second woman awarded the Eddington Medal, and in 2021 Hiranya Peiris (chapter 34) became the third.

2019: E. Margaret Burbidge was selected as the Inaugural Fellow for the newly created Fellows of the American Astronomical Society.

2020: Andrea Mia Ghez became the first woman to earn the Nobel Prize in Physics for work in astrophysics, sharing the prize with Reinhard Genzel "for the discovery of a supermassive compact object at the center of our Galaxy" and with Roger Penrose, for his work on the formation of black holes.

Notes

1. Emily Elisabeth Dobbin, Williamina S. Fleming, Lucy Ames Frost, Caroline Furness, Ida Griffiths, Antonia C. Maury, Mable C. Stevens, Sarah F. Whiting, Mary W. Whitney, Elva G. Wolffe, and Ida Woods.

2. Leah Brown Allen, Harriet W. Bigelow, Louise Brown, Mary E. Byrd, Mary Ross Calvert, Annie Cannon, Mrs. Henry Draper, Margaret Harwood, Flora E. Harpham, Ellen Hayes, Mary Murray Hopkins, Edith Kast, Eleanor A. Lamson, Jennie B. Lasby, Henrietta Swan Leavitt, Mary Proctor, Helen M. Swartz, Ida W. Whiteside, Mrs. John C. Whitin, and Anne S. Young.

3. The concept that all stars have the same chemical composition was accepted immediately, but the dominance of hydrogen in stars was only embraced much later.

4. The best modern value for this distance is 2.5 million light years (765 kpc).

5. In chronological order: Margo Friedel Aller, Andrea Knudsen Dupree, Barbara Welther, Gretchen Luft Hagen Harris, Nancy Houk, Martha Safford Hanner, Diane Reeve Moorhead, Nancy Remage Evans, Catherine Doremus Garmany, Jane Turner, Jean Warren Goad, Karen Alper Castle, Marcia Keyes Lebofsky Rieke, Judy Karpen, Karen Kwitter, Esther Hu, Bonnie Buratti, Harriet Dinerstein, Melissa McGrath, Constance Phillips Walker, John Briggs, Deborah Crocker, Edward Morgan, and Karen Meech.

6. By 1958, Williams Vyssotsky was no longer involved in astronomy; she published her last paper in 1951.

7. At the end of World War II, when the USSR pushed its borders westward, many of the Polish faculty at Vilnius University relocated westward to Torun and started a new university.

8. Williams would be followed by Mercedes T. Richards (1986, University of Toronto), Jarita C. Holbrook (1997, UC Santa Cruz), and Dara Norman (1999, University of Washington; chapter 32) as Black women earning PhDs in astronomy.

Chapter 2

Anne Pyne Cowley (PhD, 1963)

Navigating My Life with the Stars

Anne Cowley, now Professor Emerita in the School of Earth and Space Exploration at Arizona State University, is an expert on the spectroscopy and evolution of stars. In 1983 she and her two Canadian colleagues discovered the first known black hole outside of the Milky Way. In 1972 she was one of the co-founders of the AAS working group on the status of women in astronomy, which eventually grew into the CSWA. She served on numerous AAS committees and was elected to the AAS Council, as a vice president, and to Chair of the Publications Board. She also served as co-editor of the *PASP* (1998–2005). In 2020 she was selected as an AAS Legacy Fellow.

I grew up in Marblehead, a small seaside town north of Boston known for its deep harbor. Yachting and commercial fishing were major activities there. My friends and I loved swimming and playing at nearby Rocky Beach. We all walked over a mile to school every day and took our lunches. The kids whose fathers were fishermen often hid their lunches if they had only a boiled lobster. Sometimes they would ask to swap with us for our peanut butter sandwiches. Little did they realize that in a few years a lobster would be an expensive treat!

Sailing was a big activity in our family. I had my own little sailing dinghy and competed in yacht club races with it. My father owned a

FIGURE 2.1. Anne Cowley with husband, David Hartwick, having lunch on *Stella Maris.*

classic Herreshoff in which we often sailed together. For many summers I went to a girls' camp in Maine. I loved all the activities of swimming, boating, taking long canoe trips on the Saco River, climbing the Presidential Range, and many other adventures. Even today I find outdoor activities a great pleasure. My husband and I have owned our sailboat, *Stella Maris,* for decades and have taken many cruises in her while exploring the many small islands off the British Columbia coast.

When I was growing up in the 1940s and 1950s, I was totally unaware of how women were treated in the workplace. My mother and her friends never worked. They were "stay-at-home moms." All my teachers were women, as that was one of the few acceptable jobs for women. I attended an all-girls high school, then went on to a women's college (Wellesley College). I graduated in 1959 with a degree in astronomy. This meant that I had little opportunity to see how men were treated differently from women.

While at Wellesley I took an astronomy course as my required science. I immediately loved it. Dr. Sarah Hill was the head of that small department. She was a smart and interesting woman. It was from her I learned that it was possible for a woman to be a scientist. She gave me some research projects to work on, in addition to my classwork. My first refereed paper was published in the *Astronomical Journal* with her help and encouragement while I was still an undergraduate.

During my undergraduate years I held a variety of summer jobs. One was at the U.S. Naval Observatory in Washington, D.C., in 1957. Because at that time the Chief of Naval Operations lived on the beautiful Observatory grounds, we had to check in each day at a guard station, but security was simple. When we arrived, the students were interviewed to determine in which division we would work. I was sent to talk to the head of the time-service department. During my interview this man spent the whole time looking at my legs, which I guess he liked since I was assigned to work in his division. It wasn't a glamorous job, but I enjoyed working there.

When I decided to go to graduate school, my father (who worked at Harvard) arranged for me to talk with Dr. Cecilia Payne-Gaposchkin. I think my dad hoped I would go to Harvard, although at that time women actually attended Radcliffe. Instead, she suggested I apply to the University of Michigan where "her good friend Leo Goldberg was in charge." I was accepted there and moved to Ann Arbor in 1959. When I arrived, I was surprised to find that one of the outstanding astronomers there (Dr. Edith Müller) was not on the faculty but only a research scientist. After a few years Edith went back to Switzerland to take real faculty jobs there. In my opinion, that was a great loss for UofM.

While at UofM I married a fellow astronomy student. At that time women were required by the State of Michigan to take their husband's name. When I asked for an exception, I was told that after a few months and paying a fee I could go to court and change back to my maiden name. Since I had to change my driver's license, bank account, and passport to my new name, it didn't seem worth the trouble to do it all over again. Hence, any "Jane Doe" who married a "John Smith" automatically became "Mrs. John Smith," which in no way showed who she was as a person. She was just an adjunct to Mr. Smith. That really annoyed me, and it still does!

Another eye-opener happened in 1963 when my husband and I both graduated from Michigan with our PhDs. His supervisor just called up a colleague at the University of Chicago and, bingo, he had a faculty job. My advisor did nothing. Back then, there were no advertised jobs, no applications, no interviews, and no giving professional talks while seeking a job. The "old boys' network" was the way astronomers got jobs.

We moved to Yerkes Observatory in Williams Bay, Wisconsin, without a job for me. I asked around at the Observatory and was hired as a researcher by one of the faculty members. I was paid $5,000 a year, about half what the graduate students there were earning. I asked my supervisor why he was paying me so little. He said, "You have a husband who supports you, so you only need money for perfume and fur coats," neither of which I even wanted! At least it was a job where I could use the telescopes at McDonald Observatory in Texas, write astronomy papers, and apply for grants to pay for travel and publication fees.

At the first colloquium I attended, I went into the room and took an empty chair. I could see the graduate students giggling, but I didn't know what was happening. Soon Dr. Chandrasekhar came into the room and glowered at me. Obviously, I had taken his "regular" chair. At the next colloquium I chose a different place to sit. At Yerkes, faculty wives were expected to make cookies and "serve tea" before the colloquia. I was the only wife who was an astronomer. Much to the consternation of some of the staff, I didn't agree to do this. My kids would tell you it was a good thing I didn't even try as I am notorious for burning cookies. However, I mostly enjoyed working at such a historical observatory.

In 1966 we started looking around for other jobs. We were both made faculty offers at two institutions, although these were oral agreements, as that was how hiring was handled. We thought these offers were good opportunities since "the two-body problem" was already quite clear. We decided to go back to Michigan, although we had to agree to spend the first year at the solar observatory near Pontiac, Michigan. When we arrived, we found out that only my husband would be paid. He was shown his office, while I was told there wasn't an office for me. My husband told them he would work from home and I could have his office. Then the story changed; the "queen bee" in charge said that I could not have an office because "Dr. McMath never allowed pregnant women to work at the observatory." Eventually I did get a desk in a shared room. Needless to say, that year was stressful.

At the end of the year we moved to Ann Arbor. The chairman said there still wasn't money in the budget for me, even though he had promised us two paid faculty positions. I was told I could apply for grants, but I had to agree to have some faculty member be the primary applicant. After I got an NSF grant, they then decided that my (low) salary would be set by the department, rather than by the amount specified in the grant. I wasn't happy with the arrangement, but at least I had an office. The astronomy building only had enough telephone connections for two adjacent offices to share one. The man in the office next to me was delighted. He always let the phone ring until I picked it up, so if it was for him the caller would think he had a secretary.

After several years, during which no faculty position for me had materialized, I complained to the university about my position and salary. With the help of a lawyer working on women's rights in Michigan, we filed a lawsuit that went to a three-lawyer committee. They found in my favor, but the department still refused to give me the promised faculty job. When I asked one faculty member why, he said, "I don't want people making departmental decisions in bed." In a sassy way, I told him that I pitied his wife. I later learned that the astronomy department faculty had voted in 1969 to start an anti-nepotism policy to prevent family members being hired, although the university itself no longer had such a policy.

At that time, I had no idea that so many other UofM women were having similar difficulties with low salaries and lack of promotions. Women's problems at Michigan are nicely documented in a recent book by Sara Fitzgerald entitled *Conquering Heroines: How Women Fought Sex Bias at Michigan and Paved the Way for Title IX*. The University of Michigan certainly wasn't alone among major research universities in denying women equal pay and well-deserved faculty positions.

I had joined the American Astronomical Society (AAS) as a student member in 1960. This led to my becoming involved in AAS activities, not only attending the meetings but also serving on various committees. The AAS has numerous prizes and awards, the most prestigious being the Henry Norris Russell Lectureship. By 1970 no woman had ever won that award. Instead there was the Annie Jump Cannon Award, which specified that only women could win it. Hence, any excellent woman was given the less important Cannon Award rather than the Russell Lectureship.

This changed in 1971 when Dr. Margaret Burbidge was selected as the winner of the Cannon Award, but she declined to accept it. She wrote, "It's high time that discrimination in favor of, as well as against, women in professional life be removed, and a prize restricted to women is in this category." Of course, the AAS was surprised, but they quickly put together an ad hoc group to resolve this issue. Some people wanted the prize to stay unchanged, others wanted to open it to men, others said we should get rid of it. Eventually it was agreed that the award would become a grant that young women astronomers could apply for. It was handed to the American Association of University Women (AAUW) to administer, with advice from the AAS.

In 1972 the AAS set up a Working Group on the Status of Women. This later became the permanent Committee on the Status of Women in Astronomy (CSWA). Four women were appointed to serve on the Working Group: Roberta Humphreys, Beverly Lynds, Vera Rubin, and me. We carried out a very complete survey of the number of AAS women members over time, the percentages of those with academic roles, and other relevant information. Our *Report to the Council of the AAS from the Working Group on the Status of Women in Astronomy* was published in the *Bulletin of the American Astronomical Society* in 1974. It showed how few women were professors, award winners, or in other important positions. The Russell Lectureship was finally awarded to Dr. Cecilia Payne-Gaposchkin in 1976 and to Margaret Burbidge in 1984. Since that time four additional women have won it.

In the 1970s and 1980s my Canadian colleagues, David Crampton and John Hutchings, and I were studying X-ray binary star systems. The optical counterparts of these X-ray bright objects were poorly known since the X-ray satellites didn't provide very accurate positions. It took lots of observing time to find the optical identification so that we could study their physical properties. Many of these X-ray sources lie in the southern hemisphere, so we often went to Chile to make spectroscopic observations with the telescopes at Cerro Tololo Inter-American Observatory (CTIO).

Many of these X-ray sources turned out to be binary systems containing a hot massive star and a very compact neutron star orbiting each other. Our spectroscopic data on the binary system LMCX-3 showed unexpected large changes in the radial velocity (toward and away from the observer) of the hot star. We decided that during our next trip to CTIO we would obtain further observations of this X-ray bright system. Almost immediately the new data showed that the visible hot star must be orbiting something much more massive than a neutron star. We had discovered the second known black hole and the first one known outside of our Milky Way Galaxy. Our discovery received a great deal of press coverage.

In 1983, after too many years at UofM with no resolution of the "promised faculty job," I received an offer of a faculty position from Arizona State University (ASU). I was called to meet with a UofM dean. Perhaps the university administration had also noticed the press reports about LMCX-3. He asked why I wanted to leave "the Harvard of the Midwest." He also said that they would immediately make me a full professor. However, I had been fooled by them before, so I headed to Tempe, Arizona, and never looked back.

Things at ASU were much better for me. I was appointed a full professor at a fair salary. I have no idea if the male faculty members in the Department of Physics & Astronomy were paid more than I was, but I didn't care. I finally had a real position; I could teach, apply for grants, observe at the national observatories, carry out research, work with students, and publish papers in refereed journals.

I attended many astronomy meetings, both in the United States and abroad, to present our latest research and learn what others in my field were working on. Often at scientific meetings the women would meet in the "ladies' room" to chat and warn each other about which men to avoid. At a meeting in Italy I was followed into an elevator by a senior elected AAS officer whom I knew slightly. As soon as the door closed, he started sexually assaulting me. When the door finally opened, I ran down the hall to a friend's room, not wanting him to know where my room was. I never dared to complain to anyone about this since I knew he could ruin my career if I made it public. It was years before I told anyone except my closest friends about this incident.

Over the years I worked with the AAS in various capacities. I was elected to the Council (1979) and later as a vice president (1983). I served on many AAS committees, on the Publications Board, and as a scientific editor for the *Astrophysical Journal*. I also co-edited the *Publications of the Astronomical Society of the Pacific* (*PASP*) for a number of years. Later I was elected Chair of the AAS Publications Board. I worked with a wonderful group of scientists making important changes to the way papers submitted to the AAS journals are handled.

At any workplace there are always things that don't go exactly the way one would like. Overall, I felt fairly treated in the Department of Physics & Astronomy at ASU. Eventually the astronomers were moved out of the P&A Department and merged with a large group of geologists in the School of Earth and Space Exploration (SESE). When I was about a year from retirement, just before the term was about to start, I accidently discovered that I had been assigned to teach an extra course by one of the geologists. I called him from the Dominion Astrophysical Observatory (DAO) in Canada where I have long held a visiting position. I asked why he didn't talk with me about this. He said he needed someone to teach that course. I guess he thought that since I am a woman, he could just assign it to me without any discussion. I told him that he would now have to find someone to teach two extra courses as I was going to retire immediately. I'm sure he wasn't happy about having to deal with that. I retired, joined ASU's Emeritus College, and moved to a much smaller office from which I could continue to carry out research. It was an unpleasant ending to an otherwise okay part of my career.

Looking back on my career in astronomy, there has been much that I really enjoyed. I was especially pleased to be selected as one of the

FIGURE 2.2. 2014 AAS Publications Board meeting in Dallas, TX. People pictured, starting on the lower left and going clockwise, are: Jay Moody, Ed Guinan, Anne Cowley, Julie Steffen (Director of Publishing), Barb Kern (Library Representative), Roberto Conti, Eileen Friel, Dieter Hartmann, and Allen Shafter. Photo by Kevin Marvel (AAS Executive Officer).

inaugural AAS Fellows in 2020. There are other parts that are difficult to recall because of the unfairness I, and many other women, encountered. I believe that many things are better now for women in astronomy. More are in leadership roles, have appropriate jobs, and earn good salaries. However, I worry that even now not all men in our field understand how much more is expected of women than of their male counterparts. I hope progress continues and makes the field fairer for all who work as astronomers.

Chapter 3

Ann Merchant Boesgaard (PhD, 1966)

Making Things Work

Ann Boesgaard, now Professor Emerita of Astronomy at the University of Hawai'i, is an expert in the structure and evolution of stars, in particular the measurement of the lithium and beryllium content in the atmospheres of stars. She was the first woman to be awarded time on the Mount Wilson 100-inch telescope, the first tenure-track female faculty member in astronomy at Hawai'i and the IfA, and the first female President of the ASP. She was a Guggenheim Fellow, was the recipient of the Muhlmann Prize from the ASP, received the Henry Norris Russell Lectureship from the AAS in 2019, and is a Legacy Fellow of the AAS. Her asteroid is 7804 Boesgaard.

The pre-dawn night sky was clear. My husband and I were on the roof of our house in Hawai'i at 5 A.M. looking for *Apollo 8* on December 21, 1968. The launch was flawless; the first humans were leaving Earth and going to the Moon! After the spacecraft had been tracked in the Florida dawn sky, the announcer, almost as an afterthought, told the TV audience that for people in Hawai'i, it would take off for the Moon right over our heads.

This would happen before sunrise, but *Apollo 8* was at a very high altitude and would be illuminated by full sunshine. We climbed onto our roof and lay down on the sloped wooden shingles with a clear view of the full sky. We found the spacecraft trundling along in the southwestern sky. When it was over our house, it accelerated to 24,000 miles per hour to escape Earth's gravity. Watching *Apollo 8* change in speed was incredible! As it took off for the cosmos, it created a shock wave in the upper atmosphere that spread out behind the spacecraft as a huge conical yellow-orange glow. A truly exhilarating experience!

My thirteen-year-old self had wanted to go to the Moon, but my gender, eyesight, and lack of test pilot experience precluded that. Dramatic changes in social norms have taken place since the 1940s and 1950s when I was growing up. Men were breadwinners and women were homemakers and child-rearers. My father left when I was five, and my mother became a single parent long before the term was invented. After their divorce, our nuclear family consisted of me, my older sister, our mother, grandmother, and great-aunt. We lived in a five-bedroom house in a middle-class neighborhood in Rochester, New York, thanks to the generosity of my mother's mother. My mother, who was a math major at Vassar College, worked in the Controller's Division of Eastman Kodak. That household of females sent a subtle message to me that women should not depend on men to support them.

Our public grade school was two blocks from home, and we walked to school in sunshine, rain, and snow. All our teachers and the school principal were women. At that time few women worked outside the home, and those who did were primarily teachers, nurses, and secretaries.

In kindergarten, I learned the multiplication tables while rehearsing them with my sister as she was learning them in third grade; I suspect this helped train the mathematical part of my developing brain. A weekly science program on then-new FM radio in fifth, sixth, and seventh grades stimulated the science part. Half-hour programs on specific topics were accompanied by questions to which we could discover the answers while listening. In addition we were given lists of activities to do; I always did all of them.

Around the age of seven I noticed adults asked little boys what they wanted to be when they grew up, but they asked little girls how many children they wanted to have! I became a feminist then without knowing the term. I resented that girls in grade school had to take sewing and cooking and that we had to give the boys the products of our cookie-making class.

I was hooked on astronomy from a young age. My mother taught me the constellations, and the first elective badge I earned in Girl Scouts was for astronomy.

After applying to colleges, I chose Mount Holyoke, in part because of its excellent reputation in science. One sharp memory I have from my freshman year was my thrilling pre-dawn expedition to the roof of my dormitory in October 1957 in order to view the Russian Sputnik, the first orbiting artificial satellite. For my last two years in college, astronomy was taught by faculty of the Four College Astronomy Department composed of Mount Holyoke, Smith, Amherst, and the University of Massachusetts. We traveled to each other's campuses for co-ed classes. I did my senior honors thesis with Dr. Robert F. Howard at UMass on solar rotation, using observations he had made at the solar tower at Mount Wilson Observatory.

I wanted to go to graduate school in California where the big telescopes were. I accepted an offer from UC Berkeley and drove across the country, spending the summer of 1961 working for Dr. Jesse Greenstein at Caltech. There I learned about some amazing chemical elements with names that impressed me, such as dysprosium and gadolinium. We studied the spectrum of Zeta Capricorni, a yellow giant star enriched in ionized barium.

During my education and my summer work with Greenstein I became intrigued by how much information spectra could reveal about stars and galaxies. Not only was there more information to be gained from spectroscopy than from photography, but also the observing conditions were more forgiving because we could work even when the sky was a bit cloudy.

Although the Universe is composed of 98 percent hydrogen and helium, the other chemical elements are of key importance to its structure and evolution and, of course, to life. Lithium had been found in young, Sunlike stars; in fact, the presence of Li became a defining characteristic for these young stars. Strong Li lines were seen in some cool giant stars also, so I investigated the prevalence and amount of Li in red giants and supergiants during my thesis research with George Herbig.

Part of my work with Herbig involved helping align a five-mirror system, as part of the Lick Observatory 120-inch telescope, which would allow observers to study stars in the southern sky. During that process, I met Hans Boesgaard, a good-looking, charming, Danish-American engineer who was working at Lick. As fate would have it, Hawai'i soon came into the picture. He went on a sailboat race from California to Honolulu in July 1965 and I flew there to meet him. Two months later he read about a NASA telescope to be built in Hawai'i and was very interested. In May 1966, four days after the final oral exam for my PhD degree, Hans and I were married. Our lives and careers became truly intertwined.

John Jefferies, founding director of the University of Hawai'i (UH) Institute for Astronomy (IfA), remembering those early times for astronomy in Hawai'i, wrote, "I desperately needed a mechanical engineer, preferably one with experience in large telescopes [to help with the design and construction of the telescope and] to create a capability for building the instrumentation that would be needed for our burgeoning program." Hans was that engineer. He accepted an offer from Jefferies to be the chief engineer for the telescope project to build an observatory on top of Maunakea.

Like students of George Herbig before me, I applied for a Carnegie Fellowship to work as a postdoc at Mount Wilson and Palomar Observatories in Pasadena. They had never had a female postdoc and did not offer one to me either. Instead, I went to work for Greenstein at Caltech on his stellar abundance project.

Hans and I began a "commuter marriage," another term that had not yet been invented. I was based in Pasadena, he in Honolulu. He traveled to California to work with the telescope designers and manufacturers. I was able to spend half a month in Hawai'i and the other observing at Mount Wilson, so we were together more than 75 percent of the time.

Typically, I had three nights each month using the Mount Wilson 100-inch telescope, beginning in September 1966. I was the first woman assigned telescope time there in her own name. (Margaret Burbidge had done her observing with time officially allocated to her husband, Geoffrey, a theorist.) It was a given that no accommodations existed for women astronomers on Mount Wilson. The dormitory, called "the Monastery," had two floors and a bathroom on each floor, but one floor was for male solar observers and one was for male nighttime observers.

I was assigned to stay in Kapteyn Cottage, a small cabin at the summit not far from the 100-inch dome, with no heat and no hot water. For cooking, it had a wood stove called "Dudley." The cottage was used primarily during the summer for observers' families. It had one bedroom with a double bed and a narrow cot in the living room next to Dudley. Not wishing to build a fire after a night of observing, I requested an electric blanket, which was provided for the cot. The electric wiring system for the mountain was ancient, so the blanket had to be at the top level to supply warmth. And so it went, monthly, through the fall, winter, spring, and into the summer of that postdoctoral year for me. In the spring, I wrote and circulated a petition to the Observatory Committee explaining that female observers were effectively barred from some scientifically interesting projects because of the housing situation. For example, long-term, nightly monitoring projects were not possible due to the lack of proper bathing facilities. The Caltech astronomy faculty signed that petition and it was taken seriously enough to affect the female graduate students at Caltech, who were then allowed to stay in the Stewardesses' Cottage when they stayed overnight on Mount Wilson.

Traditions abounded at Mount Wilson. One was that the 100-inch observer would sit at the head of the table at dinnertime and ring the bell when the stewardesses were to come clear the plates and serve the next course. The 60-inch observer would sit at "his" right and the solar observer at "his" left. There were linen placemats and napkins; the permanent staff had individualized napkin rings. Midnight lunch was sacrosanct, so astronomers were expected to initiate long time exposures so that the night assistant could retreat to the midnight lunch shack and heat up his food. The night assistant, in turn, would keep the telescope pointed at the star under study while the astronomer left for midnight lunch, but often astronomers ate sandwiches while guiding on the star.

In the meantime, at UH Jefferies was building up an academic astronomy program and forming the IfA. He offered me an assistant professor position, which I accepted with alacrity. I wanted to actually live with my husband in our new home with its sweeping view of the Pacific Ocean. And I was getting tired of those long flights. Furthermore, I had collected considerable new data, and working with those spectra would keep me busy until the new telescope was ready. I was invited to give a talk on beryllium (Be) at the IAU General Assembly in Prague and then returned home to Hawaiʻi to start my new career of teaching and research at UH in September 1967.

Those were exciting times in Hawaiʻi, and we were pioneers. A rigorous site survey had indicated that Maunakea, at an elevation of 14,000 feet on the Island of Hawaiʻi, was superb for *nighttime* astronomy, and so a decision was made to build a permanent observatory there. Jefferies and his team made several very important decisions. For example, the telescope would *not* be put on the actual summit of Maunakea, which is the highest point in the Pacific, and the road to the top would *not* go past Lake Waiau, an alpine lake revered by Hawaiʻians and one of the highest in the United States. I was able to help in the alignment of the telescope instruments, and several of us were involved in remodeling the construction workers' barracks into sleeping quarters for astronomers. Those rooms were then named after the stellar spectral sequence known to all astronomers: O, B, A, F, G, K, M, R, N, and S, with two smaller rooms named gas and dust.

Governor John A. Burns and university president Harlan Cleveland attended the official dedication of the 88-inch telescope on June 26, 1970, held in the telescope dome. Jefferies insisted that we dress in our Sunday best for the occasion, that is, suits and ties or long muumuus and stockings. This telescope was the first computer-controlled telescope.

In order to safely observe at this high altitude, we first had to acclimatize overnight at the midlevel living quarters at 9,200 feet. This minimized the effects of high altitude on the human body (including on mental acuity). After a night of observing, we returned there—a twenty-minute drive—for breakfast and daytime sleeping before the next night.

During the 1970s, I was involved in both undergraduate and graduate teaching in our growing academic program. My research focused on the light element content of stars and the atmospheres of giant stars, with observations from the Maunakea telescopes and the UV telescope of the Copernicus satellite. I was elected to the Board of Directors of the Astronomical Society of the Pacific and served as (the first female) president in 1977–79. During 1976–78 Margaret Burbidge was elected the first woman President of the American Astronomical Society (AAS); thus, in January 1977 when the winter meeting of the AAS was held in Honolulu, the local newspaper remarked on the first female presidents of the two American astronomical societies. I later served on the AAS Council from 1978 to 1981.

For my first sabbatical leave in 1973, I sought a destination where Hans would want to come to join me, and so I spent four months in Paris at the Institut d'Astrophysique as a NATO Senior Science Fellow. By that time the French and Canadians had decided to jointly put a telescope on Maunakea, and the French wanted Hans as a consultant. So he commuted, Honolulu to Paris, spending three weeks at a time in each city. By the end of the 1970s, three large telescopes—NASA's Infrared Telescope Facility (IRTF), the Canada-France-Hawaii Telescope (CFHT), and United Kingdom Infrared Telescope (UKIRT)—had been dedicated on Maunakea.

Hans was primarily involved with the IRTF during the 1970s. In 1978 he spent two months in Lecco, Italy, where the telescope was being fabricated. So where, near there, could I go to do my research? The Nice Observatory in France was the best choice for me, albeit four hundred kilometers from Lecco. We arranged to get together every weekend, first in Nice on the French Riviera. Next, we met in the Genoa train station and visited Pisa and Florence. Then we flew to Rome to sightsee there. That was followed by a weekend exploring Venice, after which I flew to Paris for a week. And so, that next weekend Hans joined me in Paris, where he met me at our favorite café. Très romantique. Our final weekend was spent in Cannes.

My next sabbatical leave was to the Center for Astrophysics at Harvard in 1980–81. That time, Hans worked on a project at MIT. While we were there, I got a letter from the president of my alma mater, Mount Holyoke College. They were awarding me an honorary Doctor of Science degree in May 1981 and asked that I come to the graduation ceremony to receive it. I was delighted! (One of my fellow honorees was civil rights legend Rosa Parks.) In the audience at commencement was my mentor Jesse Greenstein, and during the applause he stood up and shouted, "And I hired her!"

Having finished the UH 88-inch telescope and the IRFT and participated as a consultant for both the CFHT and UKIRT, Hans next went to Berkeley to work with Jerry Nelson and his team on the Ten-Meter Telescope project. I was back to commuting, this time from Honolulu to Berkeley. The Ten-Meter Telescope project was scheduled to move to UC Santa Cruz in the spring of 1985 and so I arranged for a visiting position there for the spring semester. But, as it turned out, the headquarters did not move to Santa Cruz because Caltech became a partner in what would become the Keck I telescope. Instead, the project headquarters moved to Pasadena in July 1985, but remained in Berkeley until that time. So now I was in Santa Cruz, and commuting again—this time by car. I was at UC Santa Cruz Mondays through Thursdays and in Berkeley with Hans on the other days.

At that time I was working with spectra I had taken of 23 stars in the Hyades cluster. I had determined their Li abundances and was with Hans in his office at Lawrence Berkeley Laboratory. Although I had access to their computer, it was overused and thus very slow, so I was using graph paper and a pencil to plot Li abundances versus stellar temperature for these stars. Eventually, a stunning pattern emerged: large Li deficiencies showed up in a narrow range of stellar temperatures and masses.

The Keck project headquarters moved to Pasadena in July 1985. I knew Hans was going to be working there for three-plus years. Once again, I returned to Caltech, but now as an NSF visiting professor, a Guggenheim Fellow, and on a sabbatical leave from UH. Under those auspices I was now able to apply for time on the Palomar 200-inch telescope. There, I advanced my studies of Li in other star clusters. Because astronomers can determine the ages of star clusters, they provide special

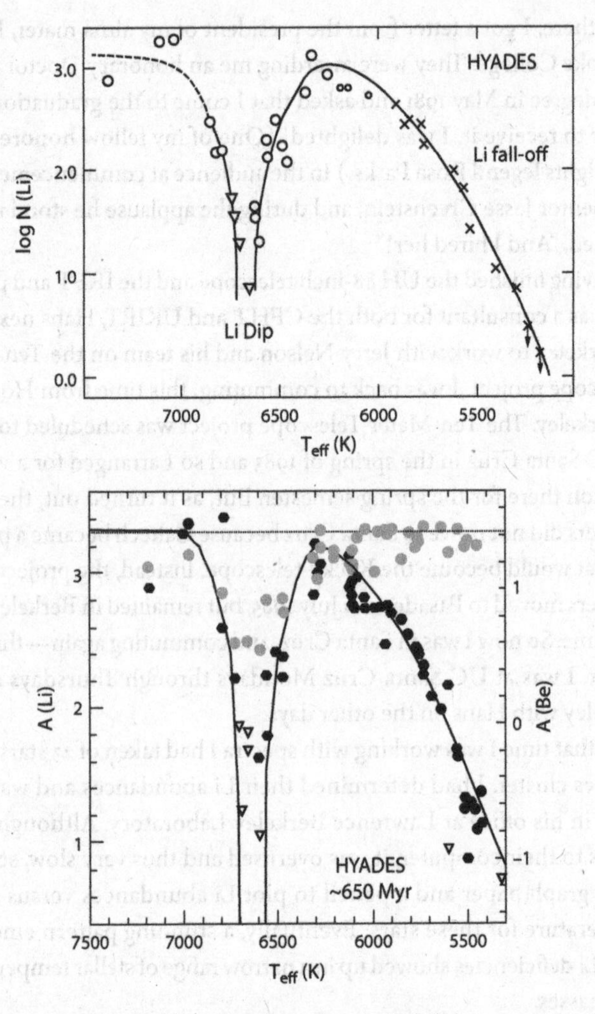

FIGURE 3.1. Top: The log of the Li abundance relative to log N(H) = 12.00 as a function of surface temperature for Hyades dwarf stars from Boesgaard and Tripicco (*Astrophysical Journal* [1986]: 302L). The large Li depletions occur in stars about 10–20 percent more massive than the Sun. Bottom: The log of the Li and Be abundances in the Hyades, scaled to their respective meteoritic abundances, from Boesgaard et al. (*Astrophysical Journal* [2016]: 830). The black symbols are Li, with the open triangles indicating upper limits on Li; the gray symbols are for Be. The Be dip is much shallower than the Li dip and there is no Be fall-off in the cooler stars. Modern stellar models can account for these results. The Li atoms are more susceptible to destruction inside stars as the surface amount indicates.

testing grounds for the study of how stars change as they age, and this is what I set out to do. At the time, I was supervising two Caltech graduate students and several summer undergrads, all of them female except one of the graduate students.

Meanwhile, Hans and I bought a five-acre lot in a new development about a twenty-minute drive from the future Keck headquarters, in Waimea, Hawai'i. We designed our new home, which had exquisite views of the Pacific Ocean, Maunakea (naturally), Maunaloa, Hualalai, and Haleakala on Maui. The house was finished in late 1988 just as the project headquarters moved to Waimea. At the IfA, our teaching responsibilities in astronomy were one course a year. The semester that I taught, I taught a Tuesday-Thursday class; I would fly to Honolulu Tuesday mornings and back to the Island of Hawai'i on Thursday afternoons.

During the 1990s and beyond I measured the abundances of the light elements Li and Be, primarily at Keck I, and boron (B) with the Hubble Space Telescope (HST). I also determined the amounts of carbon, oxygen, and a handful of other elements in an array of solar neighborhood stars. Such observations improved our understanding of the internal structure of stars and how they change with age; they also improved our understanding of the chemical evolution of our Galaxy.

Hans continued to work on the Keck II telescope through its completion in 1996. After that he was engaged in several intriguing and exciting consulting jobs using his engineering skills and telescope credentials. I was trying to push the frontiers of stellar spectroscopy, much of it my studies of Li, Be, and B, and most of it with the Keck I telescope, that Hans had worked on for ten years.

The Li measurements (very low abundances) combined with the Be results (low, but not as low) implied that our understanding of the astrophysics of how stars work was incomplete. What we learned is that as stars age, they spin more slowly, and as stellar rotation slows, Li and Be more effectively mix downward, where they are destroyed in nuclear fusion reactions.

A decade later, I continued this study with more sensitive instruments. The new data revealed that as galaxies age, the overall content of elements heavier than helium (called metals by astronomers) increases. The oldest stars are poor in metals while the younger ones are richer, having formed after previous generations of stars enriched the interstellar medium with metals. Thus the metal content of a star is an indicator of the age of a star. In our analysis we determined the amounts of Be along with iron and several other elements, such as oxygen, titanium, and magnesium. These data revealed the differences in the stellar populations that were native to our Milky Way and those that were accreted by the Milky Way from nearby systems. As expected, we found no evidence of a contribution to Be from the Big Bang. However, the data reveal that Be mostly formed in the vicinity of supernovae when the Galaxy was young and now forms by collisions of Galactic cosmic rays with carbon, nitrogen, and oxygen atoms.

Observers always hope for good weather on the nights they are assigned on the telescope. But, in 2006 on October 15 when I was scheduled to observe with Keck I on Maunakea, there was a totally unexpected, unusual loss of telescope time. At 7:07 that morning a 6.7 magnitude earthquake occurred just offshore, 30 miles west of Maunakea and 18 miles below the ocean floor. Our wooden house, only a few miles from the epicenter, kept shaking. I fled outdoors with Hans; the kittens, who were totally spooked, went under the bed. Just as we all had calmed down, almost seven minutes later, another earthquake, magnitude 6.0, hit. It occurred a few miles further north but only 12 miles deep. The first large aftershock came at 10:35 A.M. Near noon I received word that observing was canceled because all the light fixtures in the remote observing room had fallen to the floor. Then we learned that a bridge was damaged and trees were across the road impeding my way to the Keck headquarters. Still later we learned about damage to the telescopes and domes at the summit. My observing program had to wait until January 2007.

Our lives changed, but ever so slowly, after Hans had a heart attack in the fall of 2002. I continued with my research but in 2006 retired from most of the teaching aspects of my job. And after full retirement in 2009, I was still applying for and receiving time to observe with the Keck I

FIGURE 3.2. Ann Boesgaard with husband Hans in 2011.

telescope. Hans usually accompanied me on those observing runs at the Keck headquarters.

With HST observations, we were able to detect a small dip in the B content in stars in the Hyades cluster. The relative amounts of all three elements were well matched to the predictions of the theory of internal mixing of matter caused by the effects of stellar rotation.

The large size of the Keck mirror and the efficiency of the detectors made it possible to pursue many new projects. We studied the Li content of faint stars in globular clusters and in old Galactic clusters and determined the chemical composition of clusters M 71, Praesepe, and NGC 752. We were able to measure the amount of Be in the oldest stars and found that the abundance of Be increased steadily in stars that were formed more recently, with the youngest having 1,000 times more Be than the oldest. And there was a Be dispersion in stars of the same age and metal content. This illuminated issues related to both stellar and Galactic evolution.

Nearly all the oldest stars show the amount of Li produced in the Big Bang. However, a few stars are very deficient in Li. We observed the Be content in several ultra Li-deficient stars and found them to be Be-deficient as well. The depletions were greater than predicted by mixing induced by rotation but could have resulted from stellar mergers or by stars in binary systems exchanging mass.

In 2019 I was amazed and thrilled to receive the American Astronomical Society's highest award, the Henry Norris Russell Lectureship, given annually for "a lifetime of eminence in astronomical research." I delivered that lecture at the 236th AAS meeting in January 2020. Hans was with me there in spirit to celebrate.

Chapter 4

Sidney Wolff (PhD, 1966)

Changing the Landscape

Sidney Wolff has led multiple observatories, including the design and development phases for several major telescopes—WIYN, SOAR, Gemini, and the Rubin Observatory—around the world, and is the co-author of several widely used introductory astronomy textbooks. She served as President of the ASP and the AAS, was awarded the Meritorious Public Service Award from the NSF and the Education Prize by the AAS, and is an Honorary Fellow of the RAS and a Legacy Fellow of the AAS. Her asteroid is 68448 Sidneywolff.

How did I first get interested in astronomy? I often get asked this question, and it was a spelling lesson in about third grade that focused on astronomical words. I don't remember any of them, but I was fascinated by the limited amount of astronomical information in the lesson and began reading about astronomy. I remember particularly reading Fred Hoyle's books (continuous creation), *Sky and Telescope* magazine, and, when I was in high school, the classic textbook *An Introduction to Astronomy* by Robert Baker.

As I look at that book now, I am amazed how much we have learned since. In the 1950s, we knew nothing about quasars and astronomical black holes or how the elements in the periodic table were formed. We knew galaxies had young (Population I) and old (Population II) stars, but had only a very limited grasp of how stars formed. The evidence for dark energy would not be discovered for nearly half a century. Embarking on a graduate career at UC Berkeley, I could not have imagined how fortunate I would be to be able to witness—and contribute to—what has truly been a golden age of astronomy.

As I look back on nearly sixty years in astronomy, I can identify several decision points, pieces of advice, and mentors that have played important roles in shaping my career. In a way, my first mentor was my father. I think he always wished that he knew more mathematics than he did, and so when I was in high school, he insisted that I take all the math courses that were available. In those days, it was more fashionable for parents to pay their children for good grades than it is now. He promised to give me 50 cents for every "A," I had to pay back 25 cents for every "B," but the whole deal was off unless I got an "A" in math. I didn't particularly care for math and might not have persisted without his insistence. That early grounding in math was what made my scientific career possible.

For college, I chose to go to Carleton, a four-year liberal arts college in Minnesota with a strong reputation in the sciences. I chose it because it was one of the few midwestern liberal arts colleges with more than one course in astronomy. However, I was not yet committed to science, and I wanted to explore other options. My husband found a great quote, attributed to Judith Shapiro, President of Barnard, who said that a liberal arts education should be expected to make "the inside of your head to be an interesting place to spend the rest of your life."

At Carleton, I met my husband-to-be, Richard Wolff. We both applied to PhD programs at Berkeley, Harvard, and Wisconsin. One of our professors, Robert Kolenkow, agreed to drive us both to Wisconsin to check it out. We were very tempted to go there. It was close to home, we had both been offered full fellowships, and Wisconsin would certainly have prepared us for what we then wanted to do—teach in a school like Carleton. After our visit, however, Kolenkow gave us some advice that I have repeated many times to others. He said that when you are given a choice about which path to pursue in life you should choose the harder one because, otherwise, you would never know what you might have accomplished.

Berkeley was certainly the harder choice based on the experience of Carleton physics students who had already gone there, but Berkeley was also key to everything that came after for me. Professor George Preston was my thesis advisor, and we worked on stars known as "peculiar A-type stars" (or Ap stars)—A-type stars are stars that are a bit more massive than the Sun, and Ap stars are A-type stars that typically spin more slowly than the majority of A-type stars, have strong magnetic fields, and show evidence of unusual amounts of the rare earths and certain other elements in their atmospheres—at a time when the cause of their unusual abundances was not at all understood. We now know that atmospheric diffusion, not nucleosynthesis, is the explanation. This was also a time when we were just beginning to explore the magnetic field properties of Ap stars and look at the role of rotation and binary properties in determining which stars are peculiar and which are not. George was especially helpful in reading my thesis carefully. He pointed out several places where my conclusions were not clear, and those were invariably cases where I was not clear in my own mind. Science papers must be absolutely clear about which conclusions are solid, which are uncertain, and which ideas are speculative pending further observations.

I learned two different lessons about writing research papers from another of my professors, George Wallerstein, who said that no research is ever finished until it is published and that with multiple ongoing projects, one should work first on the one that is most nearly finished. I have often repeated that advice to other young scientists.

I finished my degree at the end of 1966 and spent the next nine months as a postdoc working with George Preston at the then-new campus of UC Santa Cruz. At the same time, both my husband and I were looking at options for positions after Berkeley. We considered postdocs at universities with strong astronomy programs, but we made a choice that at the time was considered very risky. Ann Boesgaard, who was a good friend all through graduate school and remains one today, had gone to the University of Hawaiʻi, where her husband Hans was to be the chief engineer in charge of building the University of Hawaiʻi 88-inch telescope on Mauna Kea. (The name was two words in those days—now Maunakea is preferred.) She talked us into considering Hawai'i. We arrived just a day after they moved into their house with a 180-degree view of the ocean. Mai Tais on their lanai probably had some small influence on our decision to go there. I remember that the wife of the director of Lick Observatory, in California, subsequently asked me why we were going to Hawai'i: "Don't you want to work anymore?"

I believe that when faced with a choice, one should go to a place where it is possible to make a difference. Getting in at the beginning of the development of Maunakea gave me opportunities that in a more established institution would have gone to much more senior people.

A few years before we went to Hawai'i, John Jefferies had established a solar program at the University of Hawai'i. The solar observatory was located on Haleakala, Maui. When John won funding from NASA to build the 88-inch telescope, he began to establish a nighttime astronomy program. In those early days, it was difficult to hire staff at such a new organization—so much so that when we visited, John spent the time trying to persuade us to come. I didn't even give a colloquium on that first visit.

In 1967 when we took up our positions at the University of Hawai'i, there were no telescopes at all on Maunakea. Construction of the 88-inch was in progress, but the choice of Maunakea was very controversial. People were skeptical that it would be possible for astronomers to work at an altitude of nearly 14,000 feet. After all, pilots of unpressurized aircraft must use oxygen continuously any time they are above 12,500 feet for more than thirty minutes. Haleakala, at 10,000 feet elevation, offered a far more convenient site because it was already developed with roads and a power line. In contrast, Maunakea had no source of power, access was by a very primitive dirt road negotiable only by four-wheel-drive vehicles, and winters were accompanied by high winds and heavy snowfall.

Comparisons of site-testing measurements of Maunakea and Haleakala showed, however, that in strictly astronomical terms Maunakea was superior in every way. John Jefferies made the bold decision to build the 88-inch telescope on Maunakea, and I think that decision changed forever our understanding of what can be accomplished with ground-based telescopes. Maunakea's outstanding image quality and infrared transparency are unmatched. It is quite simply the best observing site in the northern hemisphere.

The 88-inch telescope was not completed until 1970, and for my first three years in Hawai'i, I was unable to observe regularly. Fortunately, George Preston lent me photographic plates he had taken with the 120-inch telescope at Lick Observatory so that I could continue working on Ap stars. I visited him in Pasadena every summer, and those visits were critical in helping me maintain my scientific productivity. Those who know George have experienced his infectious enthusiasm for research and the joy he takes in so many things outside of astronomy. During my visits, he and his wife Jan often had me over to dinner, and we would listen to Schumann piano pieces afterward. It was that experience that led me to learn how to read music.

In Hawai'i, I used a 24-inch telescope, which was completed before the 88-inch, to study the photometric variability of Ap stars. I took the first scientific spectrum with the coudé spectrograph after the 88-inch went into service. The spectrum was of Vega—not exactly compelling scientifically but a major milestone nonetheless. Walter Bonsack and I built a Zeeman analyzer and continued to measure magnetic fields in Ap stars.

I also did a lot of work, partly motivated I think by aesthetics, on the properties of chemically peculiar stars that show evidence of excessive amounts of mercury and manganese in their atmospheres. The spectra of these stars are very clean with sparse, sharp lines, and I find them to be quite beautiful. In those days we used photographic plates. I still remember the magic of the moment when, after spending several minutes in total darkness to develop a plate, I could finally turn on the light and see what I had.

What I concluded from the study of these stars was that no matter how much information we accumulated about the large-scale properties of these stars—rotation, binarity, temperature, gravity, even rotation (slow rotation is a necessary but not sufficient condition)—there was no way to predict from those measured properties whether a star would have overabundances of mercury and manganese. With that conclusion,

I decided the most important contributions that stellar spectroscopy could make would be to use our deep understanding of stellar physics to probe questions related to stellar evolution and star formation, abundance gradients in galaxies, stellar activity, stellar structure, and nucleosynthesis. Collaborations with others led to research papers on a variety of topics in these areas.

By this time, however, I was devoting an increasing amount of time to administrative work. A few years after the 88-inch went into service, operation of the telescope was frustratingly unreliable. I went to John Jefferies and said that the problems had become so severe that I could no longer do my research and that if the situation did not improve, I would have to leave Hawaiʻi. Instead, John put me in charge of the 88-inch and told me to fix the problems. That assignment was a steppingstone to my future contributions to the building of telescopes. I gained additional experience by serving as acting director of the Institute for Astronomy on several occasions when John was away on sabbatical or in the interim while we searched for a new director after John left in 1983 to become director of the newly formed National Optical Astronomy Observatories (NOAO) in Tucson.

I have so many good memories of the seventeen years I spent in Hawaiʻi and especially of the dozens of nights I spent working on Maunakea. I fell in love with the mountain the first time I saw it. Its stark beauty and subtle colors are unlike anything I had ever seen before. Maunakea is most beautiful when it is blanketed with white snow under a clear blue sky, but whiteouts, high winds, and blizzards can occur in almost any month of the year. In those days, we didn't have cell phones and constant access to sources of assistance from people at sea level, and we certainly took chances that would not be permitted today.

Some of my memories:

- Seeing the Southern Cross and the glow of an eruption at Kilauea simultaneously on one of my first nights on the summit.
- Hiking up the summit cone with Walter Bonsack and tugging a Zeeman analyzer on a sled when the road was blocked by snow; I vowed never again to hike up the summit cone, only down.

- Hiking to the true summit with Gethyn Timothy; it's harder than it looks.
- Being nearly trapped in a blizzard with the day crew on the road down the summit cone; it took two hours to make our way down what is normally a five-minute drive.
- Coasting all the way to Hilo on the old Saddle Road in the middle of the night in a car that wouldn't start; there was a tricky set of turns near Hilo, and it was necessary to build up speed to get through them, but this could only be done at night when headlights would signal an oncoming car.
- Guiding at an eyepiece (no TV in those days; we had to sit out in the cold to push the guide buttons) when an earthquake occurred; the star danced in the eyepiece and made a sort of Lissajous pattern.
- Setting off Roman candles (legal on the Big Island at the time) at the summit with my husband on New Year's Eve in gently falling snow.
- Guiding at the telescope when a graduate student came in from the 88-inch catwalk and said Maunaloa was erupting; indeed it was, and we had front-row seats.
- The first time I saw all four domes open in preparation for the night's observations—the Canada-France-Hawaii Telescope (CFHT), the United Kingdom Infrared Telescope (UKIRT), NASA's Infrared Telescope Facility (IRTF), and the 88-inch—and knew that the promise of Maunakea would be realized.

It makes me incredibly sad that Maunakea has now become a source of such controversy between native Hawai'ians and astronomers, because both groups have a deep spiritual and emotional attachment to this very special place, and I wish a way could be found to share it in a mutually agreeable way.

My husband and I loved Hawai'i and planned to spend the rest of our lives there. But in 1984, the Associated Universities for Research in Astronomy Board approved my appointment as the next director of Kitt Peak National Observatory (KPNO) in Arizona. That was simply too good a professional opportunity to pass up. In Hawai'i, I experienced the thrill and the satisfaction that come from building an observatory from the ground up. I thought that transferring to the national observatory offered me the best opportunity to continue building new telescopes and new observatories.

And build we did: the Wisconsin-Indiana-Yale-NOAO Telescope (WIYN) on Kitt Peak, the Southern Astrophysical Research Telescope (SOAR) on Cerro Pachón in Chile, the two Gemini telescopes, one on Cerro Pachón, the other on Maunakea, and the Large Synoptic Survey Telescope (LSST; now renamed the Vera C. Rubin Observatory), also on Cerro Pachón, are some of the legacies of my time at NOAO. I can now claim a role in the building of seven telescopes. Besides taking the first science spectrum at the 88-inch telescope, Ann Boesgaard and I helped pick the color scheme for that facility. David Morrison and I wrote the first draft of the proposal to NASA for the IRTF, and I was acting director in Hawai'i during part of its construction. I was director of NOAO when we joined with Wisconsin, Indiana, and subsequently Yale to build the WIYN Telescope; President of the SOAR board during construction; the first director of the Gemini telescopes; and chief operating officer of the LSST Corporation during the design and development phase of what is now the Rubin Observatory.

Prior to the advent of NOAO, most of the construction at the national observatories was funded by the National Science Foundation (NSF). The telescopes that we built at NOAO required unprecedented forms of partnerships. WIYN and SOAR were both projects carried out with universities, and SOAR also involved Brazil. The Gemini telescopes were built by an international partnership, and LSST by NSF and the Department of Energy (DOE) working together. These new types of partnerships were key to expanding the capabilities offered by the national observatory.

I am proud not only of the telescopes that we built but also of the many people who came to NOAO early in their careers and gained experience at NOAO that allowed them to go on to leadership positions with other organizations. I would include in this list (in alphabetical order) Taft Armandroff, Todd Boroson, Richard Green, Buell Jannuzi, Matt Mountain, Jim Oschmann, Pat Osmer, Mark Phillips, Caty Pilachowski, Dave Silva, and Bob Williams. (I apologize to anyone that I

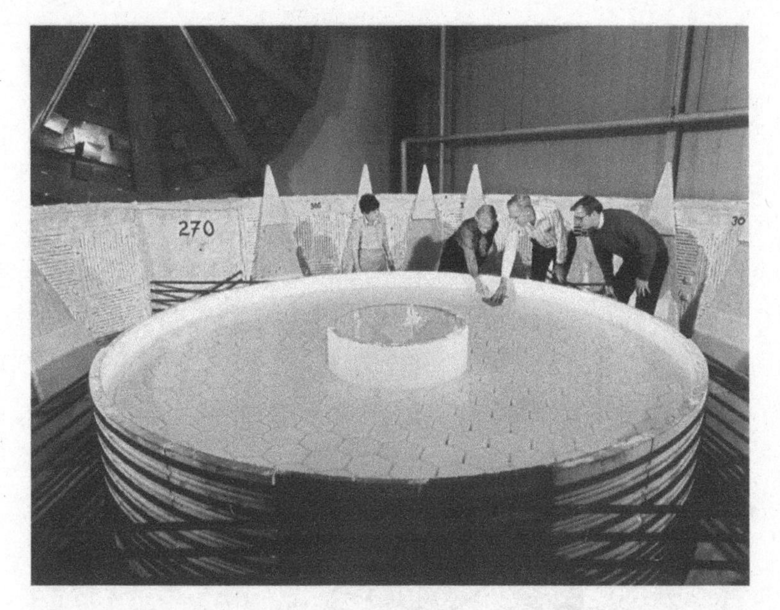

FIGURE 4.1. The WIYN 3.5-meter primary mirror seen at the University of Arizona mirror laboratory after the cooling process, in July 1989. Left to right: Caty Pilachowski, Sidney Wolff, John Richardson, Larry Stepp. *Credit*: NOIRLab/NSF/AURA.

may have left off this list.) Of course, much of our homegrown talent remains within the organization and is key to its current strength.

During my tenure as director of NOAO, I managed to build relationships between Cerro Tololo Inter-American Observatory (CTIO) and KPNO that allowed us to change our name to National Optical Astronomy Observatory (singular instead of the original plural), and the National Solar Observatory (NSO) became independent. I am very pleased to see that CTIO, KPNO, Gemini, and the Rubin Observatory have now joined together to form a true national laboratory (NOIRLab) for ground-based astronomy.

I have now retired several times—from Gemini, from NOAO, and from LSST. The third time turned out not to be the charm. After five years of true retirement, I returned in 2019 to serve as director of the AURA portion of the U.S. Extremely Large Telescope Program, which has as its goal the construction of the Thirty-Meter International

FIGURE 4.2. Vista Sidney Wolff offers views of three of the telescope projects designed under Wolff's leadership. These telescopes are located on the flat ridge line that is best viewed from this position. From left to right, the telescopes are SOAR, Gemini South, a small telescope used to make calibration measurements, and the Rubin Observatory, which was still under construction when this image was taken in October 2019. Photo by Richard Wolff.

Observatory and the Giant Magellan Telescope. NOIRLab's role in this project would be to provide end-to-end user services including the process of managing applications for observing time, a queue-scheduling system, data archiving, data pipeline processing, and visualization and other tools for working with the data. The lure of working on yet one more project to build big telescopes proved irresistible.

Chapter 5

Jocelyn Bell Burnell (PhD, 1968)

Kites Rise against the Wind

Jocelyn Bell Burnell, now Chancellor of the University of Dundee, Scotland, and Visiting Professor of Astrophysics at the University of Oxford, helped build the International Scintillation Array, which she used to detected the first pulsar, PSR B1919+21, in 1967. She is Dame Commander of the Most Excellent Order of the British Empire, and was the recipient of the Albert A. Michelson Medal from the Franklin Institute of Philadelphia in 1973, the Beatrice Tinsley Award from the AAS in 1986, the Herschel Medal from the RAS in 1989, the Grande Médaille from the French Academy of Sciences in 2018, the Special Breakthrough Prize in Fundamental Physics in 2018, and the Gold Medal by the RAS in 2021. She was the first woman to serve (2014–18) as President of the Royal Society of Edinburgh. Her asteroid is 25275 Jocelynbell.

I was born in 1943 in Belfast, Northern Ireland, but soon went home to Solitude, a sizable house surrounded by fields, a few miles from the nearest town. As the first child I joined a household of fifteen adults—my parents, extended family, family friends, two refugees (it was the middle of a major war), and domestic staff. Gender became an issue eighteen months later when a brother was born, and although my parents valued both sexes I quickly learnt that in Ireland boys were more important than girls. Later two more girls joined the family.

It was believed then that it was possible to distinguish by age eleven between those children who would have professional careers and those who would take up skilled or semiskilled jobs. Schooling post age eleven was according to which stream you fell into. Also, it was "known" that girls were only going to get married and become full-time homemakers and unlike boys would not be part of the economic workforce. Furthermore, it was known (and this I believe *is* true) that at age eleven girls are academically more advanced than boys. So "too many" girls were passing the national exam that we all had to take, keeping boys out of the academic stream. In Northern Ireland (and in some parts of England) the local authority set a higher pass mark for girls "to give the boys a chance." This discrimination persisted in Northern Ireland until 1988. Although I failed that exam I was allowed into the Grammar School, because it was planned that I would go away to boarding school in England a few years later.

Northern Ireland has a long history of sectarianism and sectarian violence (Protestant vs. Roman Catholic) and our area was one of the most polarised. As Quakers my parents tried to straddle this divide, but it was never easy, and they decided all their children should have part of their education out of Northern Ireland. So, at age thirteen I went away to a Quaker girls' boarding school in the north of England. I thrived there, but, as in many girls' schools, the maths and science teaching were patchy. I did have a very good (elderly) physics teacher, and specialised in physics and maths.

It became clear I should do a physics degree at university, but I was unclear what I would do afterwards. When my father brought home Fred Hoyle's book *Frontiers of Astronomy* from the public library, I read it. We were learning about circular motion at school, and here I am reading about the rotation of galaxies! That's it—I would become an astronomer! There was a minor glitch when I realised that astronomers work at night and I needed my sleep so I couldn't be an astronomer! Then I heard about radio astronomy and settled on that.

After school I went to Glasgow University to do a physics degree. The physics course was typical of its time, but we did have to take other subjects as well. I opted to take geology and loved it, coming top of the class without trying! I approached the head of the Geology Department to discuss changing major, but he said geology was not suitable for women, so I stuck with physics. (I subsequently learnt that women of my age at other universities were encouraged to major in geology if they showed ability!)

It was the "tradition" in that university that when a woman entered the lecture theatre all the men whistled, stamped, cat-called, and banged the (wooden) desks. Women students would congregate outside the lecture theatre and enter in a group. However, in the final two years I was the only female in the Honours Physics class (of fifty students) and had to face it alone. It was important not to blush, and I learnt to control that. My female student friends assumed I would change subject. If I hadn't been so clear about becoming a radio astronomer I might have!

In my final year I was in a group lucky enough to have a weekly tutorial with Professor Ron Drever, then one of the younger members of Glasgow's Physics Department. He was no help to us with our course material, but talked enthusiastically about developing research areas. Clearly my interests were quite close to his and I vowed to watch whatever field Ron worked in, as I judged it was likely to go places. Thus, I became an early follower of developments in gravitational wave astronomy, and one of the first in the UK.

On graduation with my BSc in 1965 I moved to Cambridge to start a PhD. I had intended to go to Jodrell Bank, but they never responded to my application. I did not expect to get into Cambridge so I reckoned I was heading to Australia to do that PhD in radio astronomy but put in an application to Cambridge just in case (Ron Drever wrote me a reference). I was very surprised to be accepted. Up till then I had lived in the north and west, the upland areas of the United Kingdom. Cambridge was close to London in the affluent, posh, southeast, and when I got there I was overawed. The students, predominantly male, were poised, well-networked, confident in their ability and their right to be there. I was female and from outside the normal catchment areas. I noted that the few women faculty referred to themselves as Miss or Mrs., not Dr. or Professor. (Was this so as not to frighten the men?) Clearly Cambridge had made a mistake admitting me; they would discover their mistake and throw me out. I decided my best policy was to work my very hardest, so that when they threw me out I would not feel guilty—I would know I had done my best and just was not good enough for Cambridge. It is a strategy I would recommend to anyone feeling overawed in a new position.

The typical PhD programme in the UK has funding for three years. I would spend two years "in the field" (with five or six others) building a radio telescope, six months as the first user debugging it and taking data, and six months completing the data analysis and writing the thesis. A few years previously quasars had been recognised as bright radio sources at considerable distances. Only about twenty were known and not much was understood about them. My project was to find many more quasars. My PhD supervisor, Tony Hewish, building on previous work by Margaret Clarke, who showed that quasars scintillated (twinkled), obtained grant funding to build a large radio telescope to search for more scintillating quasars. The telescope was built in-house. My responsibility in the construction was for all the cables, connectors, and transformers. I was spared most (but not all) of the sledgehammering of posts into the ground and became very strong and weather-beaten!

I was the first user of the telescope. After two years building, the radio telescope worked the first time it was switched on—that is probably a record! It looked due south and the beam could be swung in a north-south direction (declination) but was dependent on the earth's rotation to scan in the other direction (right ascension). With sixteen east-west rows the telescope had, in theory, sixteen beams, but we could only afford receivers for four. To detect the scintillation a short time constant had to be used.

Computers were rare—the University of Cambridge had just one (which took up a large room) and with memory comparable to that of a laptop today. Very few people had access to it and I did not! My data came out on rolls of paper chart—almost 100 feet per day, and it was my job to analyse these charts, picking out the scintillating quasars. In the six months that I observed I accumulated (and analysed!) 3.3 miles of chart paper. I quickly became used to recognising the quasars and to recognising the inevitable radio interference that the sensitive telescope also picked up.

FIGURE 5.1. Jocelyn Bell Burnell, in the late 1960s, in front of the 4-acre telescope (the Interplanetary Scintillation Array), at Mullard Radio Astronomy Observatory, Cambridge, UK. Photo by M. Burnell.

I continued to work hard and work thoroughly—still feeling the need to justify my place in Cambridge. Along with the radio interference and the scintillating quasars there was occasionally a small signal that I could not identify. It occupied about one-quarter inch on the chart paper but was so small it was not always there when observing that bit of sky. However, because I did not know what it was, it stuck in my brain. Having established that it was always from the same position in the sky, I showed the trace to my supervisor, Tony Hewish. He (rightly) observed that since it only covered one-quarter of an inch of the roll of chart paper, it was hard to see what it was—we needed an enlargement. With chart recordings that is easy—one runs the chart paper faster under the pen and all gets spread out! However, we could not leave the chart recorders permanently on that setting so I had to go out to the observatory shortly before that part of the sky was due to transit and switch to a high-speed recording. I did that for a month and got only high-speed recordings of receiver noise. The source had disappeared!

This (of course) was the grad student's fault. Finally, I caught it—a string of equally spaced pulses—about 1.3 seconds apart. I called Tony with the news and he became even more suspicious that he had a stupid grad student. However, he came out to the observatory at the appropriate time the next day and stood watching over my shoulder as I prepared for the recording and saw the pulsed signal with his own eyes. We established that the pulse period was the same as on the previous day.

We asked a colleague (Paul Scott) and his grad student (Robin Collins) who also had a telescope and receiver operating at 81.5MHz if they could see this signal, and they did! John Pilkington took two of my receivers, retuned them, and established that the signal swept down in frequency—a signature of dispersion by free electrons, most likely in interstellar space. Guessing at that electron density he came up with a distance which put the source out in the Galaxy. We were wondering how we could publish this result and be believed, when I found a second one. It pulsed at a slightly different period and was in a totally different part of the sky. A few weeks later, in early 1968, I found numbers three and four.

We could now proceed with publication of the first source—the other three would follow in a second paper. There were two likely scenarios to explain this kind of emission—a star vibrating or a star rotating. Tony, who had worked on solar flares, preferred a white dwarf star vibrating but accepted it could be a rotating object. He gave a colloquium in Cambridge a day or two before the *Nature* paper announcing the discovery came out. Every astronomer in Cambridge attended, it seemed, and Fred Hoyle sat in the front row. Tony explained what we had found and mentioned he thought it might be an oscillating star like a white dwarf. Fred spoke first at the end. Noting that this was the first he had heard of these things, he said he did not think it was a white dwarf but a neutron star. I was impressed that he could digest so much information so fast and come up with what was ultimately shown to be the right answer.

The press took much interest in the discovery, and Tony and I were interviewed many times. Tony would be asked questions about the astrophysical significance of this discovery. I was asked what my bust, waist, and hip measurements were and how many boyfriends I had!

Photographers asked me to undo more blouse buttons. I hated it and would have loved to have been very rude to those journalists and photographers, but I had still to get my PhD and did not feel I could rock the boat too much.

I got engaged to be married between discovering pulsars numbers two and three (and got married between submitting my thesis and defending it in early 1969). Many people who had not remarked on the discovery of pulsars warmly congratulated me on my engagement. Getting married was then seen as the peak of a woman's life. My husband worked in Local Government and his first job was on the south coast of England and so I looked for a job there. I got a research fellowship and moved to Southampton University. Initially I worked on ionospheric data from the *Alouette* satellite. The top-side sounder had seen an interesting trough in the electron density at mid-latitudes, but when the bulk of the data arrived after "cleaning" the trough data had been removed! I joined the gamma ray astronomy group and helped build and calibrate a gamma ray telescope to be flown from a balloon in Texas. The telescope worked in the 1–10 MeV energy range—the hardest range in which to detect gamma rays! My fellowship had run out and when one of the faculty left, his post was split in two; I got one half and started lecturing. So, I had half the pay of a faculty member, a full teaching load, and none of the perks! When I became pregnant I discovered that the university had no maternity leave or special provision! However, it was a good place in which to begin a lecturing career.

When our son was born there were few childcare facilities—mothers were discouraged from working. Also, the way my husband would get a promotion was by moving to another local authority. Neither of these helped my career! I became an editor of *The Observatory* magazine and started teaching for the Open University, which was very rewarding. Our next house move took me quite close to the Mullard Space Science Lab and I got a part-time job there—actually a support-staff job rather than an academic one, running the *Ariel V* satellite for the lab. It was an exciting time with the satellite, making discovery after discovery. The data came across my desk first for checking. Many times, there would be an additional, uncatalogued source in the field of view—a transient source had erupted! That X-ray satellite was hugely successful and kept us all jumping!

My work with the satellite was ending as my husband got a job in Scotland. We were within reach of the Royal Observatory Edinburgh and I got a part-time job there. The Observatory would be responsible for running the James Clerk Maxwell Telescope (JCMT) on Maunakea and I was recruited with that in mind, but meanwhile I had some time to learn the field. So, having started in radio astronomy and then moving to gamma ray and X-ray astronomy, I was now learning infrared and millimetre-wave astronomy. I became part of the management of the JCMT with responsibility for user liaison and for the new receiver programme. I enjoyed the role, but it had its difficulties, as the large resources I controlled were very attractive to some others.

Our marriage had broken down and our "child" was now finishing school and heading off to university. For the first time in my life I was free to go after a job because of *what* it was, not *where* it was. I won the professor of physics position at the Open University, thereby doubling the number of female professors of physics in the UK. ("Professor" in the UK was then a more exclusive, senior title than in the United States.) Here I set up an astronomy group studying energetic binary star systems at whatever wavelength was appropriate. I also had my first grad students.

For a year I was a Visiting Professor for Distinguished Teaching at Princeton and shortly after that became Dean of Science in the University of Bath for a few years. All that time I was getting many attractive invitations to travel and lecture, but being dean involved a lot of meetings—on a "good" day there were only four or five meetings, seven or eight on a "bad" day, and I wasn't going anywhere! I judged the invitations would dry up if I kept saying no, so I took early retirement and have been travelling and lecturing ever since.

Alongside all of this I have also held leadership roles: I have been President of the Royal Astronomical Society, the first female President of the Institute of Physics for the UK and Ireland, and the first female President of the Royal Society of Edinburgh, Scotland's National Academy. I have served on and chaired Research Council committees and Royal Society committees. I inherited my father's ability to explain

FIGURE 5.2. Jocelyn Bell Burnell in the Women in Science in Scotland exhibition in 2019. *Credit*: Royal Society of Edinburgh.

things clearly and have enjoyed using this skill to bring the latest developments in astronomy to a broad public. It was also useful at the International Astronomical Union General Assembly meeting in Prague in 2006, where the reclassification of Pluto was discussed. At that time, I was on the IAU's Resolutions Committee and was called in by the Executive Committee to help resolve an impasse in the General Assembly. Having offered several suggestions as to how the Executive Committee could proceed, I found myself in the midst of the exercise to craft suitable definitions of a planet and a dwarf planet. Being from another branch of astronomy, with no axe to grind, was clearly helpful, but feelings ran high. I did a lot of "shuttle diplomacy" between concerned members and the Executive Committee. There were about three thousand people in the hall for the final session when I introduced the resolutions, including about one thousand members of the press. As soon as the vote was taken the press all rushed out of the room to report the decision—Pluto had become a dwarf planet.

Chapter 6

Virginia Trimble (PhD, 1968)

Breaking through the Telescopic Glass Ceiling

Virginia Trimble, now Professor of Physics and Astronomy at the University of California, Irvine, is interested in the structure and evolution of stars, galaxies, and the Universe and of the communities of scientists who study them. She is the only person to have been president of two IAU Divisions (Galaxies and the Universe, and Union-Wide Activities). She is a Fellow of the American Association for the Advancement of Science and the APS, a Legacy Fellow of the AAS, an inaugural Fellow of Sigma Xi, and an Honorary Fellow of the RAS. She was awarded the J. Murray Luck Prize by the National Academy of Sciences in 1986, the Klopsteg Lectureship by the AAPT in 2001, the George van Biesbroeck prize in 2010 and Patron status in 2018 by the AAS, the Wm. T. Olcott Distinguished Service Award by the AAVSO in 2018, the Andrew Gemant Award by the AIP in 2019, and honorary membership in Sigma Pi Sigma. Her asteroid is 9271 Trimble.

There it was, in my mother's hand, the letter that would determine my fate for the next year, indeed, as things played out, for the rest of my life. Yes, at age twenty and finishing up a BA at UCLA, I was still living with my parents. And yes, I had applied to the graduate program in astronomy at the California Institute of Technology, even though their catalogue said "Women are admitted only under exceptional circumstances."

In fact, Caltech was the only place I had applied, just as UCLA was the only place I had applied for undergraduate study. You might almost guess that I'm a native Californian, which I am, a graduate of Toluca Lake Elementary School, Le Conte Junior High, and Hollywood High School, as well as UCLA (BA in astronomy and physics in 1964) and Caltech (MS, PhD in 1968 in astronomy), even born in Hollywood (Presbyterian Hospital, November 1943) to Virginia Frances Farmer and Lyne Starling Trimble.

What did that letter say? "Dear Miss Trimble: We have reviewed your qualifications and conclude that we cannot deny you admission to the California Institute of Technology. We think, however, you might be happier elsewhere," followed by many administrative details and signed by J. Beverly Oke, the astronomy graduate student advisor. Mother said to ignore the second sentence, so we will never know whether I would have been happier elsewhere, but I think it extremely unlikely.

A great deal of good luck is lurking in those two paragraphs. The folks married in May 1939 (at a time when there were two other young men who thought they were engaged to Mother, leaving her mercifully nonjudgmental) and had built a house in the San Fernando Valley, just over the hills from Hollywood, before the outbreak of World War II largely halted private construction. They never moved from it, and the sale of the house (I was an only child) eventually financed the endowment of the Lyne Starling Trimble lectures presented by the American Center for the History of Physics. Earlier, the sale of the house my husband, Joseph Weber, and I shared in Chevy Chase, Maryland, provided funding for the Joseph Weber Award for Astronomical Instrumentation given annually by the American Astronomical Society. Why an award? He said there were too many prizes out there, and indeed that Nobel had done more harm with his prizes than ever he did with his explosive.

Going to just one educational institution at each level made progress very smooth for me. The Los Angeles Unified School District even provided free buses over the hills because there were too many students in the new communities of North Hollywood and too few in the older communities of Hollywood. The buses continued through summer school sessions, five years of which covered requirements in history,

English, and so forth, leaving more time in the school year for extra science, music, Latin, and all the mathematics available. It helped, admittedly, that I was very good at taking traditional sorts of examinations, and I feel sorry for the younger generations coming along who will never meet "figure analogies"! More seriously, I think that pushing algebra back into eighth grade to make room for calculus in the twelfth is a mistake. And under that system I would never have had three wonderful semesters of math taught by Mitsunori Kawagoye, who remained a friend until his death in 2021 and from whom I learned of "quiz prizes" for outstanding performances. I generally give science magazines rather than candy bars, but the principle still holds.

Mother, who died in 1972 at age fifty-six of metastasized breast cancer, had the gift of tongues, which I do not. She also taught me sewing and cooking and other things that I still think every girl (and probably guy) should know how to do. Oh, yes, also typing, though I never matched her skills and never attempted shorthand.

Father was a chemist, a very good one, who held patents for color motion picture film processing and development of color on surfaces of magnetic tape. He owned and operated companies for each of these technologies, each of which eventually went bankrupt and left him working for a dozen or more other companies over the years. I carried away from this a strong desire for a stable job, which academe has provided.

One of the many things Joe Weber and I agreed upon is that it takes the child of a self-unemployed father (his was a freelance carpenter during the Depression) to appreciate tenure properly. And no one ever seemed to question that, of course, I would go to college and on to graduate school. The folks even said they could afford to pay for one year of graduate school for me if necessary. Being a UCLA undergraduate had, in those days, cost very little (mostly money for parking), and those expenses were covered by a National Merit Scholarship and various prizes from the Daughters of the American Revolution, the American Mathematical Society, Bank of America, and Coast Federal Savings. That last was most interesting: the founder worried that smart high school students were having left-wing ideas crammed into them, and so he established a Saturday morning program that paid a dozen or so of us to come to "Ninth and Hill on the Ground Floor" for many weeks to study economics from college textbooks, while getting some feel from him and other Coast Federal men (yes) how finance really works. As a result, I have never voluntarily bought or sold even one share of stock, though circumstances have dribbled some past me.

Lots of other useful lessons came from home. Comparing the piano-playing styles of Mother and Father, I concluded that I had better learn to read music. Even physical tasks will yield to the methods of rational thought, including sinking baskets, hitting a well-pitched softball, and holding up one's side at table tennis. All would perhaps have been easier if someone had noticed I was extremely nearsighted. This problem, not caught until I was in seventh grade, however, made reading very easy, and I still go through a book a day if I don't have to pay for them. Swimming? Check. Horseback riding? Check. Ice skating? Oops, never mastered, despite four tries at ages five, nine, twenty-four, and twenty-seven. Luckily, by the time I got to the University of Cambridge, the Cam never froze in winter, so my lack of skill was never revealed.

Useful skills from early schooling? Undoubtedly the most important was to agree to do whatever was expected of me. Volleyball team captain. Coaching beginning piano students. Organizing a grammar school student visit to Hollywood High auditorium. Wardrobe custodian for the choir. Captain of Spelling Bee and other similar teams. Giving the introductory talk at an IAU Symposium of 120 men (and me) when I was fresh out of graduate school. Teaching physics of music one term and general relativity the next. You get the idea, and this one I think is generation-independent.

So, it's 1968 and I'm a freshly minted PhD. What to do? In those days, there was no equivalent of the AAS job register. New PhDs these days can find out what is available, what job might suit them, and how to apply for it. We waited for our advisors to recommend us for something (though there were truly more jobs available, relative to the number of new PhDs, in 1968 than now). Perhaps because I could read and write Egyptian hieroglyphs, I sounded attractive to Smith College. (It really was that which pleased the interviewers for the Woodrow Wilson

FIGURE 6.1. Virginia Trimble at the Grant measuring machine, used for both the white dwarf redshift project and determination of motions of the filaments in the Crab Nebula.

Foundation fellowship, so that my folks did not have to pay for any of my three and a half years of graduate school. Yeah, we finished faster in those days.) But what seemed to appeal most to the president of Smith College was that I could sing second sister in the faculty production of *The Pirates of Penzance*. Thus, I settled into a teaching job my first year out of graduate school, which was a much commoner first job then than now. But what I really wanted to do was go back to Cambridge, where I had spent the summer of 1968 at the Institute of Theoretical Astronomy, run by

Fred Hoyle. That summer, I had simply shown up on their doorstep, and they gave me a sort-of job, looking after the library while the regular librarian was away and answering the phone when the secretaries were at lunch. That probably wouldn't work now. But formally applying for a NATO postdoctoral fellowship worked.

Back to Cambridge I went, which, considering the weather, was not a very good place to be an observational astronomer, immediately following an observational PhD dissertation on "Motions and Structure of the Filamentary Envelope of the Crab Nebula." What to do? Take up some theoretical tasks, like stellar structure and evolution.

Down here in the middle, where maybe you won't notice, I need to say something about the men in my life. Like one of my "classmates,"[1] I quickly noticed in both undergraduate and graduate classes, "O! Look at all the lovely men!" meaning both students and faculty. And many things happened at both UCLA and Caltech that would now get us all fired. I have, truly, no regrets about my interactions with a UCLA Egyptologist, a high school choir director, my thesis advisor, the bloke I went to Cambridge to spend time with, the other blokes I encountered there, and so on and considerably on. My year (1963–64) as Miss Twilight Zone, as the face of a publicity campaign to promote the TV show, was actually completely innocent, though very good training for answering unexpected questions, whether scientific or otherwise, because I did lots of newspaper interviews, morning radio and television wake-up shows, and so forth. And the Tuesday evenings I spent modeling for Richard Feynman (physics Nobel 1965), while he was learning to draw, yielded $5.50 an hour (when that was a lot), all the physics I could swallow, and the original of one of the drawings, which my husband framed and hung on our living room wall.

So, a year teaching at Smith. Two years as a postdoc at Cambridge, during which I began work on several odd kinds of stellar evolution calculations, using what was probably the first open-source code in astrophysics, the Paczyński stellar evolution code. One of our stars ended its life as a massive blue supergiant, but everybody, including us, had forgotten this by the time Supernova 1987A went off as a blue supergiant supernova!

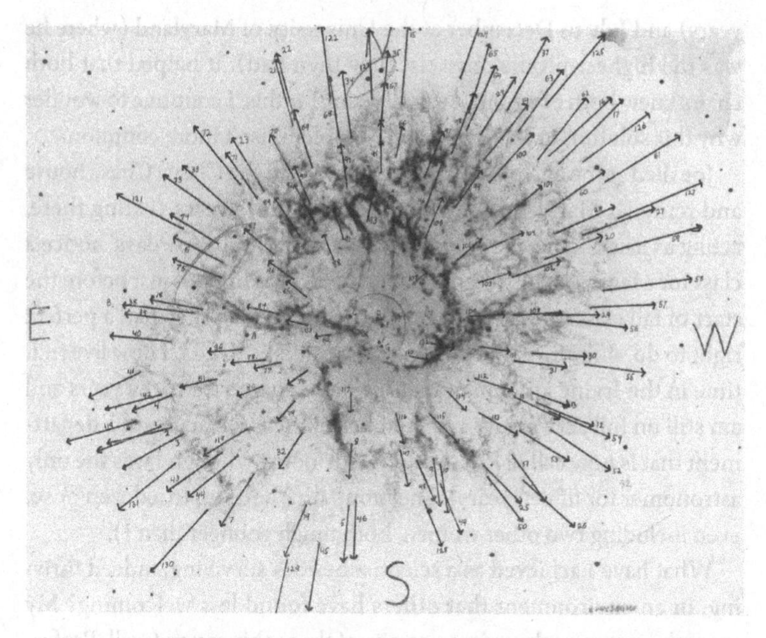

FIGURE 6.2. Photograph showing proper motions of filaments in the Crab Nebula. The arrows represent the directions and the distances materials will move in 270 years (V. Trimble, *Astronomical Journal* 73 [1968]: 535).

I had collected another engagement ring in Cambridge, returned to California engaged but unwed and enormously happy to be back in the sunshine and close to my parents, who were by then struggling with the cancer that would soon kill my mother.

A letter totally out of the blue in late January 1972 was an invitation from a distinguished, older physicist[2] for dinner a few weeks into the future, when he would be giving an endowed lecture at another Southern California university. He came back for bits of two more weekends, the second time bringing a Tiffany solitaire diamond engagement ring (not my choice, but who was going to say no?). Joe and I married in mid-March 1972, and our respective department chairs worked out an arrangement whereby we spent, each year for twenty-eight years, January to June at the University of California, Irvine (where I was, of course, the only woman in the physics department for the first fifteen years) and July to December at the University of Maryland (where he was the highest-profile physicist they then had). It helped that both chairs knew both of us, but, even allowing for this, I continue to wonder why this solution to the "two-career problem" isn't more common.

Joe died at the end of September 2000. I sold the Chevy Chase house and rented a Maryland apartment for two more years visiting there, acting as usual—that is, for instance, taking over on a few days' notice a class for a fellow faculty member who had killed himself just before the start of fall classes. Maryland then fired me (which they had a perfect right to do, since my tenure had always resided at UCI). I now live full-time in the Irvine apartment that Joe and I shared for many years and am still on full active duty as a member of the UCI faculty in a department that is now called Physics and Astronomy (though I was the only astronomer for fifteen years). The group now adds up to a dozen or so, even including two other women, both much younger than I).

What have I achieved as a scientist besides surviving, indeed thriving, in an environment that others have found less welcoming? My second-year research project was one of those things you (well, Professor Jesse Greenstein et al.) gave to a female graduate student because it was thought to be impossible and this (then) didn't matter—measuring the relativistic gravitational redshifts in the spectra of white dwarf stars. At a time when only two had been measured (one of which was drastically wrong because of contamination by scattered light from a companion), we added sixty-some to the inventory, showing the effect was there and that white dwarfs were massive enough to contribute something to the density inventory in the plane of our Milky Way Galaxy. The thesis you have heard about.

On instruction from my advisor, Guido Münch (in turn a student of S. Chandrasekhar, physics Nobel 1983), my thesis paper on the Crab Nebula was submitted to the *Astronomical Journal*, newly under the editorship of Lodewijk Woltjer,[3] rather than to the more prestigious *Astrophysical Journal* (edited by Chandra). The referee who commented on part I had been a member of the National Socialist German Workers Party. And surely there was never a graduate student who didn't feel secretly "my thesis was refereed by a Nazi!" The reviewer for part II had fought in the first World War, on the German side, but kindly caught a silly mistake I had made in one of the tables. He never told anybody but me and the editor. While at Smith, I submitted a wrong pulsar model. Well, there was a lot of it going around in those days.

From Cambridge came several Crab-related papers with a few co-authors and then the stellar evolution items, including one model for a star that might have been the companion of the X-ray source Cygnus X-1. If this had been so, the X-ray-emitting component would have been a neutron star rather than a black hole—these were not yet popular, though I had already written one paper with Kip Thorne of Caltech about looking for them. But the optically bright member of Cygnus X-1 (HDE 226868, but it won't come when you call it anyhow) was not my kind of star, and the compact component was and clearly is a black hole.

I've dabbled on and off in the statistics of the population of double or binary stars in our part of the Galaxy, including in a 2020 paper on which I am the junior author to Henri Boffin of the European Southern Observatory. Well, really, how could anyone resist having a co-author named Boffin!

One fine day, someone asked me to present the concluding remarks at a conference. Yes, I can do this (also, if needed, the hula—just take lots of notes). My first conference introductory talk goes back almost to the beginning, and I've done lots of both of those, occasionally in a sort of "tag team" with Woltjer, where for many meetings over several years we somehow alternated between those tasks.

Topics I've invented for myself have included quantitative measures of the productivity and impact of different types of telescopes (yes, big and rich is important, but so is having the right number and kinds of users who really care about what they are doing), as well as the productivity and impact of different kinds of astronomers (it paid to be a mature, prize-winning theorist, working on high-energy astrophysics or cosmology at a prestigious institution; it also paid to be male, though not very much, as long as I kept Charlotte Moore-Sitterly in the sample, even after she died.). Working with the help of Katelynn Horstman, a very gifted undergraduate from UCLA, I (oh, all right, she did all the work) looked at whether papers with women as first or corresponding authors are accepted faster, slower, or the same as papers with male skippers. The distributions of elapsed times are different, in favor of the guys, but likely not by enough to affect who gets a Nobel Prize for doing something first.

Very curiously, I think, I seem to have been the only one to look at the long-term outcomes of the decadal surveys that astronomy has been doing since 1960 to establish priorities for U.S. federal funding of large projects. The answer, summing the six to 2010, is that the U.S. government provided the money for about one-third of what the survey panels thought we wanted; one-third (like the Keck Telescope) were done through state or private funding or were done by other countries; and about one-third didn't happen at all. And many of them we would no longer want. This is, I suppose, the wisdom of small crowds, in having 20 to 200+ people involved in the various surveys, though for 2010 and 2020 anyone who wanted to could provide input in the form of a "white paper." Most ended up with many authors; there were hundreds of them, and I am last and least author of a few.

And, on another fine day, also lost in the mists of time, someone phoned from the nominating committee of some professional association and asked me to be a candidate for some office or committee. I said OK and won. Not that I have always won everything. Back at UCLA I wanted very much to be part of a team of fourteen students who would spend most of their summer in India visiting Indian college students, and I applied two years running. I was "deselected" both times. You can perhaps guess why. I did at least my fair share of the work in the preparation process, which lasted several months. But the final selection came from all of us voting for whom we would like to go with. Nicer people won. I eventually got to India in 1984, for conferences, by then a distinguished foreign visitor, just in time to have the second conference of the pair derailed by the assassination of Indira Gandhi. We visitors got to ride elephants instead. Wonderful for us; perhaps not so much so for the elephants.

But other nominating committees phoned; I always said yes, so that over the years, I have been a vice president (and many other things) for the American Astronomical Society; council member (and many other things) for the American Physical Society; chair of the commission on astrophysics of the International Union of Pure and Applied Physics; Vice President of the International Astronomical Union and president of two of its Divisions (Galaxies and the Universe, and Union-Wide Activities); committee, board, and other minor task member for the Astronomical Society of the Pacific, Phi Beta Kappa, Sigma Xi, and a few other professional organizations. And what have I been doing for some of these societies recently? In APS, membership on its Council (representing the Forum on History and Philosophy of Physics), chair of its historic sites committee, and membership on the Committee for International Freedom of Scientists (which gets to select the winner of the biennial Andrei Sakharov Prize); in AAS, membership on the Working Group on the Preservation of Astronomical Heritage and "agent" for the AAS at UCI; in the IAU, president of its Commission on Binary and Multiple Systems of Stars, membership on the Organizing Committee of the Commission on History of Astronomy, and liaison to the Commission on Astrophysics of the International Union of Pure and Applied Physics. And looking around just a little further, membership on a smattering of editorial boards and scientific organizing committees of conferences. It will strike cognoscenti that these are mostly "entry-level positions," meant for young astronomers, to which I can only say, "These jobs are generally done by those who are willing to do them." Some of the recent tasks have arisen because, with the advancing years, I find that many things I remember as "current events" (like the discoveries of quasars, pulsars, and the microwave radiation) now count as history. Thus many of my papers and talks over the past twenty or so years have been in the history of science.

It has all been, and remains, as I hope is obvious from these words, enormous fun! I think, looking back at the students with whom I went through graduate school, that the very best got the best jobs and that I ended up with at least as good a one as I deserved. I have also compiled statistics that show it pays to go to the most prestigious graduate school that will take you, while your undergraduate institution matters a good deal less, provided only that it is big enough to offer a range of classes and majors, in case you want to change your mind.

Notes

1. J. Cohen et al., "Uncle Jesse and the Seven 'Early Career' Ladies of the Night," *American Journal of Physics* 87 (2019): 778.

2. For a bit more of the back story and my take on his life, see V. Trimble, "Wired by Weber: The Story of the First Searcher and Searches for Gravitational Waves," *EPJH* 42 (2017): 261, and V. Trimble, "What are the Wild Waves Saying? Yet another Meditation on the Predictions of, Searches for, and Detection of Gravitational Waves," *International Journal of Modern Physics*, Part D Vol. 27, No. 14, ID 183009-499.

3. V. Trimble, "Lodewijk Woltjer (1930–2019)," *Bulletin of the American Astronomical Society* 52, no. 2 (2020).

Chapter 7

Roberta M. Humphreys (PhD, 1969)

Be Your Own Advocate

Roberta Humphreys, now Distinguished Professor Emerita, College of Science and Engineering and Professor Emerita of Astronomy, University of Minnesota, is one of the rare scientists to have a phenomenon named for her that emerged from her extensive research on massive stars: the Humphreys-Davidson Limit for the maximum brightnesses of stars. She was the first female hired as a faculty member in astronomy at the University of Minnesota, in 1972, and later served as Associate Dean for Academic Affairs in the College of Science and Engineering. She is an Honorary Fellow of the RAS, an Alexander von Humboldt Senior Scientist Award recipient, and a Fellow of the American Association for the Advancement of Science and the American Astronomical Society. Her asteroid is 10172 Humphreys.

I had my final oral defense for my PhD two days before Neil Armstrong walked on the Moon. I celebrated both fifty-year anniversaries in 2019.

As was the case for many of us, I was the only girl in my high school advanced science and math classes. I should give credit to my high school math teacher. Mr. Ledger was fair and encouraging, and he was interested in astronomy. His daughter, whom I met later, became a staff astronomer at the Space Telescope Science Institute. There were more women in my college classes at Indiana University. One of my classmates was Caty Garmany.

I didn't really experience overt discrimination, in the form of teasing and hazing, until graduate school at the University of Michigan. There were only a couple of women among the thirty or so grad students. But my role model and in many ways best friend was Anne Cowley, who was on the science staff at that time.

I didn't realize until near the end of graduate school that I was in a competitive environment for faculty attention and of course eventual jobs. My PhD advisor, Billy (W. P.) Bidelman, was a prominent stellar spectroscopist. I always felt that he didn't think I was serious. He had a male student, a year or so ahead of me, who was like a surrogate son. Billy never paid much attention to me so I was pretty much working on my own. Eventually the other student finished and went on to a faculty position. So I thought here is my chance for some attention. One day, Billy walks into my office, I show him my work, what I'm doing. He says, "Good work, just what I wanted to see, keep it up," and walked out.

In those days, jobs weren't advertised; they weren't open to fair application. It was indeed the "old boys' club," quite literally. By 1969, the job market had sort of just collapsed with NASA phasing back. So I asked for help in the job search. I thought Bidelman would do what he had done for his male student. But he said I couldn't expect him to stick his neck out for me. After all, I was a woman, and in a few years I'd just get married. But, he said I should write around, and if I got a nibble let him know and he'd follow up. I did get a couple. One was interested in me because I was woman and he could pay me less! This was 1969. Actually another faculty member helped me get a postdoc-like position for a year at Dyer Observatory, Vanderbilt University.

I'd always been interested in Galactic structure. My thesis was on the spatial distribution of supergiants. There were two important results. At that time there was a controversy between the distances to the spiral arms derived from optical tracers and those from 21 cm line radio measurements that traced gas clouds. I showed that when compared in velocity space (kinematics) the two agreed. The problem was with the radio distance determinations and the models that assumed circular rotation. One of the other related results was the discovery of non-circular motions between the sides of the arms, one of the expected results of the new density-wave theory.

While at Vanderbilt, I had obtained telescope time at Kitt Peak, in Arizona, and Cerro Tololo, in Chile, following up on the Michigan objective prism survey of the southern hemisphere and my work on spiral structure. While at Michigan, I had attended a summer school on Galactic structure where I met Bart and Priscilla Bok. The director at Dyer was friendly but a real misogynist. So, I wrote to Bok, told him about what I was doing, and he said, "Come! Come to Arizona." Which I happily did.

I liked Steward Observatory and Tucson. When my postdoc was ending, I told Bart that I'd like to stay. He apparently informed the faculty, and they offered me an assistant professorship. Bart was no longer the chairman/director. When the then chairman called me into his office, he said that I was "the choice of his faculty, not his choice." He said he didn't believe women belonged in astronomy, etcetera. He emphasized that the offer was only for two years, with no additional commitment. This angered one of the women faculty, and she arranged for me to speak to the dean. That was even more surprising. His reaction was, "Well, that's too bad that X is such a sexist but, you know, that's the way it is." This was 1972, post-1960s, civil rights, women's movement, and so forth. What he and the chairman said was against the law even then.

Fortunately, I had two other offers, a postdoc at Kitt Peak and a faculty position at the University of Minnesota. I opted for Minnesota. Earlier I had met Ed Ney at Cerro Tololo. Ed was a cosmic ray physicist turned infrared astronomer. These were the early days of infrared astronomy, and Ed was making some of the first infrared observations of stars in the southern hemisphere sky. I had a large collection of supergiant stars I wanted to study, and Ed was happy to observe them; many were very interesting and new types of strong infrared sources. He later invited me to Minnesota to write a paper. So when the UM group was given a couple of new faculty positions, I received an offer.

I had definitely experienced sex discrimination and met some real misogynists, but I always seemed able to move on. It wasn't until I was at Minnesota that some incidents really did affect me in a more direct, personal way.

By way of explanation, the astronomy group at Minnesota was not really a separate department from physics but part of the School of Physics and Astronomy in the College of Science & Engineering. For many years it had a separate budget for teaching, faculty lines reserved for astrophysics, and a chairman of sorts who technically reported to the head of the school. Appointments and tenure went through the school. When I arrived, five people plus the two new appointments were doing astronomy, and several faculty worked in the area of space physics/cosmic rays, which sort of made up the larger group. Physics and astronomy were part of the Institute of Technology (IT), now called the College of Science and Engineering, which included several large engineering departments and the physical sciences. Physics was very ingrown. Several of the faculty had gotten one or more of their degrees from Minnesota and were the former students or postdocs of other faculty. This was also true of the astronomy group, with four members fitting this pattern.

In 1972, I was only the fifth and the youngest woman on the faculty in IT, which then had more than four hundred members. Three women were in math, one in physics (Phyllis Freier in cosmic rays), and one, me, in astronomy. Phyllis would sort of advise me at times on the politics and personalities in the school. I've never forgotten what she told me once: "they'll let you be here, they'll let you do your science, but your role is supporting."

Initially I got on quite well with Ed and continued to work with him on infrared sources, but I had my own research and began to move off in different directions. Due to departures, some additional young faculty were hired, including Kris Davidson. The Astronomy Department was created at about this time. There were now eight or so in astronomy and Ed was the chairman of the program. Previously, I had dated some of the physics postdocs, but when Kris and I started dating, it was clear that Ed did not approve. Phyllis Freier was married to a faculty member in physics. When they waived the nepotism rules a few years before, Physics did the right thing and put her on the faculty. But my relationship with Kris was not approved of. Well, we waited till we both had been promoted to associate professor with tenure. We kept our plans confidential and didn't announce that we were getting married until about two weeks in advance. We knew the situation. But I hadn't anticipated how severe it would be.

I can sort of imagine the news spreading down the hall. One of the more senior faculty members, who was in the astronomy group, came into my office and started yelling at the top of his voice. Screaming at me, "how dare I get married," "it will ruin your career," "it will ruin the department." And I'm looking at him aghast. One of his favorite persons in astronomy was Geoff Burbidge, who was married to Margaret Burbidge, and I said, "What about Geoff and Margaret?" "That's different," he said. "What about so-and-so and so-and-so down the hall? They're all married." He says, "Yes, but they're men." This was 1976.

Of course, Ed wasn't happy. He tried one of his little tricks. Actually they weren't so little and could be quite harmful. Minnesota has a small 30-inch telescope about an hour's drive from the city. He manufactured some reason why I had to go to the telescope the night before the wedding. He was testing me. This may sound minor, but it was very stressful; we had family visiting and many wedding-related details to attend to. I didn't go.

Obviously the next few years were difficult. Kris and I were ostracized by Ed. We received very small salary raises and minimal support in other ways. We tried to maintain positive relations with all our colleagues, but the other young faculty were intimidated.

Ed was an excellent physicist, but he had a personality quirk or character weakness. He had problems with personal relations with people. He always needed to have a pet, a favorite, and would set young faculty against each other. Most of his targets were graduate students. He would set them up in a situation to see if they would fail. He was testing. If you didn't live up to Ed's expectations, then Ed would try to push you out. He had a lot of problems because of that and eventually he stepped down as chairman.

During this time (1978–79) I made what I now consider some of my most important contributions to astronomy. I had collected a very large data set on luminous stars in the Milky Way and the Magellanic Clouds. I found that the distribution of the most luminous stars across the HR Diagram revealed an empirical upper luminosity limit now known as

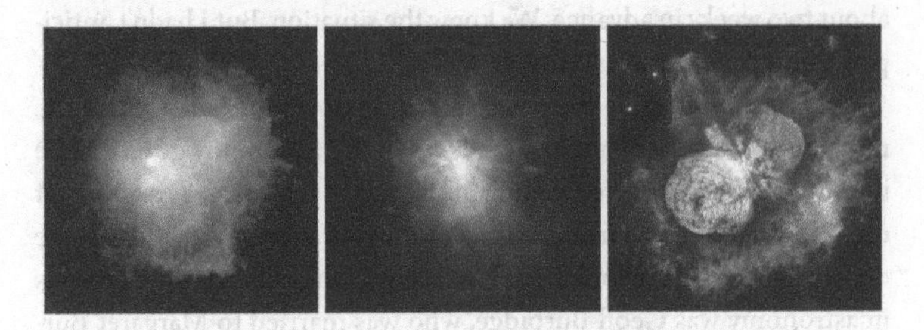

FIGURE 7.1. Hubble Space Telescope images of three massive stars with their complex circumstellar ejecta due to high mass loss events: the red supergiant VY CMa, the warm hypergiant IRC+10420, and Eta Car.

the Humphreys-Davidson Limit, which we attributed to large mass-loss events in the most massive stars due to their proximity to the Eddington Limit. This recognition has since played a major role in our understanding of massive star evolution.

At this same time I collaborated with Allan Sandage for about three years on a survey of the Triangulum Galaxy, M33. This experience was interesting but not always pleasant. Sandage had a very strong personality and could be quite charismatic at times, but he was definitely very old school. He would begin his phone calls with "Hello princess, this is your Uncle Allan." This was the late 1970s, so my first thought was of Princess Leia and Darth Vader.

Back at Minnesota, another younger person was brought in as chairman. He had a young family and seemed to be very pro-family. So Kris and I decided to start a family. In 1979–80, Minnesota had no maternity leave policy. You took your chances or, most likely, resigned. So again, we kept it quiet for a while. Eventually, I called the chairman into my office and said, "I'm expecting a baby, it's due in May." He says, "Oh, okay," and walked out. He comes back in a couple of minutes later, looks at me, and says, "This May?" (This was January, Ha!) I said, "Yes, this May." "Oh! Well, what are you going to do?" Big grin. "Are you going to resign?" Grin, grin. He's thinking open position! "No. I'm not going to resign." "Oh. Are you going to go half-time?" Grin. I said, "No." And he says, "What are you going to do?" I said, "Child care."

The baby was due in mid-May. The university was on the quarter system then and classes didn't end until early June. So I faced a few weeks of uncertainty. Earlier, a very senior member of the faculty had a heart attack, and we all covered his classes for most of a semester. So, I asked if I could be relieved of teaching for a few weeks. The answer was no, but if I double taught that term, someone could cover my classes after the baby was born. Our son was born late. There were only a few classes left and, of course, Kris covered them.

Our salaries had been kept low for several years, so I approached this same guy about getting an equity raise. His response was that I couldn't expect to get as high a salary as the men. Besides there were two salaries in my family. When I pointed out that several of my male colleagues had working spouses, his response was that he didn't have control over their salaries. Despite his interest in social engineering, this was against the rules even then.

I had a strong research record and was a candidate for promotion to full professor in the early 1980s, but Ed Ney had managed to delay it for a couple of years until Phyllis Freier went into her feminist mode and announced at a faculty meeting that there was going to be a sex discrimination case and she'd be testifying on my behalf. I later learned that Ed was simply told, "not this time." There was no problem after that.

After all of this, one may wonder why I didn't leave. With a husband and a child, it was harder than you may think. We had the famous two-body problem. In the 1980s, that was a kiss of death. It wasn't like today when a real effort is often made to solve the two-body problem. By the mid-1980s, however, there was a real push to hire women. I was asked by several places to apply, but it soon became obvious that they were just looking for a token female on their short list. Actually, one university did have two open faculty positions. We were both on the short list, were both interviewed, and were both invited back for a second visit that was serious. This opportunity looked promising, but we never heard back. Later, via the grapevine, I heard that someone had described me as an "overachiever."

The astronomy program at UM always had a problem with salaries. By the mid-1980s, I was well aware that our salaries were systematically lower than those of our peer group in physics. The problem was made worse by a couple of hires who entered with much higher salaries than those who had been there awhile. This is the well-known salary-compression problem. I pressed our then-chairman on the issue but to no avail. I later learned from a dean that he never took advantage of opportunities to raise salaries. By that time I had a very visible professional reputation. I was on committees: for the National Science Foundation (NSF), for NASA, for the Associated Universities for Research in Astronomy. I had grants and was invited to numerous international meetings. So I pressed my own case. Women's advancement was now an issue at Minnesota. The chairman actually said, "I don't give a damn what your record is." "All that matters is my perception of it, and I don't value your research." I worked on stars. I also led the Automated Plate Scanner (APS) project to digitize photographic plates, including the Palomar Observatory Sky Survey (POSS) plates. This was a very classical program for positions, proper motions, brightnesses, and colors of stars (aps.umn.edu). I always had trouble getting recognition and support from the department for my work with the APS.

A few years earlier, the University had effectively lost a big, class-action sex discrimination case. The result was the Rajender Consent Decree, with wide repercussions across the University for a decade. One consequence was a major salary adjustment for women. Because of my seniority in IT, I was in a special class of one and was therefore advised to apply for the special adjustment fund. I had to collect information for comparison with my peer group in the School. The results were as expected. I requested an equity salary adjustment with one particular faculty member. Well, I got almost twice what I requested. I guess I made my case. The year was 1989.

I continued an active research program with the APS on Galactic structure and on massive stars, including with the Hubble Space Telescope (HST). I also got active in faculty governance and was one of the faculty leaders in the "tenure wars" in 1995–96 at Minnesota. As a result I made numerous new friends and colleagues from across the University.

In about 2001, the university in its wisdom realized that its prominent women faculty were not getting the rewards and recognition they

FIGURE 7.2. Roberta Humphreys giving a talk on massive stars.
Credit: Department of Physics, University of Illinois Urbana-Champaign, courtesy of AIP Emilio Segrè Visual Archives.

deserved. The administration established a special prize just for senior women. Our chairman at that time, Len Kuhi, asked me if I'd like to be nominated. I thanked him but said that I was going to pull a Margaret Burbidge on him and said no thanks. Remember that Margaret had turned down the Annie Jump Cannon Award that was to be given by the American Astronomical Society in 1971 because it was for women only and kept them from receiving other forms of recognition. Instead, I told him I wanted the Distinguished Professorship, a college-wide award. Previously, no woman had received it. Well, I got it. At the ceremony, the dean whispered in my ear, "I'll never forget what you did." A year later, he appointed me his Associate Dean for Academic Affairs.

My position included hiring, promotion and tenure, retirements, and awards, and I was also responsible for the college's programs for women students and faculty. So you would think that I had finally escaped the Astronomy Department. Unbeknownst to me, the leaders in astronomy had decided to give the APS lab to an incoming faculty member. I was neither told nor consulted. The APS had been well supported by NSF and NASA for more than twenty years. As the APS project was concluding, Kris Davidson received an HST Treasury Program award to study Eta Carinae, so we had a couple of postdocs, students, and a technical support person in the lab, plus dedicated computers and other pieces of equipment. The APS was still there plus about three thousand POSS-sized glass plates. The space was in use. I continued my research program and was submitting a proposal to NSF for Galactic research using the APS. As part of the proposal process, the department chairman, head, and deans have to sign off on a form. It's routine. The guy who was in an acting position for just one year as Chair of Astronomy refused to approve it. He wanted the space. What was he thinking? I was in the dean's office. I saw the form. It took intervention by the head of the school and by the dean to get this guy to back down. I was thinking, does the harassment never stop? This was 2004.

I retired in 2017.

So has the academic environment changed for women? Yes. What I experienced in the 1970s and 1980s can't happen now, or at least can't happen overtly. Hiring and promotion are more transparent now. The issues are still hiring, promotion, salaries, and space. My advice is check. Check records, check salaries. A 0.2 percent difference every year may seem small, but it adds up after thirty years.

Most important, be your own advocate.

Chapter 8

Silvia Torres-Peimbert (PhD, 1969)

An Astronomer in Mexico

Silvia Torres-Peimbert, who has worked continuously at Universidad Nacional Autónoma de México throughout her career, conducts research on the chemical abundances of H II regions and planetary nebulae. She was President of the IAU and is the recipient of the Premio Nacional de Ciencias award, the Premio Universidad Nacional de Ciencias Exactas Medal of the Sociedad Mexicana de Física, the Lecture Medal of the Academy of Sciences of the World, the L'Oréal-UNESCO Awards for Women in Science in 2011 on behalf of Latin America, and the Hans Bethe Prize from the APS in 2012.

I will start from the beginning. My family was a typical middle-class example. My father was a doctor and my mother, who had been a schoolteacher, quit working when I was born. I grew up with a very traditional background. At that time, there were no expectations for women to work all their lives. The expected path was to prepare young women to be good housewives and mothers. My sisters were trained for secretarial work, since my father was very concerned that we be provided with the basic training to be able to support ourselves, if necessary. In my case, being the youngest, I was given more freedom to choose my studies. At school I stood out for my interests in mathematics and science in general.

During high school I became more acquainted with science, and chemistry caught my interest, so I decided to become a chemist. At the time I entered college my enthusiasm had shifted to physics. Therefore, I went to college at the Facultad de Ciencias of the Universidad Nacional Autónoma de México (UNAM) in Mexico City. In my country, college studies are highly concentrated on the major topic of interest, which in my case meant physics and mathematics courses. In addition, optional subjects were offered, and I decided to enroll in an introductory astrophysics course. That course captured my attention and put me on a definite road for my life: I decided to become an astronomer.

While an undergraduate I had the opportunity to join the Instituto de Astronomía (UNAM) as a research assistant. Through this activity, I became acquainted with the first computer in Mexico, an IBM 650 model. Of course, in retrospect we realize how very limited was the scope of research activities possible on this drum data-processing machine. I was hired as an assistant to Eugenio Mendoza, a photometrist, who required zenith angle tables to correct for atmospheric extinction for different coordinates and observing times.

My apprentice period revealed unexpected new avenues that were possible career paths for me. At the time the number of physicists in Mexico was very limited and the number of astronomers even more so. However, Guillermo Haro, director of the Instituto de Astronomía, had decided that young students should be exposed to other approaches and fields than the ones available locally. So he organized a program of scholarships to train graduate students abroad.

In the meantime, I had met Manuel Peimbert, who had similar interests to mine, and married him. Both of us were accepted to study astronomy at the University of California, Berkeley. We arrived at Berkeley in January 1963, where my first encounter was with Louis Henyey, who was the chairman of the department. He was very generous and offered to advise me on a small project that would allow me to obtain my undergraduate college degree (I finally obtained it in 1964). At Berkeley I also had the opportunity to meet and interact with many great astronomers, including Henyey, George Wallerstein, Hyron Spinrad, Ivan King, Karl-Heinz Böhm, and Erika Böhm-Vitense.

My graduate studies at Berkeley were very exciting and demanding. Wallerstein encouraged me to turn a class project into a research note about the presence of lithium in carbon stars. This research, which appeared in the *Astrophysical Journal* in 1964, was my first publication and a great experience! I followed this project with a study, based on observations made using the 36-inch Crossley telescope at Lick Observatory, in California, of the colors of light emitted by a group of stars rich in carbon in their atmospheres. This work was quite a challenge and another great experience. I still have visions of this venerable instrument—commissioned in 1895 as the largest reflecting telescope in the United States—which was very difficult to operate. The eyepiece had a field of view of one degree across (about twice the diameter of the Full Moon), and within that large area of the night sky we had to find the star and then manually move this heavy telescope to the required position. I envisioned continuing this research on carbon stars under the supervision of Wallerstein, but he moved from Berkeley to Seattle to establish the Astronomy Department at the University of Washington, and I was left without a project.

I again contacted Henyey, who had a group of students involved in developing a research program to understand the interiors of stars. Also at that time, based on the analysis of stellar absorption lines, Spinrad and Benjamin Taylor had proposed that there were several stars and Galactic clusters with extremely high metallicities, that is, they had anomalously high abundances of elements heavier than helium compared to most other stars; they named them "super-metal-rich stars." Therefore, I decided to compute models of stellar interiors for these anomalous stars to try to better understand their properties. The project required special opacities—parameters that characterize the effectiveness with which light is transported from the core to the surface of the star, which were kindly computed by Arthur Cox and John Stewart for my research. I then constructed my doctoral thesis on the topic of fitting the HR diagrams (plots of the luminosities versus the surface temperatures of stars) of the star clusters NGC 188, M67, and NGC 6791, all of which had been identified by Spinrad and Taylor as super-metal-rich, with my stellar interior models. For this project I had the opportunity of using Henyey's computer code, with the newly computed set of opacity tables. Of course, I was very excited about the project. I finished my computations in 1968, barely in time before the 650 computer was withdrawn from operation and a new computer with a new operating system was installed.

As part of my experiences in Berkeley, I must mention the impact of the many activities that took place in and around campus. In addition to the rich cultural activities, I followed from a distance the free speech movement, which started on campus in 1964, mainly at the South Gate and Sproul Hall. Also, I was influenced by the change in social behavior among the young population that was taking place.

At the time I was working on my thesis, Raymond Davis and his team presented the first measurements of the flux of neutrinos emitted from the core of the Sun, which come from the decay of boron-8 atoms that is part of the nuclear fusion reaction chain that powers the Sun. The solar neutrino flux that they measured was smaller by a factor of three than the stellar interior predictions made by John Bahcall. Inspired by this solar neutrino mystery, with two other student colleagues, Erik Simpson and Roger Ulrich, I decided to compute a set of possible solar interior models for comparison. Our result was that no reasonable composition for the Sun could yield the low neutrino flux observed.

Prior to finishing my thesis, I returned to Mexico City in 1968 where I was offered a job at the Instituto de Astronomía, UNAM. There I finished writing up my thesis, which was presented and accepted in 1969 by UC Berkeley. I should point out that my only employer has been Instituto de Astronomía, which has supported all my scientific activities; for this, I am most grateful. The papers from my thesis research were published in the *Boletín de los Observatorios de Tonantzintla y Tacubaya*, a journal that was published jointly by the Tonantzintla Observatory and the Instituto de Astronomía (Tacubaya).

At this point I had to rethink my career, because, although I had the stellar interior computer code from the Berkeley group that Henyey had generously shared with me, I realized I could not translate the extensive code to the local computer on my own, and I did not have enough computing time available either. As a result, I decided to change research

FIGURE 8.1. Silvia Torres-Peimbert delivering a lecture at Universidad Nacional Autónoma de México in 2018. Photo by J. C. Yustis.

fields to the study of the chemical abundances in gaseous nebulae. This field captured my interest and has held it ever since. At this time, I also started my professional association with Manuel Peimbert.

In 1972 and 1973 I observed at the Cerro Tololo Inter-American Observatory, in Chile, and had the opportunity to observe with the Image Dissector Scanner. The project was to determine chemical abundances in H II regions and planetary nebulae. Our data helped us realize that the published nebular line intensities derived from photographic plates were biased in the case of faint lines; in fact, faint line intensities had been overestimated. This result was important at the time, because great efforts had been devoted to explaining the hydrogen emission line intensities; and this effort turned out to be unnecessary. I also observed at Kitt Peak National Observatory (KPNO) in Arizona, to which I made several very fruitful observing trips from then on.

My life changed significantly in 1972 when my first child, Antonio, was born. Until that time, I had been dedicated only to my work. Soon afterward, in 1973 my daughter, Mariana, was born. Of course, I was delighted with my children, and since then I have enjoyed every minute of their presence. However, as expected, I had to reorganize my life in order to combine my family commitments with my scientific career. Fortunately, at the Institute it was possible to adjust schedules, and I was able to proceed with my projects.

In the autumn of 1975 we spent six months on sabbatical leave at KPNO. This visit was very productive, since we became better acquainted with the data analysis package called IRAF, now widely used for working with astronomical imaging data but then in the early stages of development. Then for the spring of 1976, we went to University College London at the invitation of Michael Seaton. We had a very productive visit, as we were especially interested in deriving precise abundances for elements in gaseous nebulae, and doing so required use of the excellent atomic physics data derived by Seaton's research group.

During 1976, NASA, ESA, and the United Kingdom Science Research Council were finalizing the design of the International Ultraviolet Explorer (IUE) satellite. At some point, Anne Underhill (project scientist of IUE for NASA) suggested that IUE would need and should have available only one small aperture (3 arcsec). I was very concerned about this proposed design feature and insisted that IUE should have on board the extended aperture (10 × 20 arcsec) originally planned for it, as it would be needed for studying objects other than stars. I am very glad that my suggestion was accepted, because this aperture turned out to be widely used. In particular, it allowed observers to more effectively study extended nebulae. Observations with IUE were very exciting and led to quite a change in perspective. I was granted time during several observing epochs, which allowed me to obtain very valuable ultraviolet spectra of several planetary nebulae (PNe).

In 1987 the Mexican astronomy group hosted IAU Symposium 131 on planetary nebulae, held in Mexico City. This meeting was very important and allowed us to become better known by the PNe community.

Besides my research work, from 1969 onward I became very concerned about the quality and distribution of the journal *Boletín de los Observatorios de Tonantzintla y Tacubaya,* which was edited by Guillermo Haro. I managed to get involved in different aspects of this

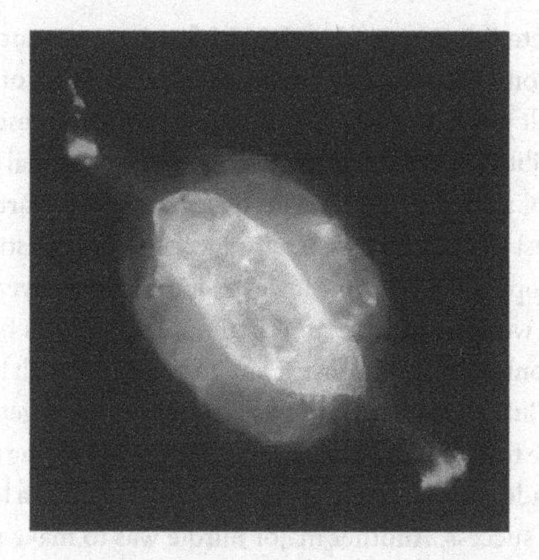

FIGURE 8.2. Hubble Space Telescope image of planetary nebula NGC 7009. *Credit*: Bruce Balick (University of Washington), Jason Alexander (University of Washington), Arsen Hajian (U.S. Naval Observatory), Yervant Terzian (Cornell University), Mario Perinotto (University of Florence, Italy), Patrizio Patriarchi (Arcetri Observatory, Italy), NASA/ESA.

publication; after a while this effort was recognized, and I became one of its editors. However, in 1972, Haro decided to break the joint publishing agreement between Tonantzintla and Tacubaya and continue this publication only from Tonantzintla. This decision frustrated many of us at our Institute, and we decided to establish a new publication, *Revista Mexicana de Astronomía y Astrofísica* (*RMxAA*), in which we would publish refereed material and proceedings of astronomical meetings held in the region. Eventually, the need to differentiate refereed papers from proceedings papers was recognized, and in 1995 we started the publication of *Revista Mexicana de Astronomía y Astrofísica Serie de Conferencias* (*RMxAC*) to publish only conference proceedings. From their starting dates until 1998 I was very involved in the trajectory of both of these journals. From 2001 onward I have been editor of *RMxAC*.

I was elected director of IA-UNAM for the 1998–2002 period, in which position I tried to carry out my duties to the best of my abilities. It is a difficult job, because we never have had enough resources to enlarge our facilities at Observatorio Astronómico Nacional at San Pedro Mártir, Baja California (OAN). This observatory was started in 1965 as a very modest installation; we finally succeeded in constructing a 2.1-meter telescope there in 1979. However, it took several years to get the telescope in working order. From then on, great efforts have been expended to continuously improve it and to provide it with better instrumentation. But the real problem is that we require a larger, more modern telescope to do cutting-edge research at OAN. During my tenure as director I made great efforts to try to find funds to build a larger facility, but without success. Another major hurdle was to make sure that the observing site would be protected and recognized formally within the Parque Nacional Sierra de San Pedro Mártir in Baja California. On the bright side, I was able to establish an agreement through which the Mexican astronomical community entered into a collaboration with the Gran Telescopio Canarias, on the island of La Palma in the Canary Islands, which was then under construction.

I have always been very dedicated to promoting the Regional Latin American Meetings of the IAU (LARIM). These meetings have been taking place since 1978. LARIM has acted as a catalyst to strengthen the Latin American astronomy community and has encouraged more collaborations. I have been very supportive of this activity, and *RMxAC* has published the proceedings of fourteen of the sixteen meetings held.

Planning for the International Year of Astronomy (IYA 2009) turned out to be very exciting. It required a lot of attention and effort from many professional and amateur astronomers and turned out to be a great success in many countries, where a whole array of different activities took place. In Mexico IYA 2009 triggered a strong interest in astronomy among the general population. It also encouraged many young people to engage in outreach activities in astronomy. I became very involved with these activities and have multiplied my outreach activities since then.

At the IAU General Assembly held in Beijing in 2012 I was designated president-elect, and since then I have thus followed closely the development of this organization. Its main mission, "to promote and safeguard the science of astronomy in all its aspects through international cooperation," continues to be pursued; nevertheless, a major change took place at the time of IYA 2009, with the adoption of the "Strategic Plan." It led to the establishment of several important new offices, namely, the Office of Astronomy for Development, followed by the Office of Astronomy Outreach, the Office for Young Astronomers, and the Office of Astronomy for Education. This enlargement of the goals of the IAU has been very important to the organization.

My overall research has focused on determining the chemical abundances of regions of ionized hydrogen (H II regions) and of PNe. PNe, clouds of gas and dust that have been expelled into space from the outer layers of dying stars, give us information about their progenitor stars and about the chemical evolution of the Galaxy, because they enrich the interstellar medium, particularly with nitrogen, carbon, and helium. From the ratio of oxygen to hydrogen atoms in PNe, I found that the chemical composition of these objects shows a spread from the very metal-poor objects (i.e., very few elements heavier than helium) to those that have the same composition as the Sun. The oxygen deficient PNe come from the least massive progenitors (less than one solar mass) while those of solar composition and with an excess of nitrogen are the result of the evolution of stars in the 2–8 solar mass range. I have studied several PNe that merited special attention, namely the variability of the spectrum of N66 in the Large Magellanic Cloud and the scattered light from M2–9.

On the topic of H II regions, I have worked on measuring abundance gradients of the gas in spiral galaxies by determining the abundances in the regions least contaminated by the matter ejected by stars in previous generations. Through this work, I have shown that stars enrich the interstellar medium not only with heavy elements but with helium as well. Through studies like these, it is possible to determine the primordial helium abundance, which provides additional information about the overall properties of the Universe. For this work Manuel and I, together, were awarded the Hans Bethe Prize of the American Physical Society in 2012.

I have seen the Mexican astronomical community grow from not more than ten to the current situation of about three hundred professional astronomers and several graduate programs. With this growth the topics of study have multiplied from stellar astronomy studies to all fields of astrophysics, with all possible observational techniques and computing simulations. I consider myself fortunate in having been able to carry out my research in my own country. I would like to think that I have helped to build our astronomical community and to excite young people on the topic of astronomy.

Each of the new results that are published in the different topics in astrophysics inspires me and leads to the confirmation that my choice of career was indeed the correct one.

Chapter 9

Neta A. Bahcall (PhD, 1970)

My Life in Astronomy

Neta Bahcall, the Eugene Higgins Professor of Astrophysics at Princeton University and Director of the Undergraduate Program in Astrophysics, has made important contributions to cosmology and our understanding of the large-scale structure of the Universe. Her work was one of the first to reveal the existence of very large-scale structure in the Universe and to show that the Universe has a low, subcritical, mass density that is insufficient to halt the cosmic expansion. She is a member of the National Academy of Sciences, a Fellow of the American Academy of Arts and Sciences and a Legacy Fellow of the AAS, and was a Vice President of the AAS. She is the winner of the de Vaucouleurs Medal from the University of Texas, presented the Bennett-McWilliams Lecture at Carnegie Mellon University and the Cecilia Payne-Gaposchkin Lecture at Harvard University, received the President's Distinguished Teaching Award at Princeton and the Distinguished Research Chair at Perimeter Institute, and was the Caroline Herschel Visitor at STScI. Her asteroid is 137166 Netabahcall.

My love of science began with my wonderful high school science teacher. He made science exciting, approachable, fun. I loved the beauty and thrill of answering basic questions of how the world works: what makes a rainbow, why is the sky red at sunset, why does water freeze, what are stars made of. The ability to understand such phenomena was awesome. In 1965, as a physics graduate student at the Weizmann Institute of Science in Israel, I met my future husband, John Bahcall, a visiting American astrophysicist. We met, fell in love, married in 1966, and moved to Caltech, where John was a junior physics professor. At Caltech—a powerhouse in astrophysics—I met leaders in the field and became fascinated with the topics they were exploring: dark matter in the Universe, cosmic expansion, quasars, distant galaxies, and so much more—and changed my research direction to astrophysics. I have worked in this field ever since and loved every minute. After completing my PhD I moved with John and our young kids to Princeton, where I continued my research in astrophysics, exploring the mysteries of our cosmos. My awe at how much we have learned about our Universe has never waned. It still amazes me; it still inspires me.

I grew up in Israel during a turbulent time, from before Israel's 1948 Independence through the young country's infancy and the ensuing wars for its survival. This was a life-forming experience—living through tough times of war and meager resources, learning to become independent, self-sufficient, resilient, and to appreciate the values of life, peace, family, and education. I was born in Tel Aviv in 1942. My mother emigrated to Israel from Russia just before World War II; she became a nurse and worked in a hospital her entire life. My father escaped Nazi Europe and arrived in Israel from Vienna during the war; most of his large family did not survive the Holocaust. My father, a lawyer, established his law office in Tel Aviv, where I frequently worked over the summers in my youth. He wanted me to become a lawyer; my own desire was to become a medical doctor. I saw doctors every time I visited my mother at work, and I was impressed with their dedicated work to save people's lives. There were no openings available at the only medical school in Israel in 1960; so I turned to my second love—science.

I loved the beauty, clarity, and excitement of science. I was greatly influenced by my science teacher who made science so exciting. I majored in physics and math at the Hebrew University in Jerusalem. Originally, I was planning to become a high school science teacher, like my favorite role model. At the University, I was one of only a couple of female students in a large class of over one hundred physics and math students. I was not distracted by that fact; I simply did what I enjoyed doing and thrived as a student. After graduating, I decided to postpone my teaching career and, instead, pursue a master's degree in nuclear physics at the Weizmann Institute, which I completed in 1966. These were exciting early days in physics at Weizmann, with much progress in nuclear physics. I greatly enjoyed my work at Weizmann and still regard the place as a second home. Even more special, I met my future husband there. John had come for a brief visit from the United States to Israel in 1965 and gave lectures at Weizmann. We met; he asked my advisor to introduce us; I showed him my lab; he asked me out for dinner; I said I was too busy; he asked me again and again; I finally said yes. For our first date we went to the Old Opera in Tel Aviv by the Mediterranean Sea. After the performance, we walked along the beach in the beautiful moonlit summer evening—I can still vividly see it in my mind—and we fell in love on that first date. John did not speak Hebrew, and my English was not good, but that did not stop us. We talked about physics, about his early work on solar neutrinos, about my work in nuclear physics, and about life in general. We married a year later and never stopped loving each other. It was a match made in heaven.

After we married, I moved to Caltech where John worked. These were exciting times for us—a young couple starting their life together and a new place and new culture for me to adjust to. And I had to improve my English. I loved it all. I greatly enjoyed Caltech, Pasadena, and our colleagues who became lifelong friends—including Wal and Anneila Sargent, Pete and Sue Goldreich, Gerry and Naomi Wasserburg, Marshall and Shirley Cohen, Maarten Schmidt, William Fowler, and many more. I worked on my PhD in nuclear astrophysics, with Fowler (a physics Nobel Prize winner) as my amazing and inspirational advisor. This work allowed me to combine my nuclear physics background with the astrophysics topics being explored at Caltech. Understanding nuclear reactions in stars was a major scientific frontier at that time, following the seminal 1957 paper by Burbidge, Burbidge, Fowler, and Hoyle and the outstanding nuclear physics experiments being done at Caltech. My thesis focused on determining the effect of excited nuclear states on stellar reaction rates and related topics. I was privileged to be in such an outstanding place and work with Fowler, the world's leader in the field. We remained close colleagues and friends till his passing in 1995. Those days were also a gold mine for astronomy at Caltech, with Schmidt's discovery of distant quasars and Fritz Zwicky's work on clusters of galaxies, dark matter, and more. I loved learning about these exciting discoveries. I was delighted to work with Schmidt on identifying quasars on the three-color image plates he obtained at Palomar Observatory; my role was to identify quasars on the plates and explore the relationship between quasars and galaxies. We published a couple of papers together (the quasars we discovered were named, following Schmidt's suggestion, NAB—####, after my initials); I greatly enjoyed the work. I also had wonderful discussions with Zwicky, who was extremely nice to me, though I was only a young graduate student then. We talked about dark matter and clusters of galaxies, which impressed me and influenced my future work. (He was the first scientist to discover, in the 1930s, dark matter in clusters of galaxies.) This began my love with astrophysics and my remarkable, enjoyable, and productive journey of research, exploration, and exciting discoveries. And it all began in those early days at Caltech.

During my years as a graduate student, from 1966 to 1971, John and I enjoyed our new life together. Our two boys, Safi and Dan, were born then; our daughter, Orli, was born after we moved to Princeton. I was busy—working on a PhD in astrophysics, starting a family, and raising two small children. John's dedication to our family life—always being a supportive, helpful, and loving husband and father—was pivotal. We were also fortunate to have great help from my mother, who came from Israel to help us for several months after each child was born. Friends frequently ask if these were difficult times for me; my answer is "No," these were times full of excitement, happiness, and joy. Caltech, too, remains my second home.

There were only a few women physicists or astrophysicists at Caltech at that time (just as at other places). I was one of several women graduate students, including my friends Virginia Trimble, Anneila Sargent, and Donna Weistrop. Margaret Burbidge visited frequently to work with Fowler. I always enjoyed talking with her; she was a singular role model at the time. As before, I mostly focused on what I enjoyed doing—research, studying, interacting with my colleagues, and of course spending time with my family, which has always been my top priority.

After I completed my PhD, John was offered a prestigious position as professor at the Institute for Advanced Study (IAS) in Princeton, New Jersey. It was a highly desirable position (Einstein was famously the first professor at IAS). John was recruited by Freeman Dyson and other IAS faculty to start the Astrophysics Program at this distinguished institute. John was only thirty-five years old then, very young for such a high-level position. I was offered a postdoctoral fellowship in the Astronomy Department at Princeton University. I was hesitant about moving to Princeton—I loved Caltech and Pasadena and did not particularly want to live in a small town and in cold weather, or what I imagined (incorrectly) was a formal East Coast town—but we decided to move and give it a try. I vividly recall John's friend Marshall Rosenbluth—a world leader in plasma physics and an IAS professor then—convincing me to relocate to Princeton with this simple and wise piece of advice: "Don't think of it as a permanent decision; come for a couple of years; if you don't like it you can move." That logic persuaded me. We moved to Princeton in 1971, fell in love with the place, and never left. We enjoyed the Princeton community, the outstanding science at the IAS and University, and our many colleagues and friends.

When I started my postdoctoral work in the Astronomy Department we had two little boys at home; a babysitter helped during the day (no daycare programs were available then) and my mother would come from Israel to help for long periods of time. She was a wonderful help. It was a great time for our young, growing family; we enjoyed and thrived in Princeton. At the IAS, John created one of the top theoretical astrophysics groups in the country, which continues today. He mentored more than three hundred postdocs and young scientists, many of whom became leading scientists in the field, including in Israel. John helped bring astrophysics to Israel and helped train many of their leading astrophysicists. The IAS was a perfect fit for John, for his initiative and creativity, his high scientific standards, and his devotion to his

FIGURE 9.1. Neta Bahcall in her office at Princeton University in 1993.
Credit: Princeton University.

postdocs and young scientists. They loved him like their father, they said. The IAS has been a most supportive community throughout, with outstanding colleagues, a peaceful environment, and complete freedom to explore science. John thrived at the IAS. He was awarded the National Medal of Science from President Clinton, all the top awards from the American Astronomical Society (AAS), and numerous additional awards, medals, and honorary degrees for his pioneering work on solar neutrinos (for which he was expected to share the Nobel Prize) and his important research in Galactic and extragalactic work. He served as President of the AAS, President-elect of the American Physical Society, Chair of the Astronomy Section of the National Academy of Sciences (NAS), and Chair of the 1991 Decadal Survey in Astronomy, and along with Lyman Spitzer and Robert O'Dell was a founding father of the Hubble Space Telescope (HST). In honor of John's work on Hubble, the astronaut John Grunsfeld took John's and my wedding rings up to the Hubble during the last HST servicing mission in 2009. I proudly wear the rings on my necklace every day.

My postdoctoral years at the University went well. The Astronomy Department was small but outstanding; it included luminaries such as Lyman Spitzer, Martin Schwarzschild, and Jerry Ostriker, but no women at that time (the 1970s).

Initially, I did research on clusters of galaxies, determining their structural and physical properties; I also studied the large-scale structure of the Universe, as well as quasars and related topics. Over time, I was promoted to senior research scientist—but not to a faculty position. It did not seem to bother me too much, as I enjoyed what I was doing. In the mid-1970s I was offered a faculty position at a good, nearby university. I considered it. I was hesitant about the daily commute, with two small kids at home. I consulted with several colleagues, and of course with John, who was always extremely supportive, helpful, and encouraging. One senior colleague, whom I liked and admired, made a comment that stunned me: he said, "Isn't it enough to have one science career in a family?" It took me a while to grasp what he meant. I realized then that my cultural experiences in Israel, where women could be and do anything—women serve and fight in the military, they pioneered building the country, and one famously served as prime minister—showed me that cultural differences exist and that his comment not only was bad but could have devastating effects on young women scientists. I ignored his misguided advice, of course, but I learned a lesson I always share with young women: You can do anything you want—don't let anyone tell you otherwise. Eventually I decided not to accept that faculty position—mainly because I enjoyed working at Princeton and did not want to commute. It was a good decision, and soon our third child, daughter Orli, was born.

In the early 1980s, when the Space Telescope Science Institute (STScI) opened in Baltimore to run the science operations of the future HST, I was recruited by the institute's first director, Riccardo Giacconi, to serve as head of the General Observer Branch. My job was to plan how best to select the most important HST science and establish the science policies for this soon-to-be-famous telescope. I had no faculty position at Princeton at that time; I mostly obtained grants from the National Science Foundation to help support my research. After consulting with John and others I decided to accept the position; it was an exciting opportunity to work on the long-awaited HST and help create its science legacy. I started in 1983 when STScI was in its infancy, with only a dozen people. It was an exciting, pioneering experience. I commuted from Princeton to Baltimore weekly, rented an apartment near STScI to stay a few nights per week, worked hard while I was there, and developed with my team the policies and procedures for selecting and conducting the best science with HST. I am proud that the policies and procedures we developed, including the introduction of Key and Large Projects, the science selection panels, the funding programs for General Observers, the open access to archival data for the community of users, and more, have all been highly successful and have now been routinely adopted by all space observatories and many ground-based telescopes. I have continued to advise NASA on their policies for science missions, including for the Spitzer and the James Webb Space Telescopes. I greatly enjoyed my time at STScI, and I loved working with Riccardo—he was inspirational and a dear friend, and we remained close friends throughout his life. It was a busy and demanding period at STScI—but also a most exciting experience in helping create the HST legacy. John and the kids provided vital support throughout; they were terrific. I became accustomed to the commute routine and kept my focus on the critical work to be done while at STScI and on my family when in Princeton.

After more than six years at STScI I decided it was time to end the commuting routine. By then, I had established all the science policies and selection procedures for Hubble, and the system was operational and ready. John and I were fortunate to receive several outstanding offers at that time. Princeton Astrophysics, headed by my colleague and friend Jerry Ostriker, was one of them. We decided on Princeton, where John continued at the IAS and I became, in 1989, the first woman full professor in the department. I have remained there ever since. I feel fortunate to be in a place like Princeton, one of the top astrophysics departments in the world, with outstanding colleagues, excellent students, a friendly and supportive atmosphere, and a close and warm community. I have thrived in Princeton, and I love it here.

Shortly after my return to Princeton I was appointed as director of the Undergraduate Program in Astrophysics. In this capacity I direct

FIGURE 9.2. The Bahcall family: Neta and John (center), Safi, Dan, and Orli, in Tel Aviv in 2003, at John's Dan David Prize award ceremony.

and mentor all our undergraduate majors, establish our teaching program, and oversee the academic development and excellence of the program. I have been serving as director for nearly thirty years, from the early 1990s to the present. During this time I have expanded our undergraduate astro majors class from a tiny program of typically ~5 students to a large and vibrant program of ~20 students. Our astro major is regarded as one of the most attractive and most supportive on campus. The enrollment in our courses has similarly increased dramatically; approximately one-third of all Princeton undergraduates take astronomy courses. I am proud of these achievements. I am also proud of mentoring hundreds of undergraduate majors and helping direct them toward their goals, and proud that about half of our astro majors are female; they thrive in our supportive department. Just as my high school teacher influenced me decades ago, I try to do the same for my students. I find it fulfilling and delightful to see my students succeed.

I have served in numerous science administration functions. I was director of the Council of Science and Technology of Princeton University and academic advisor at Forbes College at Princeton. I served on numerous national committees, including Chair of the Astronomy Section of the NAS; editorial board member of the Publications of the NAS; member of several Astronomy Decadal Survey Panels; member of several National Astronomy Advisory committees; member of several NASA and STScI Advisory Committees; and Vice President and Council member of the AAS.

The combination of doing research, teaching, mentoring, attending science conferences, giving lectures, working in science administration, and serving in national advisory and leadership activities has kept me busy and been educational, rewarding, and fun. I have enjoyed and benefited from them all.

My research has focused on observational cosmology, trying to learn more about dark matter and the mass density of the Universe, dark energy, the large-scale structure of the Universe, clusters of galaxies, the formation and evolution of structure, and quasars and their environments. My work addresses questions such as: What is the large-scale structure of our Universe? How did galaxies and large-scale structure form and evolve? What are the properties of clusters of galaxies (the most massive systems in the Universe), and how best can we use them as tools in cosmology? How much dark matter exists in the Universe and how is it distributed? What is the fate of our Universe—will it expand forever or just collapse? I use different methods and a variety of observational tracers to answer these questions, including studies of galaxies, clusters of galaxies, superclusters, and quasars. I combine observational data from large-scale surveys and other observations to determine the structure in the Universe and its properties and compare them with those expected from cosmological computer simulations of how these structures might have formed. My work with students and colleagues provided the first determination of how clusters are distributed in space ("the cluster correlation function," in 1983), the number-density of clusters as a function of their mass and redshift ("the cluster mass function," in 1992 and beyond), the distribution of mass as compared to the distribution of light or stars, the geometrical shapes of clusters, and large-scale structures. I have used these tools to provide powerful constraints on cosmology, including one of the first determinations of the mass density of the Universe, showing that it is only ~25 percent of the critical density needed to halt the cosmic expansion. This result is now part of the standard model of cosmology. Our determination of the distribution of clusters of galaxies[1] found that clusters were, unexpectedly, twenty times more clumped together in space than individual galaxies (relative to their mean separation). It revealed, for the first time, the existence of very-large-scale structure (~100 Mpc) and opened the door to new models of the Universe. Our work has been cited in more than three hundred scientific publications.

My life in astronomy has been an exciting and fulfilling journey. It has greatly exceeded my expectations since I started as a young student in Israel. My main advice to students, and especially to young women, is what I have learned over time:

- Do what you love; don't let anyone tell you that you cannot do it.
- Focus on your research and studies. This is where the fun and rewards are.
- Persistence; Persistence; Persistence.
- As my close friend Vera Rubin said: "Half of all brains in the world are in women."

My final words are those of gratitude and thanks to my many colleagues and friends along the cosmic journey. Most especially, I thank my wonderful children, Safi, Dan, and Orli—all scientists—for your love and support throughout; I am so proud of all of you; you are the stars that light my way. And last but not least, to my wonderful husband, my best friend, my inspiration, the love of my life, who passed away in 2005, after forty loving years together. He was, and is, the sunshine of my life, the wind beneath my wings.

Notes

1. N. Bahcall and R. M. Soneira, "The Spatial Correlation Function of Rich Clusters of Galaxies," *Astrophysical Journal* 277 (1984): 27–37.

Chapter 10

Catherine Cesarsky (PhD, 1971)

Equations, Satellites, and Telescopes

Catherine Cesarsky, currently High-Level Science Advisor at Commissariat à l'Energie Atomique and Chair of the SKA Council, conducted research in high-energy astrophysics and was the principal investigator for the infrared camera ISOCAM on ESA's ISO satellite. She served as director general of ESO, High Commissioner for Atomic Energy in France, editor in chief of *A&A*, and President of the IAU. She received COSPAR's Space Science Award in 1998, the Prix Janssen de la Société Astronomique de France in 2009, and the Tate Medal from the American Institute of Physics in 2020. The French government named her Grand Officier de la Légion d'Honneur. She is a member of the Academia Europaea, of the Académie des Sciences de l'Institut de France, and of the International Academy of Astronautics. She is a Foreign Associate of the National Academy of Sciences, a Foreign Member of the Royal Swedish Academy of Sciences, a Foreign Member of the Royal Society of London, and a Fellow of the American Astronomical Society.

Looking back at my life, I realize how incredibly happy and fulfilling it has been, and I owe much of this satisfaction to astronomy and astrophysics. I have had a successful career, have enjoyed doing research, and have held or now hold positions of power in science, much beyond any of my expectations. Also, very important for me, I have always enjoyed a good family life.

Trying to summarize aspects of my life, the first thing that comes to my mind is that I have been lucky. At the same time, I am wary because for some time now I have been paying attention to comments by men and women about their careers, for instance, as recipients of prizes or decorations. I noted that more often than not, successful women who talk about their careers say that they were lucky, while men are more inclined to assert their merit.

Never mind, I reiterate that I have been lucky and I am going to prove it. My first streak of luck was to have loving, intelligent, and nurturing parents and two remarkable older sisters. They were French but moved to Argentina at the end of World War II, and I was raised there, fully immersed in French culture.

At the French High School of Buenos Aires, I was lucky to have two excellent teachers, one man and one woman, in scientific subjects. Without these professors, I would never have thought of pursuing a career in science. They were at that time my one and only contact with the world of hard sciences, very different from the world of arts and humanities to which my parents gravitated. I was generally a good pupil, but by far my favorite subject was mathematics; I never had enough of it. We had relatively little physics, studying that subject only during the last three years. I found it challenging, but compelling. I was deeply attracted to nature, plants, trees, animals, rocks, the ocean, the Sun and Moon, and the beautiful starry sky of the southern hemisphere, which I watched all night while on camping expeditions. My goal in life became "putting nature into equations."

I was lucky again, after having decided to pursue physics at the University of Buenos Aires, to have studied there during one of the rare intervals in the second half of the twentieth century when Argentina enjoyed a democratic government. The professors at the University of Buenos Aires were incredibly motivated by their teaching duties. The competition was tough and there were many dropouts along the way, but the students who graduated had a high international standing. Luck also, at the university, led me to meet Diego Cesarsky, my accomplice, my husband, and my partner for life, and to have, later on, two beautiful sons with him. Lucky finally to witness, in my fourth year, the arrival of a new professor, Carlos Varsavsky, who was the first, in my university, to teach astrophysics. Diego and I, and several of our friends, joined his group, did master's theses with him, and assembled with our hands the first Argentine radio telescope, still in use. In 1965 and 1966, with a fellow student, Zulema Abraham, I developed stellar models with an incredibly primitive computer. Carlos Varsavsky held a PhD from Harvard, had worked at the University of Cambridge, and maintained excellent international connections. Thanks to him, we were able to meet in Argentina a Harvard professor, David Layzer, who helped Diego and me obtain PhD scholarships at the Harvard College Observatory. On the evening of our departure to the United States, after a military government had taken power in Argentina, the police went to our school and beat up professors and students. Carlos had his skull cracked.

Once in Cambridge, Massachusetts, as I enjoyed my graduate studies at Harvard, I continued the work begun in Argentina on stellar models with a professor from MIT, Icko Iben. With the superior computers I had access to there, my models of main-sequence stars ran in no time and facilitated the exploration of second-order effects. I also did some bold modeling of solar flares with David Layzer but lacked sufficient confidence in this work to publish it. I wished to do a theory-based thesis, but the professors at Harvard offered observational projects.

I had the opportunity to meet, at an astronomy meeting, an astrophysicist and plasma physicist from Princeton, Russell Kulsrud, who offered me a summer job studying cosmic rays. My fresh perspective allowed me to unveil new facets that strongly changed the conclusions of his work. In 1971, we published together an article that was well received. I was very impressed with Russell's scientific breadth and clarity of thinking, as well as his deep knowledge of plasma physics; I chose him as my thesis advisor even though he was far away in Princeton. I was then leading exactly the kind of life I had dreamt of. I was a theorist, I worked at night, my head full of ideas, and my tools were essentially a block of paper, a pencil, and an eraser. From time to time I sent a handwritten letter to Russell and every three or four months I would go and see him for a couple of days. This allowed me to develop a model for the propagation of cosmic rays in the Galaxy that

FIGURE 10.1. Catherine Cesarsky in 2016.

was original. Parts of it, published in 1980, and my subsequent work are still used today.

I will admit that not luck but our own merit, or the resonance of the work done during our PhD theses, led Diego and me to being awarded postdoctoral contracts at Caltech. I had three years of total immersion in astrophysics, with access to Peter Goldreich, one of the greatest astrophysicists of the twentieth century. That is really where I ended up acquiring the passion for astrophysics that has never left me and that has illuminated my life.

We had planned to return to Argentina, but the political situation there was again unpalatable. We had met several French astrophysicists in the United States, and unrequested job offers arrived from my native country. I was hired in 1974 to work at Saclay, close to Paris, by the French atomic energy commission, CEA, which has a strong sector of activities in fundamental sciences.

For the first ten years, I did theory work on the origin and propagation of cosmic rays and on the interstellar medium, and I achieved high recognition in my field of work. I was working as I had wished, alone or in small groups, supervising graduate students, having great moments of inspiration in the middle of the night and at dawn cracking problems that had kept me awake. I led the small theory group of the Astrophysics Department, about ten people. Our department was very strong in building hardware for space experiments, and I became very ambitious for it and in general for European space astronomy and started acquiring influence in France and Europe. I became interested in infrared astronomy, which could provide unique complementary data for my studies. I promoted the construction by the European Space Agency (ESA) of ISO, the Infrared Space Observatory, judging that it was a unique opportunity for European science in space to be in the front line. To consolidate this effort, I was compelled to organize the development of the ISO camera, ISOCAM. I ended up taking the lead of the ISOCAM consortium at the request of the other participants. I then accepted the leadership of the Astrophysics Department at the CEA (about 150 people), which previously had not been involved with infrared astronomy, in order to consolidate the laboratory around the project. From then on, I have been involved in several space projects and have played important advisory roles for the French space agency CNES and for ESA.

As ISOCAM Principal Investigator, in 1985, I was the first woman PI at ESA. I hired an excellent engineer to manage the construction, Danièle Imbault, and our principal contractor also hired a woman, Danièle Auternaud, to run the industrial team. Almost ten years of intense collective work among the two Danièles, Francois Sybille, Laurent Vigroux, my husband, Diego, and 150 other people from various countries, laboratories, and industry allowed the delivery of ISOCAM to ESA. After ISO's launch in 1995, ISOCAM exhibited excellent performance, and for the first time we could see the shape of a galaxy in the mid-infrared domain. It was a unique moment in my life. In my wallet, next to the pictures of my children, I have always kept the thumbnail bearing this blessed image.

Seeing is good; understanding is better. The period of reflection concerning the results of observations of distant galaxies, as we tried to define the impact of these results on the general picture that we had then of the evolution of the Universe, provided perhaps the most interesting of all our efforts. The ISOCAM data from distant galaxies were particularly difficult to analyze, and I needed an excellent team, working hard for many months, to reduce the data properly. In the end, we were able to demonstrate that the rate of star formation in galaxies had been much higher in the past than it is now. We could look back to eras when the Universe had about half the age that it does today. Later satellite observatories and ground facilities have extended this work to cover most of the life of the Universe. Many other topics in astrophysics were enhanced with ISOCAM data.

In the meantime, the CEA Directorate promoted me to the position of Director of Basic Research in Physics, Chemistry, and Universe Sciences, leading a team of about 1,800 scientists and engineers. I had to familiarize myself with the work going on in nuclear fusion, particle and nuclear physics, chemistry, lasers, condensed matter, climate studies, and other subjects. I found myself in charge of numerous laboratories, each one seemingly doing more exciting studies than the other. My membership on the CEA's board of directors allowed me to learn more about its missions, activities, and potential. I became extremely interested in the very large European Intergovernmental Research Organizations (EIROs) around large instruments, the ESRF synchrotron source, the ILL neutron source, the JET fusion device, and the accelerators at CERN. At some point in my life, either then or later, I have been involved in all of these.

In 1999, another great turning point arrived in my life: I became the director general of an EIRO, the European Southern Observatory (ESO), the foremost European organization for ground-based astronomy. My parents had just died, my children had grown up, and my husband could easily pursue his career in Munich, near the location of ESO headquarters. This gave me a double return to my roots, first in astrophysics, second in the Latin America so dear to my heart, since the ESO observatories are in Chile. Running ESO proved a beautiful adventure for me. I was lucky to arrive just in time to supervise the exploitation of the Very Large Telescope (VLT), whose construction had occurred under my predecessor, the Nobel Prize winner Riccardo Giacconi.

ESO's discoveries followed one another in rapid succession, as all the sophisticated instruments constructed and installed for the four 8-meter telescopes of the VLT, as well as the interferometric systems that allow their use as a single instrument to achieve even greater accuracy, yielded excellent results. Particularly celebrated among astronomers has been the accurate follow-up over more than a decade of the motions of stars that orbit the central regions of our Galaxy, the Milky Way. This work, led at ESO by Reinhard Genzel, allowed him to demonstrate the presence of a black hole with a mass of about four million times the Sun's at the center of our Galaxy, a discovery that earned him 2020's Nobel Prize in Physics, shared with Andrea Ghez, who did similar work in Hawai'i with the Keck Telescope.

At the same time, we began the construction of a world-wide project in millimeter and submillimeter astronomy, ALMA, dizzying in scope from the ambition of its scientific objectives to the complexity of its creation, which produced 66 highly accurate dishes installed at an altitude of 5,000 meters in the Atacama Desert. ALMA is the only large-scale scientific project in which Europe and North America are equal partners, with the strong involvement of East Asia. Its harvest of results is totally extraordinary and touches on every field of astrophysics. Particularly striking is the sharp image of the protoplanetary disc around the star HL Tauri showing planets in formation within the disk-like structure.

I also launched the studies for a future European Extremely Large Telescope (ELT) project, with a diameter of 39 meters, now under construction close to the VLT. As the time of my return to France approached, the French Research Minister Valérie Pécresse called me in Munich to invite me to participate in the implementation of her new plans for French universities. Thanks to her, I began again to forge ties

FIGURE 10.2. Catherine Cesarsky at the VLT on Mount Paranal. *Credit*: ESO.

with the French scientific world, which I found to be in the midst of profound change.

In 2009, I returned to France, and, as President of the International Astronomical Union, was feverishly busy preparing the actions linked to the International Year of Astronomy 2009, a year-long celebration of astronomy geared to the public. At the end our tally estimated that we had reached 815 million people world-wide.

Some time after my return to France, there was a new turning point in my life: I was appointed by the French government as High Commissioner for Atomic Energy, a position at the subminister level of advisor to the government on science and energy. As a result, I was lucky enough to find myself immersed in one of the most interesting issues in today's world, that of energy. I am very concerned about the future of our planet, in the face of multiple threats, including climate change. I have been interested in climate studies for a long time; I had set up a CEA-CNRS laboratory on this subject, which is performing spectacularly. During my tenure as high commissioner, the CEA has evolved to become the French Atomic Energy and Alternative Energies Commission.

Since 2017, I have been devoting much of my time to another large astronomy project, complementary to the ELT, ALMA, and space experiments: the establishment of the world's largest radio telescope, SKA, part in Australia and part in South Africa. I chair the council of the new intergovernmental organization, SKAO, running this project, which started construction in 2021.

In this book devoted to women astrophysicists, I am inclined to comment on my experience of being a female scientist in a period when we were a small minority. At the French High School of Buenos Aires, the last year was one of specialization. Of the eleven students of the math and physics section, I was the youngest, sixteen years old, and the only girl, the first one in the history of the school. It was a great year; I loved the contents of the courses; I had a lot of fraternity and complicity with my classmates. At the end of the year, at the final exam, I came well ahead of the rest of the class. *Once and for all, I learnt that men are not necessarily better than women in math and science.* But after the announcement of the results my fellow students shunned me and I experienced strong discomfort. As a result, later, at the University of Buenos Aires, I wanted to be among the good or even excellent students, but not *the* best, a wish that was easy to fulfill because I had very brilliant classmates of both sexes. I was part of a small group of top students; we discovered together the wonders of math and physics with glee, and overall had an excellent time.

At Harvard, where every year there was only one woman among the ten new PhD students in astronomy, we were two in 1966. This provided me with strong legitimacy in my own eyes: we were not token women, as we had really been selected on our merits, and this gave me the self-confidence needed to succeed. My fellow student, Sandra Faber, is today a very high-level astrophysicist in California. Like me, she has raised two children.

As a postdoc at Caltech from 1971 to 1974, I was confronted with a world in which professional women were extremely scarce. In fact, the department chair from Caltech met with me while I was still at Harvard finishing my thesis work, but after I'd been offered the postdoctoral appointment at Caltech, to tell me that he didn't believe in women in the sciences and to warn me in advance not to have children while working at Caltech. Despite his not-so-elegant warning, I was eager to launch a family and soon was pregnant; he then threatened to terminate my contract. Luckily the professors who had hired me, Peter Goldreich and Wal Sargent, defended me well. Of course, I was not entitled to maternity leave. I worked until the day before the birth and took just my annual month's vacation after the birth, which took place in early August 1972.

At Caltech, I worked with two other young women—Anneila Sargent, who was working as an assistant to the department chair, and a chemist, the only other female postdoc on campus—to survey by questionnaire all the Caltech women, plus a representative sample of men, about attitudes toward women on campus. In 1973, we also invited the young female students (who were pioneers because Caltech had just opened up to women undergraduates) to attend meetings in my apartment. Caltech's meager female population was made up exclusively of handpicked individuals. Yet the ambitions and expectations of the young female students were far lower than those of their male colleagues at the same level. Was it realism (there was only one female professor in the university), a real lack of ambition, or rather the fear of losing femininity by leaving the traditional position of women? At the time, this was much talked about and labeled "fear of success." I have the impression that it is far less prevalent in young women these days. Personally, I had a good dose of self-confidence, as I mentioned. But I have always been prey to what is now called impostor syndrome, something common in women, in men too but to a lesser extent. For example, when I obtained the prestigious COSPAR Space Science Award in 1998 I felt compelled to say in my acceptance speech: "I don't deserve it." This was the first time this award, given every two years, went to women (I shared it with Marcia Neugebauer). Since then, it has gone to only one more woman, Janet Luhman, in 2012.

Once in France, at the CEA, I felt accepted and recognized by all, with courtesy and simplicity. The director of the Astrophysics Department who hired me was a woman, Lydie Koch, but the group included really very few women. In twenty years, I went through all levels before becoming the head of the Directorate of Basic Research in Physics, Chemistry, and Universe Sciences in 1994. This is when I considered I had broken the glass ceiling. I was the first woman at this level in the structure of CEA, and the second, who has even greater responsibilities, started only in 2019. I am the only woman to have been DG of ESO. In 2016 CERN was the second EIRO to be led by a woman, Fabiola Gianotti, who is now in her second term of office. In 2006 I became the first woman President of the International Astronomical Union, but it is a great pleasure to see that I have already had several female successors: Silvia Torres-Peimbert, Ewine van Dishoeck, and Debra Elmegreen. When I returned to France in 2008, I became the first and only, up to now, female high commissioner.

Being a woman in a male-dominated environment helps a person emerge quickly from anonymity. I sincerely consider that as far as my career is concerned, the advantages and disadvantages of my gender have exactly balanced each other out. I have never made any requests; I have never leaned on a protector or a mentor. I have attributed to myself the right to be a woman, a researcher, a mother, and a leader, probably drawing from a background of willpower, good health, a good dose of work power, and a passion for my profession. I think I have given up nothing important in life, except sometimes the chance to take a rest.

Chapter 11

Judith (Judy) Gamora Cohen (PhD, 1971)

A Long and Winding Road

Judy Cohen, the Kate van Nuys Page Professor of Astronomy at Caltech, Emeritus, is an expert on the structure of the outer halo of the Milky Way and has identified some of the oldest stars in the Milky Way. She also played an important role in the design and construction of the Keck Telescope and helped design and commission the LRIS for the Keck Observatory. She was one of the first women to observe at Palomar Mountain and Mount Wilson Observatories and was appointed in 1979 as the first woman on the Caltech Astronomy faculty. She is a member of the National Academy of Sciences and has presented the Cecilia Payne-Gaposchkin Lecture at Harvard University.

My grandparents fled to the United States from Kiev in the Ukraine and a small shtetl in Poland, to escape from Russian pogroms against Jews. They arrived in the U.S. with very little in terms of both money and worldly goods and settled in or near New York City in heavily Jewish neighborhoods.

My father was an accountant with his own practice. He graduated from New York University, which at that time charged only nominal tuition. My mother was a nurse who went back to school to get her bachelor's degree at the age of forty-five and later became the head of nursing at a large retirement home.

I was born in 1946, the oldest child of four girls. We lived in Brooklyn. I went to public schools, graduating as valedictorian of a class of more than five hundred students from Midwood High School. I also went to the Workmen's Circle (a descendent of the Bund in Eastern Europe) schools, both their local elementary and their high schools (the only ones in the city), which met on Sundays in lower Manhattan.

I became involved in the Junior Astronomy Club (JAC), which met near the Jewish high school. There are a number of people in the American Astronomical Society (AAS) who also were members of the JAC at that time, including Joel Levine. We went on observing trips with the small telescopes of members, held at the Woodlawn Cemetery in the Bronx, the nearest relatively dark site. An indicator of those times is that my parents let me go there alone on the subway at night, although they refused to let me go to the Apollo Theatre in Harlem where many truly great musicians played.

That left me with no time for friends, boys, or dates. But it didn't matter because none ever asked me out, although the bookworm in me had and sometimes still has feelings that hurt about these issues.

I got the money to go to Harvard (i.e., Radcliffe at that time for women) by winning a National Merit Scholarship. I was a very awkward, socially inexperienced young adult. Until after my second year at Harvard, when I spent a summer at Kitt Peak National Observatory in Tucson, I had never been west of Buffalo, east of Long Island, north of Maine, or south of New Jersey.

At Harvard, I wanted to have a social life as well as excel academically. I rejected math—the class I took was full of arcane proofs—and I hated blood, which ruled out biology. I ended up at the Smithsonian Astrophysical Observatory talking to a young staff member, Steve Strom, who offered me a part-time research opportunity; I wrote several papers with him, his wife, Karen Strom, and graduate student Bob Kurucz.

In my sophomore year I met a nice Jewish boy at Harvard from a wealthy family living in Manhattan. His parents liked me, and my parents were ecstatic. His parents took us out to dinner at places where the women were given menus with no prices. But he decided, without ever consulting me, to go to graduate school in Pittsburgh, and I had no desire to live there. That was the end of my first serious relationship.

A year later I met a Harvard astronomy major who was one year ahead of me. We became formally engaged. He was not Jewish, and my parents were not happy, but they put a good face on the situation and arranged (with his mother) a large engagement party. He decided to go to graduate school in astronomy at Caltech. Without saying anything to me over that year while I was finishing at Harvard, he met someone else in Pasadena, came back just before my graduation, and broke off the engagement. My parents had to return all the presents. Meanwhile I had been accepted for grad school at Caltech and had notified them I would come. One more note from that time: I also applied to Princeton Astrophysics but received a letter from that program saying that they did not accept women.

So I showed up at Caltech in the summer of 1967, totally bereft. People all thought that I was with my former fiancé, so I put a notice on a bulletin in the lobby of Robinson Lab (then the astronomy building) saying (approximate quote) "abandoned mistress seeks new friends and companions."

Caltech professor Jesse Greenstein had offered me a summer job. When I told him that my wedding plans had fallen apart, he suggested that I go to Europe for the summer to recover, but I could not afford this, so I worked analyzing some of his spectra with the techniques I had learned from Bob Kurucz and Steve Strom. Jesse was kind to me and invited me to his house several times for dinner.

I had also asked Caltech to give me a small house to set up a co-op for women only, which they did. But this living situation did not work for me, and I was the first person to move out, as almost all the other women were foreigners who were totally immersed in their work and not interested in any kind of social life.

I shared an office in the back of the Astronomy library in Robinson Laboratory with the very beautiful, young Virginia Trimble. Many people came to our office to talk to her, including Richard Feynman and Jim Gunn. As a result, I felt even more awkward, jealous, and unhappy.

Then I met a postdoc in astronomy. We spent many nights together in the small apartment I had rented close to campus. But he did not want to get married. I and he (at my request) went to the Caltech Counseling Center, but he was firm that he did not want to marry me, although he was happy to have good times together. Along the way, I suggested that the Counseling Center add a woman psychiatrist to their staff (part-time), which they did.

At this point I gave up on ever finding at Caltech or anyplace else an intelligent, educated man who would want to have a serious relationship with me. I had some relatives in Berkeley and shifted into a mode of working very hard for a month or two, then going to visit them in Berkeley. I even hitchhiked from Pasadena to Berkeley once to go to one of the big peace marches, this being the era of the Vietnam War. My cousin, who later was elected to the Berkeley City Council, took me to a party there but neglected to warn me that the punch was spiked with LSD; that was my introduction to drugs.

Meanwhile in Pasadena I was hard at work and had started observing at Mount Wilson and also at Palomar. Mount Wilson had some amusing traditions; an example: your seat at the dining table was determined by the telescope you were using, with the 100-inch telescope observer at the head of the table. When I first observed at Mount Wilson, women were not allowed in the dormitory. Instead, I stayed at the rustic Kapteyn Cottage, which had been built early in the twentieth century for J. C. Kapteyn and his wife Elise to stay in during their annual visits to Mount Wilson, was subsequently used for astronomers who brought their families with them, and was quite far from the Monastery (the name of the dining and sleeping area normally used by visiting astronomers). This cottage had, as its only source of heat, a wood-burning stove—I had never tried to use one before. After a while I was allowed to stay in the cook's cottage, which was close to the Monastery, then eventually in the Monastery itself. I believe this was finally achieved by Caltech pressuring the Carnegie Institution, which owned the Observatory.

At Palomar, women were already being accommodated even when I first arrived and had progressed from sleeping in the roped-off second floor of the dormitory (again, called the Monastery) to the normal bedrooms downstairs, where women were allocated both of two adjacent rooms that shared a single bathroom.

My PhD advisor, Guido Münch, was born in Mexico and did a PhD with Subrahmanyan Chandrasekar. He had many women in his life but was always more or less professional to me. My thesis was on the lithium isotope ratio in stars. My observational data relied on a then novel Fabry-Perot interferometer that he had built. The line profile had to be built up one wavelength bin at a time (i.e., it was a single channel measurement), and shifting the wavelength was also done manually, so it was quite time-consuming. Nevertheless, one eventually ended up with a very high resolution spectral line profile from which one could fit models that would yield the isotopic ratio for Li.

After getting my PhD, and taking a great month-long road trip across the western United States with my youngest sister to celebrate, I took up a Miller Postdoctoral Fellowship at UC Berkeley. That was an exciting time to live in Berkeley, in large part because of the protests at People's Park over the Vietnam War. I worked hard, and I played hard too.

The first meeting of the AAS that I attended was held in Seattle in April 1972. The dinner for the meeting was an outdoor barbecue on a small island close to Seattle. At this event, a senior astronomer who was a vice president of the AAS, whom I had never met before, perhaps got a bit too drunk and accosted me. I did not tell or complain to anyone, but just shrugged it off.

Having lost access to Palomar because I was no longer at Caltech, I started observing at the Cerro Tololo Inter-American Observatory (CTIO) near La Serena, Chile, which was built by the United States and accessible to all U.S.-based astronomers. I met Gaston Araya there. He worked as a support astronomer and mechanical technician at CTIO, and prior to that worked fixing trains and train parts for the Chilean National Railroad. Although he spoke little English at that time, I spoke semi-reasonable Spanish. He struck me as a brilliant man who grew up very poor and did not have a chance to get a good education. On every observing trip to Chile, I would stay an extra day or two to be with him.

I knew he was married, but he said it was over, and we discussed him coming to the United States and us getting married. I think this happened because he was the first really intelligent man I had met in years who was pleased rather than turned off by the fact I was very intelligent. After about two years of long-distance romance, we decided that he would come to the United States and we would get married. You can imagine my parents' very negative reaction.

After my three years as a Miller Fellow at UC Berkeley, I accepted a position as an assistant astronomer with an initial three-year contract at Kitt Peak National Observatory (KPNO) in Tucson. I was short of money and, given my situation with Gaston, did not want to ask my parents for help. Steve and Karen Strom graciously let me stay in their guest house for a few months. I needed to buy a house for us; I ended up buying one in the Black part of Tucson; the price was right, and I thought, stupidly, that he, being brown, not lily white, would be more comfortable there than in a white neighborhood.

Several months later, Gaston came to the United States as planned but explained to me that Chile, being a Catholic country, had no divorce. One had to obtain an annulment, and that took longer and was harder to do than he had expected. This led to a lot of problems, because he obviously needed a job, spoke very little English, and basically was illegally in the United States overstaying his visitor's visa. He and I had a huge mess to clean up. He got a job as a precision machinist through someone I knew. This shop made dewars (double-walled, vacuum-sealed containers used in astronomy to hold and cool detectors with liquid nitrogen and liquid helium) for astronomer Frank Low.

Meanwhile I was trying to establish myself at work, while tremendously worried about both our legal and financial problems, since Gaston was initially earning below the minimum wage, although that soon changed. We/I wrote letters to the senators from Arizona. All of this time and worry about homelife issues affected my performance at work, and after my two-year review, I was told that my contract at NOAO would not be renewed. So I decided to study civil engineering, which I viewed as a very portable job, unlike astronomy. I signed up for classes at the University of Arizona—they gave me lots of credits from my

Harvard classes. The classes were relatively easy for me, so this was not too big a deal.

Also, we were having problems at home. Although all the neighbors seemed to like us, we were robbed numerous times (not that we had much to steal). After our Nth report to the Tucson police, two policemen came to our house and said that the neighbors did not want us living there as we were not Black, that this would not change, and we needed to move. By then we had enough money saved to buy a small house in a non-minority neighborhood, which we did.

Finally after a substantial effort, all our legal problems were straightened out and we married. I could then put all my attention into my job, and I blossomed. I began to explore the scientific importance of infrared observations of galaxies. I had met Jay Frogel, a support astronomer at CTIO, and started working with him and his collaborator Eric Persson, from Carnegie and Caltech. They were exploiting the rapidly improving new infrared instrumentation to obtain broadband near-infrared photometry for a large sample of stars along the giant branches of globular clusters. They produced the observational data, and I joined this collaboration as the modeler, as I had previously worked on predicting stellar parameters, using model atmospheres from Kurucz. This turned into a very successful and long-term collaboration; we wrote a series of several papers together. At that time, I also wrote a series of papers on elemental abundances in Milky Way globular clusters.

At Kitt Peak I also helped in the commissioning of the new echelle spectrograph for the 84-inch telescope and wrote some manuals and a guide to the spectra of the calibration lamps.

As I recall things, my contract with NOAO was extended. About two years later, in 1978, Peter Goldreich phoned me and suggested that I apply for a faculty position at Caltech. I was very surprised but I did so and was invited to come for a visit. Subsequently I was offered a faculty position, which I accepted. Idiotically, I felt so honored (after all, I was a Caltech PhD and knew the quality of the faculty), and also felt so beat down by being told I was going to be terminated at KPNO, that I accepted the position of associate professor without tenure. I was so stupid that I did not even ask for start-up money. (NOAO was a government lab, and there the subject of money, be it for operations or for science, was never discussed among the trolls.)

This was the biggest mistake I made in my entire life. I found out much later that if I had asked for tenure, I would have gotten it. Ditto for a reasonable start-up package. Instead I was back in the prove-yourself mode, which I did. But I did not get tenure until 1984, at which time I was thirty-eight years old, too old and too set in my ways to have a child, especially since I was the principal breadwinner.

At Caltech, things went reasonably well. I wrote a paper with Ken Freeman, which suggested the opportunities that better spatial resolution observations might bring to astronomy. But my major effort at that time switched to instrumentation. I became involved in the design and building of the Keck 10-meter telescope, resuming my friendship with Jerry Nelson, now based at UC Santa Cruz, which began when I was a postdoc at Berkeley. Early in the overall design phase, I showed my husband some rough drawings for the proposed Keck design. He was intrigued and from these drawings made a wooden model of Keck that rotated and tilted. Ed Stone (the PI at Caltech for the Keck project and former director of the Jet Propulsion Laboratory in Pasadena) saw the model in my office and said he would like to purchase it. Gaston then made a slightly improved version of this model, which was presented by the president of Caltech to Mr. Keck at a formal dinner and was subsequently displayed in the Keck Foundation Headquarters, while we kept the original version in our house.

J. Beverly Oke (a professor of astronomy at Caltech and a great instrument designer) and I worked together to build LRIS (the Low Resolution Imaging Spectrograph), one of the three first-light Keck instruments. That project was extremely difficult, money was very tight, especially toward the end of construction, and there were many disputes between the project manager (whose job was to produce the telescope on schedule and on budget) and the scientists, as represented by Jerry Nelson, whose goal was the best performance possible. Commissioning was brutal because there was a pack of hungry astronomers wanting to get going with observing, but getting all the parts of the telescope and the instruments to play nicely together took a lot of

FIGURE 11.1. The first high-speed link from the U.S. mainland to the Maunakea summit was achieved in 1995 by a project led by Larry Bergman of JPL and Judy Cohen from Caltech, with Patrick Shopbell of Caltech as a postdoc. Information was routed from JPL to Honolulu via a NASA satellite, then to Waimea via an underwater cable, then several microwave antennae to reach the summit of Maunakea. Photo by Gaston Araya.

engineering effort and time. As an illustration of our desperation over money, I was leading the charge for installation and commissioning for LRIS at the 14,000-foot summit of Maunakea, as Bev Oke had heart problems and could not go to such a high site. I had rented a truck at the Kona airport to take the things we had shipped from Pasadena up to the telescope. We were to be at the summit for about a week. So I immediately returned the truck to the airport to save money (maybe $1,000 in total) and decided to save another few dollars by hitchhiking to the Keck headquarters in Waimea from the airport, there being no mode of public transportation beyond the big hotels at Waikaloa.

Somewhat later, George Djorgovski complained at a Caltech astronomy faculty meeting that I was getting all the dark nights at the beginning of scheduled observing. The fact was that the engineering and instrument calibration was going very slowly, it was still impossible to align a slitmask, and we had used up all the engineering time allocated to LRIS, so I was using my hard-earned nights awarded to me to do my own science project to, instead, do integration and systems engineering that benefited the entire community. Jim Westphal defended me at that meeting. It was a crazy scene, getting the telescope to work and the instruments to work, with insufficient commissioning time and an only partially debugged telescope.

I never got any substantial reward for all those years of work. Bev and I got about $20,000 from Caltech, but there was no official reward policy from Keck for these major endeavors. This issue becomes more crucial as instruments become bigger and more complex, construction lasts longer, and a person's scientific productivity is interrupted for years. It is particularly bad at Caltech in comparison to the UC campuses also involved with Keck; Bev and I did not have a project manager for LRIS, while UC has many more support people to help the PI.

This issue is also a serious one for the future of astronomy, in terms of both big instruments and big software efforts. I note that one reviewer of the first NSF proposal to support my science that I submitted after LRIS was at least sort of working wrote that this woman had not done any science in a decade and should not be awarded any NSF funding. I have had the least money for science of any of the astronomy faculty at Caltech for years and have long felt that this impacted my ability to carry out major science projects, as I have had trouble obtaining NSF funding to pay for graduate students or postdocs to work with me ever since.

Once Keck was finished and LRIS was working well, I explored various aspects of stellar evolution, the globular clusters of the Milky Way, and related astrophysics projects.

When I joined the Caltech faculty there were about five other women in total across all fields. One of them, Jenijoy La Belle, professor of English, had been denied tenure; she sued Caltech and won (in 1979). I started and hosted intermittent lunches for the faculty women at the Athenaeum (the faculty club); the Office of the Faculty agreed to pay for this. Since we were so few in number, this was very useful in discussing our concerns and institute-wide issues.

FIGURE 11.2. Jerry Nelson, the father of the Keck Telescopes, and Judy Cohen, in 2009. Photo by Mike Bolte, UC Santa Cruz.

At that time, only tenured faculty attended faculty meetings. After I received tenure, I went to my first meeting. I sat down in the back and listened as the division chair Charles Peck said, "Gentlemen, let's get started." That was a harbinger of things to come.

Once I got tenure, and even before, I spent a fair amount of time fixing things. A female grad student came to see me to say that the student health plan did not cover abortions; I fixed that. Then another graduate student came to say that he had cancer and the student health insurance maximum coverage was too low and he had substantial bills that he and his family could not pay. So I joined the Student Health Committee and was able to improve the health insurance coverage. Several years ago, I was the first person from the Caltech faculty or administration consulted by the grad student who raised a successful sexual harassment complaint against Caltech.

One of the biggest issues, at least to me, that arose over the years was the equity or lack thereof of women's versus men's faculty salaries. After some of us suspected that this might be an issue, an investigation followed. This was non-trivial as Caltech faculty salaries are confidential, and there were at that time not more than fifteen women on the faculty. I had decided that if the differential was large enough I would sue and urge the other women to do so; in the end the difference was just below the margin to justify a lawsuit. I did, however, get a substantial raise of over $40,000/year as a result; all the senior women were raised to the same salary, I believe.

My distinguished colleague Vera Rubin once told me that my website was not very good and I should fix it up. I told her I was much too busy to worry about that, but in hindsight I think what was really going on was that she wanted to nominate me for something, but without a good website that people who are not familiar with your work can look at, she couldn't do it.

I was elected to the National Academy of Sciences in 2017; by then I had fixed up my website.

So here is my advice to all women in astronomy: do not underestimate yourself, check out all offers of positions very carefully to find out what you should ask for, and maintain a good web page.

Chapter 12

Judith Lynn Pipher (PhD, 1971)

Taking Advantage of Opportunity

Judy Pipher, now Professor of Astronomy Emerita at the University of Rochester, is one of the pioneers in developing the field of infrared astronomy and is an expert in building astronomical IR detector systems, including those for the KAO, the SST, and the future NEO Surveyor. She is often called the "Mother of Infrared Astronomy." She was hired in 1972 as the first female, tenure-stream astronomy faculty member at the University of Rochester and directed the C.E.K. Mees Observatory in Bristol, New York, from 1979 to 1994. She is a Legacy Fellow of the AAS, the winner of the de Vaucouleurs Medal from the University of Texas, and was inducted into the National Women's Hall of Fame. Her asteroid is 306128 Pipher.

The hallmark of a curious person is that any number of possibilities can change one's direction in life. Looking back at my career, I see that taking advantage of a series of unplanned-for circumstances has heavily influenced my professional life. For example, I did not always plan to be an astronomer, particularly one who spent much of her career working on enabling technology for astronomy. Lucky circumstances led me there, and I reacted positively to the opportunities presented.

My first astronomical experience, if you can call it that, occurred in the early 1950s—I decided to try to find constellations on a very dark night, lying on the beach at our summer cottage on Georgian Bay on

Lake Huron. Having been raised in Toronto, where even back then, the sky was never very dark, I had never been tempted before! That experience was enjoyable, but not challenging, so did not attract me into astronomy.

The so-called hard sciences and math were my academic interests, and when I entered the Mathematics, Physics and Chemistry option at the University of Toronto in fall 1958 I planned a career in chemistry. However, university physics and calculus proved to be my favorite subjects. Exposure to astronomy courses didn't occur until my sophomore year—then I was hooked! Professor Helen Sawyer Hogg hired me that summer as a research assistant to work on her catalog describing the properties of the approximately one hundred known globular clusters in our Galaxy: globular clusters are roughly spherical, compact entities containing up to a million stars. Four undergraduate students traveled each day by several buses to the David Dunlap Observatory in Richmond Hill, fifteen miles north of home, to work with our professors on their projects. While our research projects were fairly pedestrian, Professor Hogg encouraged me to spend as much time in the Observatory library as I wished, reading various journal articles.

One of our senior year astronomy courses had evening observing components at the Observatory, with reduction techniques learned in the afternoon. With such an auspicious beginning, one would think that I would have immediately leapt into graduate study. For various personal reasons, however, I decided to work in laboratories to improve my experimental abilities at Cornell University (where my new husband was a student). After several years, I left to teach at the high school and then the college level, initially entering the astronomy graduate program at Cornell on a part-time basis while continuing to teach. Graduate studies became my full-time pursuit in 1966, and I completed my PhD in 1971 in the very new discipline of rocket infrared astronomy under the guidance of Professor Martin Harwit and his postdoctoral associate Dr. Jim Houck. My student days, both at Toronto and then Cornell, provided exceptional opportunities.

At my BSc graduation, my dad was questioned as to why his daughter was "taking away professional positions from her male counterparts"—this attitude did not affect me, partially because of my parents' support, and partially because I had Helen Hogg as an example and a mentor. When working with Martin and Jim in graduate school, no one paid any attention to my gender, even though I was in a definite minority as a woman among the mostly male Cornell graduate students in astronomy. By always behaving as if I expected to be treated equally with my male peers, I found that usually I was.

It was natural not to draw attention to my gender, because the intensity of graduate school classwork and research shaped our common emphases.

In those very early days of astronomy done at infrared wavelengths, well beyond the reddest part of the visible spectrum, we graduate students hand-built our own single pixel sensors, and even our specific wavelength-defining filters, because they were not commercially available. These sensors did not work at room temperature: they needed to be cooled to temperatures only a few degrees above absolute zero. With fellow graduate student Tom Soifer, we spent many hours testing our devices in the rocket telescope dewar (an evacuated container built to hold liquid cryogens, the telescope, sensors, and electronics), conceived of and designed by our advisors. The sensor focal plane was cooled by liquid helium to a temperature of 4 degrees Kelvin.

Technology issues dominated our lives, and Jim Houck, a solid-state experimental physicist turned astronomer, was an amazing technology instructor. Martin, who had conceived of the cryogenically cooled rocket-borne telescope experiment, also educated the group on the astronomical mysteries awaiting us. When our group gathered to educate ourselves further in journal club, there was a definite paucity of infrared astronomy papers to study, particularly at the longer infrared wavelengths for which we were preparing the rocket payload.

Rocket launches of the infrared telescope experiments took place in New Mexico at White Sands Missile Range—and the military personnel had unusual rules for females on base. Despite the fact we were often working on the ground while making measurements or modifications to the equipment, females had to wear skirts. Tom and I became part of a very limited community of far infrared astronomers—we started at the beginning of a fairly new field. Not a bad way to begin a career! We obtained (in the five minutes of observing time each flight!) observations of dust emission from scans across the Galactic center, from regions filled with ionized hydrogen, and from the Galactic and zodiacal planes. Many years later at an event celebrating Jim Houck's sixtieth birthday, I showed that our very early rocket results were confirmed by the much later, more sophisticated COBE (COsmic Background Explorer) space mission results.

Post-PhD, I was lucky again—I was hired directly out of graduate school as the first woman faculty member in a department of approximately thirty to start an infrared astronomy research program at the University of Rochester. I was still relatively shy, and not particularly assertive. Professional life changed that as various subgroups in the department competed for students and other resources. My arrival at Rochester was before the days of large start-up packages for new faculty. I was given a lab, the equipment left behind when the cosmic ray group there dissolved, and a grand total of $500 in start-up funds. Martin Harwit graciously provided me with a small liquid helium test dewar to get my lab work started (he called it my dowry), as well as a few sensors. But even then, liquid nitrogen and helium were expensive, and while postdoctoral associates in condensed matter labs occasionally loaned me cryogens, that could not always happen. So, I began ground-based observing at the Kitt Peak National Observatory outside Tucson, Arizona, using their equipment on the 50-inch and 84-inch telescopes and began to write both observing and research proposals to gain funding for my research.

Our ground-based "maps" of the dust emission associated with extended ionized hydrogen sources in massive star-formation regions led to an ancillary radio interferometry program using the then Green Bank Very Large Array three-element interferometer to characterize the ionized gas emission of those same sources. Although most astronomy faculty paid no attention to my gender, on one trip to Green Bank, West Virginia, with an older male faculty member as well as with an infrared-astronomy colleague from an external institution, the faculty member went to unusual lengths to open doors and so forth for me. He was being polite in the old-world sense but overdoing it. My infrared-astronomy colleague offered to intervene—as we approached the door to the main building at Green Bank, he rushed ahead of the other faculty member, opened the door, and entered first. That not-so-subtle hint was taken!

I do not recall many overt instances of gender discrimination in my own department. An exception was a time when the then department chair wanted to deny the female half of married graduate students a stipend because that would mean two stipends to one family. I told the chair (before I had tenure) that if he tried that I would go to the university lawyers with a complaint concerning his plan. That chair later brought me in and intimated that I might not get tenure and that I should consider applying elsewhere. Happily, the female graduate student received her stipend, and I did receive tenure. I was informed that my gender never came up in the faculty discussions on my case. Most of my fellow physics and astronomy faculty were supportive and gender blind.

Writing research proposals then, as now, was necessary. In one interesting experience, a National Science Foundation proposal of mine was rejected, but the program officer offered instead to fund me to work in the lab of a well-known infrared physicist. I declined the offer, knowing how another woman had fared in that lab (and also believing the offer to be an inappropriate response to my proposal). However, my proposal to NASA to build a far infrared interferometer using a lamellar grating (comprised of interleaved fixed and movable panels acting as a beam divider) to fly on the Kuiper Airborne Observatory (KAO) proved successful—and once again, another instance of luck affecting my career took place. As I found out later, the inventor of the lamellar grating interferometer, John Strong, was a member of the proposal review committee! In fact, the only reason I had ever heard of lamellar grating interferometers was that a Cornell postdoctoral associate, Barrie Jones, built one for lab measurements that measured the performance of our far infrared detectors. I was lucky enough to have used his instrument as a student, so knew its basic properties well. My KAO instrument was unusual in that it had the lamella built on a sphere in order to allow focusing onto the sensor. We were able to obtain far infrared spectra of cold dust emission in molecular clouds associated with regions of star

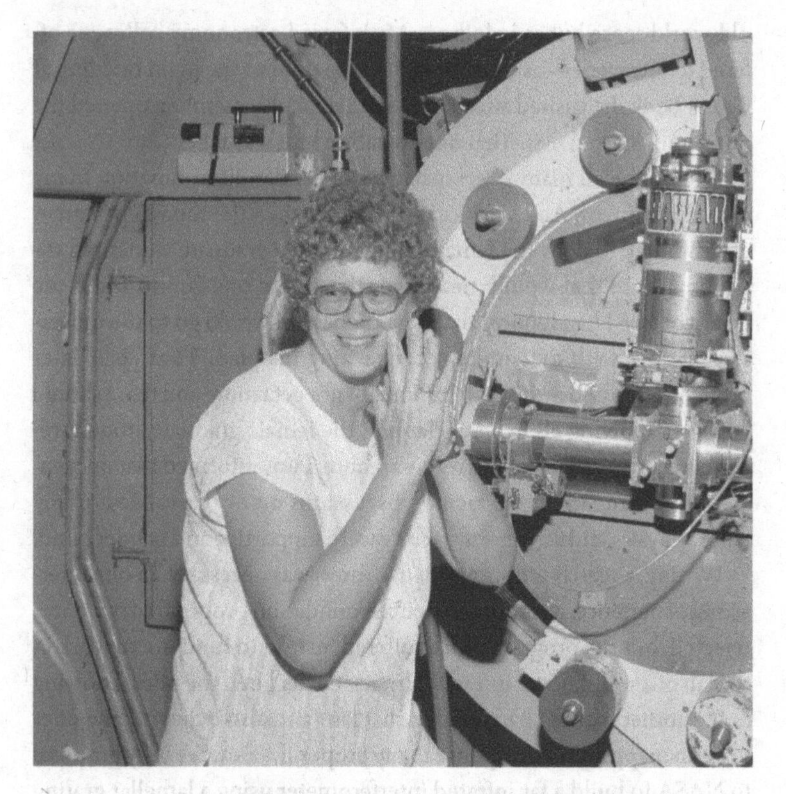

FIGURE 12.1. Judy Pipher on the Kuiper Airborne Observatory plane with the Cornell Spectrometer, sometime in the late 1970s or early 1980s.

formation. Around the same time, I also began working on a program to determine abundances in ionized hydrogen regions as a function of galactocentric radius, using infrared line emission of several species: this work employed mid-infrared spectrometers on the KAO, as well as spectrometers at ground-based national and other observatories. The infrared line emission method avoided the issues that visible wavelength astronomers had to face, namely the temperature sensitivity of visible wavelength line emission.

During my first sabbatical in the late 1970s, spent at Cornell working on yet another rocket launch, perhaps the most auspicious aspect of my career took place. A phone call from a former Rochester physics graduate student, Alan Hoffman, who was working at Santa Barbara Research Center (SBRC) in Goleta, California, changed the course of my professional life. Alan offered to lend me an infrared 32 × 32 pixel sensor CCD array (the light-sensing element of a 1 Kilopixel camera). At that time there were no infrared cameras for astronomy, and this device had in fact been developed as a research and development project aimed toward the military. Alas, it was not ideal for making measurements in an environment in which the brightness of the astronomical background and sources is up to a billion times less than that to which the military was accustomed. That meant changes had to be made in the operation of the CCD.

Rochester was able to convince Bill Forrest to join our faculty, and, with our students, Bill and I worked to get our first camera operational and then began to acquire the first infrared astronomical images with this very small-format camera. To put this effort into context, because the effort was then poorly funded, we had an undergraduate build an LSI-11 Heathkit computer to operate the camera (the LSI-11 was the poor sister of the then-common PDP-11 computer), hired an engineer friend at Ithaca College to build a CCD controller, built associated electronics in the lab, and hired a computer engineer to program in FORTH to operate and read out the CCD (FORTH was a language for the exceedingly memory-poor computers of that era). This was one of the first infrared astronomical cameras in the U.S. that gathered tantalizing data. The other group developing an infrared camera for astronomy at about the same time had the advantage of scientists and resources of two NASA labs (Ames Research Center and Goddard Space Flight Center). The first astronomical images from both groups were published in the astronomical literature at about the same time.

In 1983, Giovanni Fazio of Smithsonian Astrophysical Observatory asked Bill and me to join his proposal for the InfraRed Array Camera (IRAC) instrument on what became, twenty years later, the Spitzer Space Telescope following its launch in 2003. In the intervening years, with SBRC (later named Raytheon), we developed far more sensitive and larger infrared sensor arrays for Spitzer and tested them thoroughly at ground-based observatories as we worked on a variety of infrared observational studies. Our two shorter-wavelength (3.6- and 4.5-micron) IRAC cameras continued to obtain important data from the August 26, 2003, launch until Spitzer was turned off on January 30, 2020.

Amazing astronomical observations have been acquired with these two cameras, from the discovery and characterization of exoplanets (planets around other suns) to measurements of the high redshift cosmos: the IRAC cameras we worked on have played a role in enabling an outstanding observational legacy. Part of their strength has been the fact that unlike the two longer-wavelength (5.8- and 8.0-micron) IRAC camera sensors, the MIPS camera (24-micron), the MIPS longer-wavelength photometers, and the IRS camera and spectrometer (5-micron to 40-micron) onboard Spitzer, our two camera sensors continued to work after the liquid helium in Spitzer ran out in 2009. IRAC's focal plane temperature, initially 13K, increased to 26K once the cryogen ran out, a temperature achieved through balancing the heat output of the electronics against the input of the cold cosmic radiation.

This balancing concept—passive cooling in space—has driven our current and recent past work on the development of longer-wavelength infrared sensor array cameras for future space missions. Passive cooling of cameras in space would obviate the need for cryogens or cryocoolers to cool the sensor arrays. As already noted, the other longer-wavelength infrared cameras on Spitzer failed to operate once the cryogen ran out. My major scientific interest with Spitzer concentrated on star formation—particularly understanding protostellar evolution from the earliest stages, through the formation of circumstellar disks around the young stellar objects, and the disks' ultimate dissolution. Those studies required both our two short-wavelength cameras and the two or three long-wavelength cameras to characterize these star-forming systems.

Our next goal was to develop longer-wavelength cameras that could be passively cooled in space in the same way that our two short-wavelength IRAC cameras were. We anticipate that the long-wavelength (10-micron) 4 Megapixel cameras using the sensor material HgCdTe, which we have developed with Teledyne Imaging Sensors, Jet

Propulsion Lab (JPL), and the University of Arizona, will fly on the Near-Earth Object Surveyor (NEO Surveyor) when it is launched (planned for 2025). They will perform at a passively cooled focal plane temperature near 40K. Near Earth Objects (NEOs) include comets but are primarily asteroids whose orbits around the Sun have been modified by other planets to come very near to or cross the Earth's orbit, leading to the possibility of an impact with Earth.

Since 2005, when the congressional George E. Brown Jr. Near Earth Object Survey Act specified that orbits and velocities of 90 percent of NEOs larger than 140 meters in diameter be identified within five years, the NEO Surveyor and its precursor NEOCam have been shown by several panels, including a National Academy panel, to be the type of mission best suited to this task, since ground-based visible-wave efforts add data extremely slowly to what is known. In addition to identifying these potentially hazardous objects, sites for future asteroid exploration and exploitation will be identified, and important NEO data, combined with ancillary data obtained elsewhere, will lead to improved scientific understanding of the evolution of our solar system. In addition to these devices, we at Rochester have developed with Teledyne even longer wavelength sensor arrays. These most recent arrays have attracted the interest of other scientists; for example, they could be exploited for space missions such as the proposed flagship ORIGINS Space Telescope mission, one facet of which plans exoplanet spectroscopy in a search for biosignatures of life. Detecting very strong CO_2 line absorption in the atmospheres of such exoplanets at a wavelength of 15 microns is one driver for our longer-wavelength work.

How has luck or opportunity entered into this latest career path direction? In a much more indirect way. Spitzer's thermal design was extremely good, and it was executed to guarantee that the liquid helium cooled phase of operation would last as long as possible (designed for a goal of five years, this phase lasted from 2003 launch to 2009). That good thermal design, when taken together with the low power output of our IRAC sensor arrays, guaranteed the new focal plane temperature of 26K. Because our IRAC sensor arrays continued to perform admirably, we became convinced that passive cooling was the future, echoing

FIGURE 12.2. The sensitive element of a NEO Surveyor camera long-wavelength focal plane is a set of four 4 Megapixel 10-micron sensor arrays, produced by Teledyne Imaging Sensors with guidance by the University of Rochester, JPL, and the University of Arizona. One of those sensor arrays is shown in the photograph.

the "Cheaper, Better, Faster" paradigm of Dan Goldin, the NASA Administrator from 1992 to 2001.

We knew that the HgCdTe sensor material, which is related to the HgCdTe material for the shorter-wavelength arrays flown on HST, WISE, and OCO-2, among other missions, could be developed for even longer wavelength arrays if we were able to solve the many problems that the 10-micron NEO Surveyor devices presented. It took years of effort to overcome these problems with limited support until the NEOCam (now NEO Surveyor) design specified these devices as critical to mission success. Our group at the University of Rochester, with the Jet Propulsion Lab and the University of Arizona, took advantage of the opportunity, has largely succeeded, and is looking forward to the 2025 NEO Surveyor launch and beyond. Our current experimental efforts for NEO Surveyor focus on unusual effects of potential flight arrays that could plague calibration unless identified and characterized.

Astronomy has even affected my personal life indirectly. My husband, Bob, designed and built a beautiful forty-foot trimaran of western red cedar. It was a labor of love, built with the help of friends and my four stepchildren. I suggested that it be named *CYGNUS* after the constellation of that name, because of its graceful swan-like appearance. Sailing *CYGNUS* gave us many years of pleasure.

When I started out to obtain a BSc degree in physics and astronomy more than six decades ago, it never would have occurred to me that my career path would travel in the directions it did. Looking back, it seems like a completely natural path. I learned quickly that when opportunities present themselves, it is important to take advantage of them. Sometimes those opportunities arise because of your own efforts—but more often because of the good fortune of where you are and what has gone before. I also give enormous credit to the opportunity of working with exceptional, caring scientists, including a series of very talented postdoctoral associates, engineers, and graduate students at the University of Rochester.

Chapter 13

Gillian (Jill) Knapp (PhD, 1972)

Princeton 1984

Jill Knapp, now Professor Emerita of Astrophysical Sciences at Princeton University, has spent her career studying the life cycles of stars and the interstellar medium from which they are born. She made important measurements that established the distance from the Sun to the center of the Milky Way and showed that elliptical galaxies contain far more dust and gas than previously thought. She also has worked on planning and implementing the SDSS. She was the first tenured female faculty member in astrophysics at Princeton. She also has worked with the Pace Center to teach college-credit courses in the New Jersey prison system. She received the Presidential Award for Excellence in Science, Mathematics and Engineering Mentoring. Her asteroid is 159826 Knapp.

I was born on October 10, 1944, in Widnes, up the River Mersey from Liverpool. My father, born in 1916 in Sheffield, worked in a pharmaceutical company while my mother, born in 1920 in Cudworth, was a nurse. They married in 1940 but spent most of the first years apart, for Dad's profession, the manufacture of opium-derived drugs, was crucial to the war effort and he spent most of that time in London, "cowering," he used to say, "in the Underground." My mother told stories of sitting in wards with patients too ill to be moved to shelter as the bombs fell outside; of pushing incendiaries off the hospital roof; of walking home after the night shift through unrecognizable streets.

My mother was an only child, orphaned and homeless at fourteen, and survived by signing on as a student nurse in Sheffield. Dad's father worked in one of Sheffield's fine-steel companies and had eight children. His family had the working-class respect for education, owned encyclopedias, and took trips to art galleries. Two of my aunts were nurses, one uncle was a very successful artist, and Dad, the youngest, was the first member of his family to get a university degree, a BSc in chemistry. Like the grandparents, my parents were very invested in their children's education, so I was well launched, and much more than that: I grew up in the 1950s, when the Labour government was expanding access to university education beyond its traditional upper-class boundaries.

I attended the local schools, did a BSc in physics at the University of Edinburgh and a PhD in astronomy at the University of Maryland; was on the research staff at Caltech; then research staff followed by faculty in the Department of Astrophysical Sciences, Princeton University. High school was the make-or-break point. The teachers were terrific (thanks, especially, "Teeny" Burns, "Hen" Brown, and "Abe" Angus, maths and science teachers). I was enthralled by physics but it seemed, with the single exception of Marie Curie, that there was no overlap between being female and doing physics. But then I came across John Stuart Mill's "The Subjection of Women," and Mary Brück, astronomer at the Royal Observatory, Edinburgh, came to talk to our physics class when I was about fifteen, and everything was OK. Up front, I'd like to express gratitude beyond measure to the above, to innumerable colleagues and friends (especially the graduate school friends from Maryland), to my PhD thesis adviser Frank Kerr and department chair Gart Westerhout, to Princeton astrophysics chair Jerry Ostriker, and to my parents and my brothers and sister, Stevie, Hil, and Johnny. The humanity and decency of these folks is the reason I got to live the life I've lived.

This story will focus on the science I've been fortunate to work on and some of the difficulties getting there, given both the class systems and the received wisdom that the participation of women in physics is unseemly, we aren't capable of it, and we aren't interested anyway. The fit role in academia for an educated woman was, in that epoch, the supportive faculty wife, and I can't resist an anecdote from 1984. An invitation to a party with the Princeton president arrived: thick paper, gold lettering, addressed to "Professor and Mrs. James E. Gunn." "The h*** with that," I said. "I'm not rearranging my observing time to go to a party as Mrs. anybody." For I was observing regularly at Bell Labs in Holmdel, New Jersey, where the terrific group (Arno Penzias, Rich Linke, Tony Stark, Bob Wilson, and John Bally) had given me time on their 7-meter millimeter telescope on weekends. So Jim attended the party alone and reported, hugely amused, that its purpose had been to welcome new Princeton faculty, and I was the invitee and he the spouse.

At the University of Maryland Astronomy Program, much of the research was led by two eminent radio astronomers, Frank Kerr and Gart Westerhout. There was growing interest in what were known as dark clouds. These had large dust column densities, but did not show up in observations of atomic hydrogen (H I). During a class in Elske Smith's solar physics course on reversals in the atmospheric CaII H and K lines (these are dips in the spectral line profile formed by colder gas in front of hot gas), it occurred to me that the H I in these clouds was likely to be cold, with a correspondingly small velocity dispersion, and absorbing the background line radiation, that is, one might be able to study dark clouds via H I self-absorption. Most H I work to that point had been done with spectral resolutions of several km/sec, but there was a new autocorrelation receiver in use at the NRAO's 140-meter telescope at Green Bank. Frank Kerr (whom I was lucky enough to have as thesis adviser) and Bill Howard gave permission to take fifteen minutes out of a Maryland telescope run to observe a high-resolution H I profile towards the Taurus dark cloud, and there it was, a lovely self-absorption profile and the beginning of my thesis.

There are lessons here: science is driven by advances in instrumentation, and it's useful to (sometimes) let your mind wander in class. And a historical note: like Elske Smith, Mary Brück was a solar astronomer. She told me that if you wanted to be an astronomer in those days and happened to be female, it was solar astronomy for you, to avoid spending the night with men you weren't married to. Yet another piece of

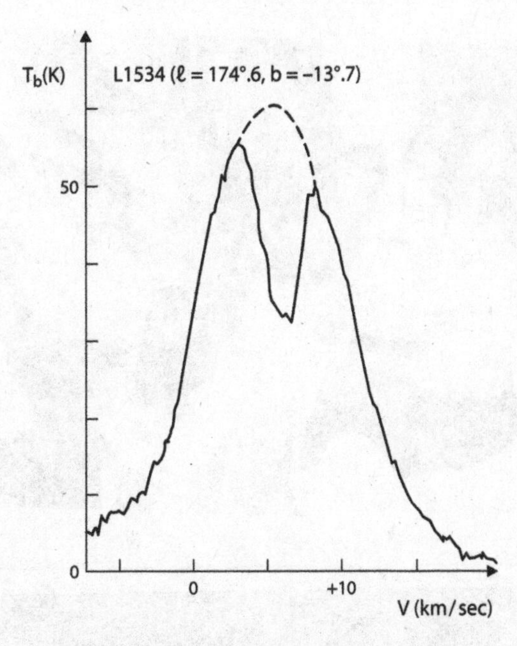

FIGURE 13.1. H I line profile towards the center of the Taurus dark cloud, observed with the NRAO 140-ft telescope at Green Bank (Knapp, *Astronomical Journal* 79 [1974]: 527).

unpredictable good fortune: radio astronomy is pretty oblivious to night and day and this issue never arose for me.

While I was the sole female astronomy graduate student for the first two years, this changed rapidly and the department, something of a pioneer, admitted a growing number of both Black and female students and was at the forefront of such efforts across the campus thanks to Gart Westerhout, a remarkably fair-minded, imaginative, and socially responsible individual as well as a great scientist. I worked with this effort (Intensive Educational Development) for a couple of years.

Astronomy was growing by leaps and bounds in the 1960s, but like many rapidly growing populations it hit a limit in the early 1970s and positions became very tight. (I don't think it's ever been quite as bad since, but the current pandemic is likely to make that time look like a

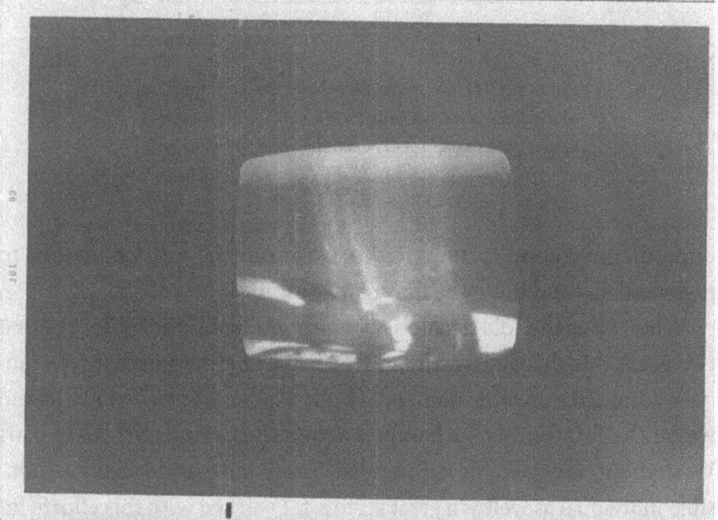

FIGURE 13.2. 10:56 P.M., EDT, July 21, 1969: Jill Knapp watching TV with fellow graduate students as Neil Armstrong steps onto the Moon. Hard to realize it's been more than fifty years. Photo by Woody Sullivan.

picnic.) Many other big changes were happening: thanks in large part to the American Astronomical Society, hiring and graduate school admissions were moving to open applications (a huge positive change from the "old boys' network" method), the number of postdoctoral positions was growing rapidly, and departments were facing, thanks to the issues raised by the civil rights and women's liberation movements, pressure to make their hiring and admissions less discriminatory. There was, of course, widespread resistance to the last, the justification du jour being the very tight employment market, in which one could not "take a risk" on a woman or Black scientist.

A couple of anecdotes give the flavour. An eminent astronomer was overheard complaining that "now we are forced to read those applications from women" instead of, presumably, dropping them in the circular file. And at an AAS meeting, one of my contemporaries rushed over, exclaiming: "You stole my job!" Apparently, the letter turning him down explained that "though we'd love to hire you, our administration has required that we appoint a woman." The job in question had, it goes without saying, gone to another young white man, as my friend knew perfectly well.

But a lovely thing happened around that time; I became close friends with Beatrice Tinsley and Sandy Faber. Both were superb scientists and, rather to my surprise, approved of me, something that held me together during any number of difficult times.

The next thirty years or so were spent continuing and expanding work on the interstellar medium, first at Maryland, then at Caltech in Al Moffet's radio group, with many observing trips to Caltech's Owens Valley Radio Observatory and Tom Phillips's Caltech Submillimeter Observatory on Maunakea, and then at Princeton using these and many other telescopes. This took several directions: molecular clouds and star-formation regions, mass loss from evolved stars, and cold gas in early-type galaxies.

In 1980, I moved to Princeton with Jim Gunn, whom I married in 1982. It was, of course, said at the time that I had taken up with the legendary Jim to further my career. When we arrived, there were and had been no tenured women faculty on our side of the tracks, that's

Washington Road to you Princetonians—all of engineering, physics, plasma physics, astrophysics, and mathematics, and as it turned out I was to be the first. Indeed, Princeton had only recently become coeducational; you could count the faculty women in all subjects on one hand, and there was huge nervousness around the whole topic. I had faculty offers elsewhere, and the informal deal was that first I would be on the research staff and then be proposed for the faculty. The astrophysics department was much smaller then, though its close ties with John Bahcall's group at the Institute for Advanced Study and with the Department of Physics resulted in a sizable astronomical community that was, however, almost entirely devoid of women. (Neta Bahcall, who had been on the research staff and should have been on the faculty, had moved to the Space Telescope Science Institute and was to return, as professor, in 1989.) Not only was I a woman but my research was observational when the department was largely focused on theory, and it was a difficult time, though made much more bearable by working with the Bell Labs group and by Princeton Astrophysics' unfailing collegiality, courtesy, and supportive culture.

After something of a saga, I was appointed to the astrophysics faculty as tenured associate professor and director of graduate studies (DGS) in 1984. But I learned very soon after, via drafting an NSF proposal, that I was to receive salary for only half of the academic year and was expected to raise the rest. I enquired of the university administration and was told, "Well, we really want to have you on the faculty but it is so difficult that this is the only way." (Princeton University was as fabulously wealthy then as it is now.) This did, eventually, get sorted out and I became fully-paid like everyone else on the Astrophysics faculty. Despite the drama, I snapped up the DGS job, which I held till retirement, because of the opportunity to help change things.

Over the years, I worked on President Shapiro's university-wide effort to increase the numbers of women and minority undergraduate and graduate students, postdocs and faculty, with strong support from astrophysics chair, Jerry Ostriker; expanded the scope and depth of the department's graduate program; set up and ran for a while the postdoctoral fellowship program; added to the graduate program a postbaccalaureate/bridge component; and set up and ran for a while the summer undergraduate research program, including participation of students from the National Astronomy Consortium. As well as this and teaching, I continued studying the interstellar medium with many telescopes and am pleased to have about 15 undergraduate and 35 graduate student co-authors on the resulting publications. The department is currently 25 to 40 percent women from undergraduate to professor, including on its wonderful faculty Neta Bahcall, Jenny Greene, Eve Ostriker, and Jo Dunkley. The department also did quite well, relatively speaking (but in absolute terms terribly), with regard to other marginalized groups, but not remotely as well as it could have done. Now, recent dreadful events have woken people up. Perhaps, finally, we will be serious about this fundamental issue.

And so to the Sloan Digital Sky Survey. As the project was getting underway, I was continuing interstellar medium work and was involved long-term in what was becoming ALMA (the Atacama Large Millimeter/submillimeter Array on the Atacama Desert in northern Chile). But SDSS was pretty completely underscoped, there was far too much to do, Jim was drowning, and we had young children at home: something had to give. So I took on some of the bigger organizational tasks: writing the funding proposals and organizing big pieces of the science planning, including the development of data policies and processing software. These were actually a lot of fun, because they played a big part in setting up, and demonstrating the value of, SDSS's famous openness and cooperation, and the breadth, rigour, and sharpness of the working discussions were phenomenal. In the process, SDSS evolved from its initial hierarchical, dysfunctional organization to one much more flat, open, egalitarian, and effective.

The science was and continues to be amazing: the luminous-red-galaxy correlation function, which clearly showed the baryon acoustic oscillation feature; high-redshift quasars and the Gunn-Peterson effect; stellar populations and galaxy evolution; asteroid colour families; metal-contaminated white dwarfs; stars with escape speed; field brown dwarfs; AGNs and supermassive black holes, to mention but a few. But it all took far longer than projected, and re-establishing observational work on the interstellar medium in the absence of a local group (the Bell Labs group had long since shut down) was more than I was capable of doing.

For the next several years I worked on a new venture, direct imaging of disks and planets around nearby stars using adaptive optics and the Subaru 8.4-meter telescope, first as part of SEEDS (Strategic Exploration of Exoplanets and Disks with Subaru; PI Motohide Tamura) and later with CHARIS (Coronagraphic High Angular Resolution Imaging Spectrograph), built in Princeton's Department of Mechanical and Aerospace Engineering. I'm so grateful to the folks who took me under their wing and involved me in this work. It produced some lovely results: detections of several giant planets at large distances from their stars and the measurement of symmetric and non-symmetric structures in disks.

In 2005, Mark Krumholz arrived from Berkeley as a postdoc and began a program to contribute to community-college teaching in the New Jersey state prisons. The department fully endorsed this and has been enormously supportive since. Mark, Rebecca Anderson, Rob Crockett, and I taught the first course (Algebra), and soon Mark put in place further algebra and English I, well supported by Kiki Jamieson and Andrew Nurkin at Princeton's Pace Center. But Mark was leaving for a faculty position and letting this useful effort die wasn't an option, given racism and its manifestation in the entwined disasters of terrible high school education and incarceration. New Jersey has the highest "over-incarceration" rate of Black citizens in the country, an eye-watering factor of 12.2, as just one dreadful statistic.

So the growing body of volunteers, teaching with Mercer County and Raritan Valley Community Colleges and Rutgers University, continued, greatly expanding the number of courses, the number of students, and the number of prisons. My personal commitment, apart from running things, was to set up and teach mathematics, science, and laboratory science: preparation math plus seven college math and eight college science courses. It was great to work with the other teachers from many departments and learn about the wider scholarship at Princeton in by far the most effective cross-disciplinary effort I've ever encountered. And I was privileged to work for many years with Jenny Greene, involved since the beginning, Rev. Toby Sanders, Drs. Robin Shore, Beth Knight, Ellen Benowitz, Sheila Meiman, and Mary van Doren, and Rutgers history professor Don Roden, who set up the organization to admit students to Rutgers upon their return to society. Don and his wife, Chie, are gentle, infinitely kind, untiring, patient, and totally committed to other people and a decent society. The results include hundreds of former students with degrees, families, and good jobs, forming a tight and supportive community and the beginnings of a much better world.

I'd like to close by reflecting on personal experience of the kind of treatment routinely handed out to women in the field. The first example is the event relevant to this volume, becoming a member of the Princeton faculty. I would not have been at Princeton were it not for being married to Jim, and the institution's message was clear: we think we have to do this but you don't belong and aren't good enough. But Princeton got it backwards: these shenanigans were the doings of the university, not me, and the event was a rite of passage for the university, not for me.

I've emphasized humorous incidents, but destructive behavior is anything but funny. Many times (including quite recently, and by members of my own department) I've been formally ignored at meetings—the usual excuse is that someone thought you were staff, but is not a staff person also entitled to respect and "inclusion"? It's been assumed that male colleagues did all the work/had all the ideas; I've been called a vile epithet by a fellow physicist; been, more than once, the only faculty member in my department not invited to a local astrophysics meeting; been told any number of times that I "don't fit in" or got a job/grant/postdoc "because you're a woman" or am "taking a job someone else needs" or am not of the calibre for this distinguished institution; been listed as an investigator to tick off a box in proposals on topics I was not involved with; even experienced the classic challenges from campus police; and much else, including the circumstances of the Princeton appointment. But the treatment is immeasurably worse for our colleagues in other excluded groups, as a reading of #BlackintheIvory will show, and what choice do those at the receiving end have but to endure and gain strength despite it, as Neil deGrasse Tyson emphasizes?[1] The whole situation is a stunning indictment of the scientific community and our claimed commitment to truth, honesty, and scientific merit. It's way past time to eradicate this wretched behaviour, to embrace with honesty our common humanity, and for science to become the human endeavour that it truly is. At this awful time, the worst that most of us have ever experienced, the changes that are beginning to happen in our field are a candle of light and hope shining in the dark.

Notes

1. N. D. Tyson, "Reflections on the Color of My Skin," 2020, haydenplanetarium.org/tyson /commentary/ 2020-06-03-reflections-on-the-color-of-my-skin.php.

Chapter 14

Patricia Ann Whitelock (PhD, 1976)

The Southern Half of the Sky

Patricia Whitelock has spent almost her entire career at the SAAO in Cape Town, first as Astronomer, then as Head of Division: Science Awareness and Education, then as deputy director and acting director, then Head of Division: Astronomy, and finally as director. In 1987, she made some of the first observations of Supernova 1987A. She started the National Astrophysics and Space Science Programme in South Africa, is a Fellow of the Royal Society of South Africa and of the TWAS, is a member of the Academy of Sciences of South Africa, is an Honorary Fellow of the RAS, and served on the IAU Editorial Board and as President of the South African Institute of Physics, which awarded her the De Beers Gold Medal in 2018.

I was born in Tynemouth, Northumberland, in the northeast of England, on 22 February 1951. From there the family moved to West Wickham, where my brother, Peter, was born, and then to Seascale in Cumberland followed by Stroud.

My parents, Marie and Phillip, had a profound influence on my life. My father was an engineer and worked on nuclear power stations. He talked a lot about science, how important it was and how it would improve the world. I remember him describing the structure of the atom and the importance of nuclear fusion and fission. It all seemed so exciting and filled with possibility, but frustrating that I had to wait so many years to learn about science at school, and then what I learned seemed so mundane.

My mother was extremely ill with polio as a child. Her parents were told that she would not live, and later that she would never have children. Although she had no university education, she was strong and tenacious and lovingly devoted herself to her family. I learned a great deal from her, much of which I recognized only long after leaving home. I gather that we were poor in the early days; no fancy birthday presents, and holidays in Whitley Bay with father's mother. It must have been hard on my "housewife" mother, but I was oblivious to financial problems. Even when a truck driver thought my father's car was two bicycles and tried to drive between them and he was hospitalized, my mother managed to deal with it and not greatly upset Peter or me.

My first clear memories are of Stroud in Gloucestershire, where we moved in 1957. We lived in a very old house that Charles I was said to have visited. The walls were so thick that the windowsills made perfect seats for two or three children, and there was a tall gnarled mulberry tree that I loved to climb. It was mostly a happy place, and I have fond memories of playing in the snow with neighbouring children, of long visits of uncles, aunts, and cousins over Christmas, of us all eating, laughing, and arguing together, and of playing cards when it rained.

My first memories of school were not so happy. I was extremely shy, and although I got on with others I did not easily make close friends. But most of all I struggled to read, a huge obstacle when you really want to understand and know! To this day I still think that learning to read was the hardest thing I ever did.

At first, I went to a state school within walking distance. Years afterwards, my mother told me that after I had been there for a while, the headmistress told them I was mentally defective and would never be able to live a normal life. I must have been around nine at the time, and, although I had learned to read, spelling remained a struggle. My parents were horrified and took me to a psychologist, who made me do all sorts of puzzles and tests. He told my parents that I was very intelligent, but had an obvious problem with reading and spelling, which we now call dyslexia.

In 2008 I visited Harvard, where Matthew Schneps was making a study of astronomers with dyslexia. I was pleasantly surprised to learn that I was not unique, and there was evidence that those with dyslexia were better at recognizing patterns, seeing the big picture, and multitasking.[1]

My parents moved me to a fee-paying school, which could not have been easy for them. Things got better, but I still failed the eleven-plus. This exam, given at the age of about eleven, tested English, arithmetic, and IQ. If you passed, you could go to a grammar school and from there to university. Otherwise you went to a secondary modern school, from which you might go on to learn a trade or become a housewife, as of course most women did at that time.

My failing the eleven-plus and having problems with spelling fixed in my mind that I was not as clever as other children and that if I wanted to succeed I would need to work very hard. I desperately wanted to succeed! I didn't care about the same things as most girls my age, and, although I loved my mother dearly, I did not want to be a housewife. I wanted to visit the stars and planets, explore the Universe, and if this were not possible, then I would be an astronomer. My parents were told that if I wanted to be an astronomer I'd have to go to university. Not knowing I was listening, they replied that they didn't think I was capable.

In 1964 we moved to Welwyn Garden City in Hertfordshire and I went to Sherrardswood School, where things improved. I learned physics and chemistry from Tony Rook, who would rather have been an archaeologist, but who thought like a scientist and taught me to do the same. Maureen Bickley taught mathematics (which I loved) and needlework (which I hated). She showed no resentment when I rebelled and became the first girl in the school to do woodwork instead of needlework, which was an outrageous thing to do at the time. My English teacher, Victor Watson, was particularly patient and gave me extra lessons that got me through English language O-level, at my second attempt. In English Literature, we studied Chaucer's *Franklin's Tale*, which includes some references to astrology. Victor Watson thought the class should hear the modern astronomical view—and asked me to present it, giving me my first chance to "teach" astronomy!

Having obtained A-levels in mathematics, physics, and chemistry in 1970, I went to University College London to read astronomy. Our year comprised eight students: seven men and me. It was my first experience of living away from home, and I struggled with homesickness and the ever-present sense of not being clever enough—my solution was to work hard and make frequent visits home. In practice, I was incredibly lucky. My costs were covered by a means-tested grant that my parents supplemented to the full amount. I got a place in hall for the first two years, breakfast and dinner provided, with my own room and a bathroom that I shared with Eleanor Onon, a Mongolian American studying chemistry. We got on well, and I was grateful that she got up in time for breakfast and often kept some for me when I slept in after doing astronomy practicals the night before at the Mill Hill Observatory.

I loved the practical work, despite the fact that stars were rarely visible from north London. In addition to telescopes and laboratory spectrographs, we learned to use theodolites and sextants (both for making precise measurements of angles). One exercise involved working out our position using a sextant. My measurements put us in the middle of the North Sea, which gave me a lasting respect for sailors observing from the moving decks of sailing ships.

Our year worked well together and I would not have managed without my fellow students. I was extremely fortunate to have friends and colleagues, including Phil Charles, Keith Mason, Clive Chapman, and Steve Sykes. Keith and Phil became professional astronomers, and we met up from time to time throughout our careers. We were told that we were a particularly good year, but I think we were also rather difficult. Staff-student meetings started during our time, and I represented our group for about two years. We got quite heated about the old-fashioned nature of the coursework (using sextants was definitely not modern astronomy) and insisted that modern be taught. An X-ray astronomy course was introduced, sadly starting only the year after we finished. In 1972 I graduated with first class honours, much to my surprise.

For my PhD I joined the astronomy group led by Jim Ring, at what was then the Imperial College of Science and Technology; they were developing a variety of new techniques and instruments. Everyone else was technical, and as the only astronomer my job was to decide what to observe, which was rather daunting. So, I was delighted when I was asked to help a different team over the summer, making infrared observations from the Imperial College Flux Collector (later Telescopio Carlos Sánchez) on Tenerife. My first job on arriving at the observatory was to fish the dead, bloated lizards out of the drinking water tank. The Imperial College hut, where we lived, had only a tiny bedroom with bunk beds that I happily left for the men, while I slept on the floor of the storeroom. I was, of course, used to being the only woman. Despite the hardships, I loved every minute of it and at the end of the summer started a new PhD topic, on infrared astronomy, working with Mike Selby.

I spent almost six months in Tenerife on various three- to four-week trips to observe or support others. I learned a lot about getting equipment out of customs and vacuum technology and a little about astronomy. I met and worked with numerous interesting people, including Carlos Sánchez and Gerry Neugebauer. But most interesting of all was John Menzies, an Australian postdoc from Oxford, whom I would eventually marry.

Writing my thesis, spelling was again a problem, but by then I was engaged to John, who painstakingly "translated" my spelling, while advising on the science. Before graduating, in 1976, I went back to University College London as a research assistant for Bob Wilson, working on very blue objects. During this time John had moved from Oxford to Preston Polytechnic (soon to be the University of Central Lancashire), 230 miles from London, and we were seriously thinking how we might get jobs closer to each other. I knew that John had spent time in Pretoria as Radcliffe Travelling Fellow, but I said I would go anywhere except South Africa. Nevertheless, I took up an opportunity to observe at the South African Astronomical Observatory (SAAO) and was impressed by the people I met.

John and I married in 1977, and, in 1978, after a brief visit to Australia where I met John's family for the first time, we took up jobs at SAAO in Cape Town. Apartheid was in full force, and we did not expect to stay long, but things rarely turn out the way you expect. Somewhat to my surprise, many white people were anti-apartheid. Michael Feast, the director at SAAO, had no sympathy for the government, and his wife, Connie, was a strong activist. As a foreigner, although I joined Connie in various protest activities, I felt ineffective. I eventually found a role helping Muriel Crewe digitise Black Sash court records. Members of the Black Sash attended most political trials and did what they could to support those being prosecuted. I was occasionally used as a Sunday relief driver, transporting families of political prisoners to the departure point for Robben Island for their rare visits.

At SAAO there was no shortage of observing time. We were encouraged to do long-term projects, allowing us to tackle problems that were largely impossible elsewhere, such as monitoring very long period variables, for example, Eta Carinae—a massive binary system with an orbital period of 2,022 days. I enjoyed observing at Sutherland and had no problem with long nights alone, although we worked outside, in the unheated dome, and it was sometimes very cold. Again, I was fortunate with my colleagues and particularly enjoyed working with Robin Catchpole and Michael Feast and was able to build on work they had started on red variable stars. I learned about infrared equipment from Ian Glass and about instrument control software from Luis Balona. After deconstructing and copying bits of Luis's code I wrote control software for our infrared equipment, first in assembly language and later in C. I enjoyed coding and even thought I might make a living at it if I had to leave astronomy.

Observing Supernova 1987A, which erupted in the Large Magellanic Cloud, a nearby galaxy, was definitely the most exciting astronomical event of my career. After being discovered from Chile, we had sufficient notice for John Menzies to get the first spectrum. From then on, for as long as we could, all four telescopes at Sutherland were dedicated to observing 1987A. At the end of each night the photometric results were telexed to Cape Town and plotted on a graph kept in the library for everyone to see. They were also regularly sent to various theorists, in particular to Stan Woosley in the United States and Ken Nomoto in Japan. In return they told us what they thought was happening and made predictions; their telexes were up to a meter long! Almost everyone at SAAO

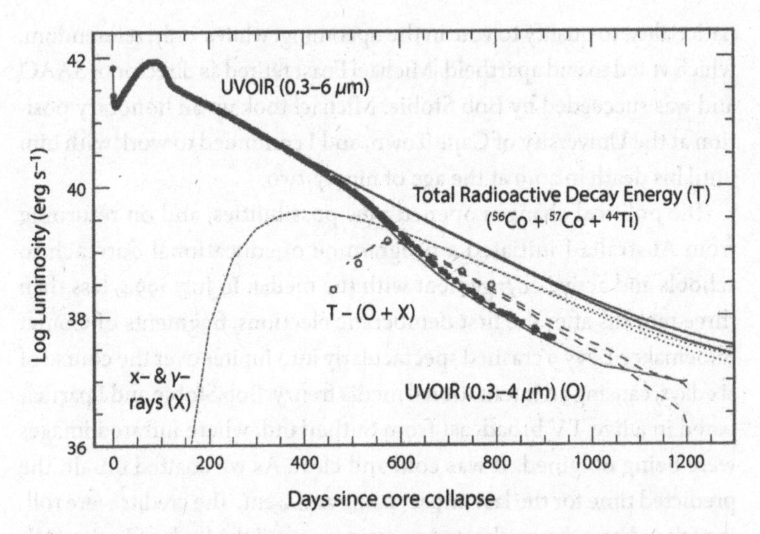

FIGURE 14.1. The changing total luminosity of SN 1987A, showing the various contributions from different wavelengths (Whitelock, Menzies, and Caldwell, *PASA* 9 [1991]: 105).

contributed, although Michael Feast, John Menzies, Robin Catchpole, and I took the lead in strategizing and writing the papers. In 1989 Robin and I attended a major conference on SN 1987A in California, where I gave the opening talk, describing how the supernova had formed dust and how the optical radiation was reprocessed into the infrared. It always struck me as curious that my paper was moved to second place in the published proceedings.

By 1989, SN 1987A was too faint for our small telescopes, and I was lucky enough to get a job offer from the Anglo Australian Observatory (AAO) and a leave of absence from SAAO. I moved to Sydney for about eighteen months and continued observing the supernova. Working at the AAO was my first experience of women colleagues, and I enjoyed the change. AAO was next door to the headquarters of the Australia Telescope, and so I also spent a considerable amount of time with the radio astronomers.

By the time I returned to South Africa, in 1990, Nelson Mandela was free and we were on the path to democracy. In 1992 I took South African nationality, to qualify to vote in the upcoming, whites-only, referendum, which voted to end apartheid. Michael Feast retired as director of SAAO and was succeeded by Bob Stobie. Michael took up an honorary position at the University of Cape Town, and I continued to work with him until his death in 2019 at the age of ninety-two.

The political changes opened new possibilities, and on returning from Australia I initiated a programme of educational outreach to schools and active engagement with the media. In July 1994, less than three months after the first democratic elections, fragments of Comet Shoemaker-Levy 9 crashed spectacularly into Jupiter over the course of six days, causing an international media frenzy. Bob Stobie and I participated in a live TV broadcast from Sutherland, where infrared images were being obtained. It was cold and clear. As we chatted on-air, the predicted time for the last impact came and went. The credits were rolling when I saw the crash site appearing around the limb of Jupiter. We went back on-air for just a few more minutes. Years later, I was told that the public interest in that event was a crucial influence in the government's decision to fund the Southern African Large Telescope (SALT). It demonstrated that the South African public, particularly young people, could be interested in science.

We were fortunate that the new government recognized the potential of astronomy to attract people into science and to help develop the technical skills our society desperately needed. In 1996 Khotso Mokhele became President of the Foundation for Research Development, the parent organization of the SAAO. Under pressure from Mokhele to focus on plans for the new telescope, Bob Stobie agreed to appoint a deputy director to run SAAO. I was offered the position in 1998. However, Bob made it clear that he really did not want me as deputy and that Khotso had persuaded the appointments committee to take me rather than his preferred candidate. He said he was only going to let me run the administration and education side of the observatory, while he kept control of the science and technology. This was not a good basis for a working relationship and problems soon arose.

The construction of a 10-meter class telescope was approved by Cabinet in 1998 pending 50 percent international investment. Khotso

FIGURE 14.2. Patricia Whitelock with her nephew Bob Whitelock. Quarters 7, SAAO, Cape Town, April 1993. Photo from Peter Whitelock and Maria Leedham.

decided that South African astronomy should undergo a strategic planning exercise to prepare for SALT and decide what should come afterwards. I was happy to lead this process. We identified the biggest threat to the success of SALT as the lack of South African, and, in particular, Black astronomers and established the National Astrophysics and Space

Science Programme (NASSP), a postgraduate training programme that took physics graduates from the many and scattered astronomical centres in South Africa. I chaired the NASSP Steering Committee from its inception in 2001 until 2015.

I took over as acting director of the SAAO in 2002 for a year after Bob Stobie's untimely death and as director for another eighteen months in 2011, by which time I had a professorial joint appointment with the University of Cape Town.

I first heard about the Square Kilometre Array (SKA) from Ron Ekers in Australia in 1989. It was clear to me that, if South Africa was to have a future in radio astronomy, we would have to become part of SKA. So, this went into the plan, and, in 2002 when I was President of the South African Institute of Physics, we invited Ron Ekers to be guest of honour at our annual meeting and ensured that he had ample opportunity to meet the scientific and political role players. South Africa and Australia were ultimately selected to co-host SKA.

I have greatly enjoyed being an astronomer. I have been privileged to be in the right place at the right time to make contributions. I am grateful to have used astronomy to do something for South Africa, and to enable South Africa to do something for astronomy. What I have facilitated has largely been through persuading very able people that they should collaborate rather than compete. I encountered some sexism which, although extremely upsetting at the time, was of no lasting consequence. My primary strength has been my persistence! This, together with my wonderful family, teachers, friends, and colleagues, has seen me through interesting times.

Notes

1. M. H. Schneps, "The Advantages of Dyslexia," *Scientific American*, August 19, 2014, http://www.scientificamerican.com/article/the-advantages-of-dyslexia/.

Chapter 15

Anneila I. Sargent (PhD, 1977)

A Long Way for a Wee Lassie

Anneila Sargent, the Ira S. Bowen Professor of Astronomy Emeritus at Caltech, studies how stars and other planetary systems form and evolve in the cores of dense molecular clouds of dust and gas in the Milky Way and other galaxies. She was Director of OVRO from 1996 to 2007, was start-up Director of CARMA from 2003 to 2007, and chaired the international board of ALMA in 2006–7 and 2009–11. She has also chaired NASA's Space Science Advisory Committee and the NRC Board of Physics and Astronomy, and was appointed to the NSB by President Obama. She served as President of the AAS and is a Legacy Fellow. A distinguished alumna of both Caltech and the University of Edinburgh, she is a member of the National Academy of Sciences and a Fellow of the American Academy of Arts and Sciences and of the Royal Society of Edinburgh. She was awarded NASA's Public Service Medal in 1998. Her asteroid is 18244 Anneila.

Years ago, when I was flying all too frequently for work, I chatted a little to a man seated across the aisle. Like me, he was from Scotland and had left to make his life elsewhere. He seemed surprised not only that I was an astronomer but also that this had taken my life in such unexpected directions. As we deplaned, he remarked somewhat bemusedly, "You've come a long way for a wee lassie from Burntisland." To me, it has always felt both such a long way and such a natural path.

The journey that took me from a small town in Scotland to a life in astrophysics began with educational opportunities. Growing up in Burntisland, a small town on Scotland's Fife coast, just north of Edinburgh, I did well in math and science at the nearby high school. I was lucky to have the support of a particularly good physics teacher, who gave me the right foundational background to go on to study physics at the University of Edinburgh. In my second year I took an introductory astronomy class that was much more demanding than I had anticipated. Luckily, the faculty, astronomers at the Royal Observatory Edinburgh (ROE), paid unusual attention to student learning. This led me to an astronomy summer school at the UK's Royal Greenwich Observatory (RGO) in Herstmonceux and to a senior year astrophysics class at Edinburgh. Neither I nor my one female classmate, British astronomer Margaret Evans Penston, recalls any real gender bias. We shrugged off the "heavenly twins" moniker assigned us by the Observatory staff.

I was well into my career before I realized that I became an astronomer as a direct result of gender bias. After graduation I intended to continue in nuclear physics. I hoped to avoid the constraints that UK funding practices then placed on women graduate students by taking a government laboratory position that would support me financially through PhD research, but that fell through. Nearly fifteen years later, in a chance meeting with my interviewer, I learned that his recommendation to hire me was denied because "she'll get married and be of no use to us, or she won't get married and then she'll cause trouble in the lab." Instead, I became a research assistant at the RGO. That led me to meeting and subsequently marrying another astronomer. My professional life did not come to a halt.

Wallace (Wal) Sargent and I met at the RGO, where he had taken a research position following a postdoctoral fellowship at the California Institute of Technology (Caltech). Wal was English, a theoretical astrophysicist from the University of Manchester, who had returned to Britain and the RGO expecting opportunities to use British telescopes in South Africa. When these hopes were dashed, he looked for faculty positions back in the U.S. and was already considering offers when we met. It became a family joke that his proposal of marriage came in the disguise of a suggestion that I apply to graduate school at the relevant American universities. In 1964 we set off for La Jolla and the University of California, San Diego (UCSD), where Wal would work with Geoff and Margaret Burbidge and observe at UC's Lick Observatory. I began graduate school in astrophysics with Geoff as my advisor. Although she taught in the physics department, Margaret could not formally take on this role; due to nepotism rules, she had been hired as a professor of chemistry.

Graduate school at UCSD was my first experience living abroad, my first experience of American academia, and the first time I became aware of the discouraging climate for women scientists. It was well-illustrated by the histories of two exceptional women who taught me, the distinguished astronomer Margaret Burbidge and physics Nobel Laureate Maria Goeppert-Mayer. Their professorships, and the respect and admiration they now enjoyed, had been hard-won despite their remarkable research contributions. Both remained positive and charming, and both had successfully combined their careers with marriage and children, as I saw firsthand through the close and lasting friendship that Wal and I forged with Margaret, Geoff, and their daughter, Sarah.

Midway through our second year in La Jolla, Wal accepted an invitation to return to Caltech as an assistant professor. I transferred to graduate school in astronomy at Caltech but soon decided to drop out with a master's degree. I no longer recall all the contributing factors. On the positive side, the astronomy department was small, and both graduate students and faculty were welcoming. Indeed, seven women were part of Caltech's astronomy graduate program around that time. Virginia Trimble has written of this elsewhere. My grades were good; my physics credits from UCSD had been accepted and I was taking mostly astronomy classes. But the question hovered: Was I admitted to ensure Wal's appointment? Much later, I became aware of the prevalence of "imposter syndrome." I now think I had a bad case of that. I remained at Caltech as a research assistant to Jesse Greenstein, the founder of the department and definitively its leader. Jesse was largely responsible for the number of women graduate students in astronomy and I now recall how strongly he argued that I should not give up. But somehow doubt prevailed, and I focused instead on the next logical step for a young married woman: motherhood.

During my years away from graduate school, Wal and I became the parents of two wonderful daughters, Lindsay and Alison. I have always been glad that they were born relatively early in my life and presented us with four grandsons before we even thought of the possibility. Life with an infant and a toddler, however, made it very clear to me that I missed the thrill of research that I had experienced while working with Jesse. My first significant publication was as co-author of his paper on the spectra of faint blue stars in the Galactic halo. Motherhood clarified for me that my answer to the marriage or career question was "both." I applied to re-enter Caltech as a student, gaining considerable confidence merely by being successful—this time around, I knew admission had relied on my own performance.

I've always been grateful to Caltech for that second chance. I'm equally grateful to my PhD advisor, Peter Goldreich, for giving me the luxury of a summer reading period to select a thesis topic before I formally began graduate school. I was soon hooked on the question of how the formation of stars propagates through molecular clouds. I could investigate this with large-scale millimeter-wavelength observations of the parent clouds of OB associations, loose aggregates of high-mass stars segregated by age, and perhaps complement these with infrared observations of any detected protostellar cores. At Caltech, Robert (Bob) Leighton, aware of early results from single millimeter-wave telescopes, was planning a three-element, millimeter-wave interferometer for Caltech's Owens Valley Radio Observatory (OVRO). At the same time, Gerry Neugebauer and his group were pioneering infrared research observations from Mount Wilson and Palomar. It seemed to be the right time and the right place to venture into star-formation studies that relied on both kinds of measurements. For me, millimeter-wave astronomy had the added advantage of separating Wal's and my observational interests. I really wanted to make my own way in the field.

Acquiring the necessary observations might have posed a problem, but Dr. Eugene Epstein offered extensive access to his recently constructed millimeter-wave telescope at the Aerospace Company in El

Segundo, California. I began mapping out cloud structure in the Cepheus and Perseus star-forming regions early in 1975 and completed the project within a few months, thanks to a generous allocation of observing time enhanced by the superb millimeter-wave detection conditions produced by a severe California drought. The location of my sources on the sky at that time of year meant I could commute between El Segundo and Pasadena every day, fitting my family around my observing schedule in an ongoing balancing act between the two. I presented preliminary results at the 1976 IAU Symposium on Star Formation, held in Geneva, Switzerland. At this, my first meeting with the broad international community of star-formation researchers, I was relieved to finally see my own research in a broader context and confirm its relevance. Of course, much of my ease with the give-and-take of this conference derived from my interactions with the lively group of infrared and millimeter-wave astrophysics graduate students and postdoctoral fellows at Caltech. The accompanying photograph of some of us at a star-formation workshop in Ensenada, Mexico, in 1976 captures this camaraderie very well.

I completed my PhD in late 1977 and became a Caltech postdoctoral fellow for the next three years. In addition to molecular cloud mapping, I undertook more detailed cloud observations with various other U.S. millimeter-wave telescopes. And realizing that the pre-stellar cores I was finding were often sufficiently obscured by dust to be undetectable at the near-infrared wavelengths accessible from Mount Wilson, I began a series of far-infrared balloon observations of clouds, in collaboration with Reinder van Duinen's Space Research group from the University of Groningen. At Caltech, I became progressively more involved with the millimeter-wave interferometer as it moved towards operations. By now, my husband was well-established at the Institute and did not seem likely to leave for anywhere that would be better for my research. In addition, our daughters were happily ensconced in a school conveniently close to Caltech. I saw a number of advantages in having a less-pressured job while they were growing up and so, as my fellowship drew to a close, I made no effort to look for a position elsewhere and joined Caltech's professional staff as part of the OVRO array support group.

FIGURE 15.1. Caltech students and postdocs relaxing at a conference in Ensenada, Mexico, in 1996. Clockwise around the table from the front left: Steve Beckwith, Steve Wilner, Eric Becklin, Fred Lo, Anneila Sargent, Mike Werner, France Córdova, Cathy and Ian Gatley, Neil Evans, Dan and Pirio Gezari, Dick Joyce.

The lack of pressure enabled our family to continue to return to the UK each summer and spend time with the grandparents, a tradition that started early when Wal, along with the Burbidges, Willie Fowler, and his Caltech postdocs, descended on Fred Hoyle's Institute of Theoretical Astronomy (IOTA) in Cambridge. Beginning in the late 1980s, thanks largely to Wal's connections, we began to visit research institutions in other parts of Europe. Our daughters quickly formed "insider" views on the culture and food of a number of countries that they still recall fondly. For me, these visits often led to new and rewarding collaborations. Our Paris summer sojourn in 1984 had surprisingly long-term effects. While Wal collaborated with colleagues at the Institute d'Astrophysique, I joined the astrophysics section led by Catherine Cesarsky at Centre d'Etudes Nucléaires de Saclay as they were putting together a proposal to the European Space Agency (ESA) to build the camera for the

Infrared Space Observatory (ISO). Following the proposal's success, I was invited to be part of the science team planning the program for ISOCAM guaranteed time. My first grant was from NASA and provided the necessary support for my participation. The family summers had opened wide a new professional avenue.

That first grant led to NASA committee work and, in the longer term, to a wide variety of scientific policy or organizational committees. I admit freely that I became fascinated by the ways in which science is focused and funded in the United States. By the 1990s, I was involved in a rather broad range of boards and advisory committees, spanning NASA, the National Science Foundation (NSF), and the National Research Council (NRC). I especially enjoyed chairing the NASA Space Science Advisory Committee and the NRC's Board on Physics and Astronomy. My long stints on the Board of Trustees of Associated Universities, Inc. (AUI), which manages the National Radio Astronomy Observatory (NRAO) on behalf of NSF, and on the Atacama Large Millimeter-wave Array Board were memorable. All of this work gave me a wider scientific perspective and proved to be excellent training for aspects of my later roles as Director of OVRO and Vice President for Student Affairs at Caltech. The serendipity of my choice to begin in a support staff position and remain for quite a long time in a research position emerged only over time, but I certainly could not have taken on all these community commitments as a teaching faculty member with associated university obligations.

With the arrival of Nicholas (Nick) Scoville at Caltech in 1984, as a professor and OVRO director, the millimeter array was completed successfully, enabling advances in a broad swathe of astronomical areas ranging from planetary science and star formation to extragalactic studies. I was particularly excited about the study of a very young Sun-like star, HL Tauri, that I undertook with Steven (Steve) Beckwith, once a fellow Caltech graduate student but by then a faculty member at Cornell. We suggested that our detection of circumstellar molecular gas and dust around this object indicated a forming planetary system. Our suggestion stimulated considerable follow-up both by us and others. I found myself often invited as a colloquium speaker or giving the lead-off presentation at a conference. This professional success coincided with our daughters becoming independent and leaving for college. In 1988, I asked for, and eventually was given, the opportunity to move from the professional staff to the more prestigious world of the research faculty track.

Over the next few years, I continued a variety of protoplanetary disk studies with Steve, my students, and postdocs. Among other projects, I ventured into interferometer investigations of the origins of the stellar mass distribution and the role of bipolar outflows in star formation with postdocs Leonardo Testi (ESO) and Héctor Arce (Yale), respectively, Eventually, I became OVRO associate director for millimeter operations. From time to time, I became frustrated by the professional limitations of the research track and made up my mind to leave the Institute. Each time I let this be known, attractive outside offers appeared. But I stayed on. Caltech's counteroffers and research opportunities remained a compelling option. By the mid-1990s, the OVRO array had been expanded to include six telescopes and continued to produce exciting scientific results. Nevertheless, it was recognized that continuing cutting-edge investigations would soon require further technical enhancements, perhaps even a move to a higher-altitude site. When, in 1995, Nick Scoville announced his intention to step down as OVRO director, I believed that I had the background knowledge and experience to take the Observatory forward and threw my hat into the ring. I served as OVRO's executive director for two years before the department recommended my appointment but, in 1998, I became both OVRO director and professor of astronomy at Caltech.

After that, with teaching, research, and expanded OVRO duties, my life was a whirlwind. An immediate concern was how to upgrade the millimeter array. Increased support from the NSF seemed unlikely, as the community now sought to build a large national millimeter-wave interferometer. While this would take years, combining the OVRO and BIMA array telescopes on a higher site, with improved receivers and electronics, could produce an enhanced array relatively soon. BIMA, the Berkeley-Illinois-Maryland Array at UC's Hat Creek Radio Observatory, and the OVRO array had been stimulating competitors since their inception. Both groups now agreed on the wisdom of at last joining forces. The devil was, as always, in the details, and the partnering of sometime rivals had inevitable issues. Nevertheless, the desire for better science was strong and the Combined Array for Research in Millimeter-wave Astronomy (CARMA), funded equally by the university partners and NSF, became a reality in the early 2000s. In the United States, its scientific contributions were an obvious driver for the international Atacama Large Millimeter-wave Array (ALMA). For me, its construction was a highlight in a career that saw a lot of rather unexpected twists and turns. I enjoyed that enormously, even on the bad days. Often, as I stood at the new site in the Inyo Mountains above Owens Valley, watching trenches being dug and cables laid and telescopes arriving, I thought about the hopeful young woman who had left Scotland. This was beyond her wildest dreams.

I remain proud not only of CARMA's scientific accomplishments but also of the numbers of young people, graduate students and postdocs, who came, learned, and went on to successful careers at academic institutions in the United States and around the world. The effort of bringing the array about certainly cut into my research time, but I was fortunate to have colleagues like John Carpenter (ALMA) and Andrea Isella (Rice University) and students like Laura Perez (University of Chile, Santiago) and Stuart Corder (ALMA), who kept me involved in the acquisition and analysis of the much-improved protoplanetary disk images. The first high-resolution ALMA image of HL Tauri, showing circumstellar rings of material that might indicate the presence of planets enchanted me, but by then I had moved on to be Caltech's Vice President for Student Affairs. And that is another story.

As I look back, the theme that emerges is that each of my varied experiences turned out in the end to be valuable and to open doors I could not have anticipated. And I have been remarkably fortunate in my friends and colleagues. I am still amazed to say that today I am professor emeritus at the California Institute of Technology. Reading my résumé, you can follow a more or less neat path as I re-enter Caltech as a graduate student in astronomy in 1974 and move though the postdoctoral fellowship, professional staff, and research faculty ladders, eventually

FIGURE 15.2. Anneila Sargent at the Owens Valley Radio Observatory millimeter-wave array circa 1998. *Credit*: Corbis/Corbis Collection via Getty Images.

attaining a named professorship. Along the way I was, at various times, Director of OVRO, Director of CARMA, and Caltech's Vice President for Student Affairs. As I lived through the experience, there was nothing neat or tidy about it and I could not have persevered without the support of my late husband, Wal Sargent. Wal, in his own way, always supported my aspirations and made it possible for me to follow both the science and my instincts to make choices that were interesting rather than conventional.

Chapter 16

Martha P. Haynes (PhD, 1978)

Hands-on Adventures with Telescopes: From the Backyard to Cerro Chajnantor

Martha Haynes, Distinguished Professor of Arts and Sciences in Astronomy at Cornell University, studies the large-scale distribution of galaxies in the local Universe and, in particular, has identified the filamentary nature of the Pisces-Perseus galaxy supercluster. Her accomplishments have been recognized with the Henry Draper Medal of the National Academy of Sciences in 1989, the Catherine Wolfe Bruce Gold Medal of the ASP in 2019, and the Karl Jansky Lectureship from the NRAO; she is a member of the National Academy of Sciences and of the American Academy of Arts and Sciences, a Fellow of the American Association for the Advancement of Science, and a Legacy Fellow of the AAS. Her asteroid is 26744 Marthahaynes.

How did I get here? Frankly, my life story and research career have not involved much strategic planning. I was lucky to grow up in an era when we didn't worry so much about eventual employability. For me, the key elements of my career (and life) have been the support and inspiration of my family, friends, colleagues, and (yes, definitely) students; the thrill of observing and making things work; the reward of sharing the Universe with students of all ages; and lastly, the satisfaction of contributing to the development of new "stuff": software, databases, facilities, and instruments. For me, research, teaching, mentoring, and service are all exciting, and I get the most satisfaction from a mix of them. Following my own path has always been important. Some luck, a hard head, a thick skin, and a sense of humor have also helped.

I don't come from an academic family, and as the youngest of five children I was always trying to keep up with my four amazing and smart older siblings. In truth, they blazed my trails.[1] My father was a self-employed manufacturer, and my mother was a first-grade teacher before she married.[2] They valued education highly and encouraged us to pursue our passions. In retrospect, the fact that neither of them ever said "you cannot do that" had a huge impact. I often have had doubts about myself, particularly as the only female in the room, but I have always had especially my mother's and my sisters' voices in my ear telling me to follow my dreams wherever they led.

I never knew how tight money was in my family when I was growing up. We helped with chores, took jobs to earn spending money, and put creative effort into designing homemade personalized gifts. As we grew older, all of us worked summers in the family business. My job (from high school through college) was as the shipper, in retrospect, a primitive form of database manager. Frankly, it was incredibly boring, but I never had to worry about finding a job, and I learned a lot of organizational skills. It certainly made me appreciate even more the opportunities afforded me to pursue my education.

On a fateful day during my second year of graduate school at Indiana University, I crossed paths with Riccardo Giovanelli. As the famous quote from *Casablanca* goes, it was the "beginning of a beautiful friendship." Most of the big projects in my career have been undertaken in partnership with Riccardo. Beyond the personal relationship, we have enormous respect for each other as scientists and human beings. Having a supportive and inspiring partner has certainly made my journey easier, not to mention a lot more fun.

I grew up less than ten miles from downtown Boston, definitely not a dark-sky site. I don't remember being particularly interested in astronomy, but I was an outdoor kid, participated in numerous team sports,[3] and earned lots of nature-related Girl Scout badges. When I was about thirteen, my brother convinced me to give him $30 of my babysitting money so he could buy us a three-inch reflecting telescope. At the time, he was busy with college, his car, and his girlfriend (later my sister-in-law). So, I became the one fascinated with what the night sky held. I remember showing the rings of Saturn and the moons of Jupiter to a couple of passing police officers one night. Explaining to them what they were seeing gave me the same thrill then that explaining cosmology to "students" in Cornell's Adult University (CAU) does today.

That little telescope led me to Wellesley College with its Whitin Observatory. Just as I was beginning my senior honors thesis, a new spectrograph arrived, and my task included getting it to work properly. To observe, I had to round up one or two friends to help me replace the viewing eyepiece with the spectrograph and then rebalance the telescope, so I learned to be ready to make the best use of the few clear nights. By good luck, I managed to catch the outburst spectrum of a variable star. While not the most brilliant piece of observational work ever done, my project taught me a lot about astrophysics, about the workings of astronomical instrumentation, and, indeed, the thrill of discovering something. Hands-on observational astronomy was definitely for me.

Just as I graduated from Wellesley, I was offered a summer internship at the Arecibo Observatory. There, my project involved calibrating the telescope's sensitivity in support of radar studies of the Earth's upper atmosphere. I knew nothing about radio astronomy, electronics, programming, or the ionosphere, but being at the Observatory with scientists talking science large fractions of the time was exhilarating. The next summer, I was off to the National Radio Astronomy Observatory (NRAO) in Charlottesville where I learned the basics of interferometry and worked on a project that introduced me to "normal," as opposed to "radio," galaxies. This second experience, coinciding with the end of a first year of graduate school, focused on stellar astrophysics, reaffirmed my interest in extragalactic radio astronomy, and as a result, I charted a path toward the galaxies, not the stars.

On a trip to the Green Bank Observatory in West Virginia in the early summer of 1975, I met Mort Roberts in the lounge of the residence hall.

FIGURE 16.1. Martha Haynes taking measurements at the CCAT Observatory site, future home of the Fred Young Submillimeter Telescope, at 18,400 feet (5,600 meters) elevation on Cerro Chajnantor in the Atacama region of northern Chile in April 2010. Her backpack contains the requisite oxygen tank. "Being involved in using telescopes and planning for new ones has, for me, always added adventure to the excitement of scientific discovery."

I remember that Mort asked me what I wanted to do with my life. I believe I expressed interest both in galaxies and in using radio telescopes to try new techniques. Mort suggested that I think about what observations of neutral atomic hydrogen at a wavelength of 21 centimeters (known as "the hydrogen line") might be interesting to pursue and write him again in a few months with a thesis proposal. I read a lot of papers that summer. Apparently, my idea to look for atomic hydrogen gas in the spaces between galaxies intrigued Mort because soon a plan was in place for me to spend two years as a graduate student at NRAO under his mentorship.

Arriving at NRAO, I found myself in a very different working environment, with almost no talk of classes, homework, or exams. For me, this time was special, filled with lots of seminars and discussions about both radio techniques and science. As one of only a few in-residence graduate students, I felt like part of the staff and shared an office with Mort's data analyst.[4] Mort was the perfect advisor for me: amazingly knowledgeable and ever-available, but never hovering and always challenging. I learned so much about so many things during those two years. And opportunities indeed arose to make observations that pushed the technology to its limits. As a result of my youthful shipper years, I was the fastest card puncher in the group so I got to see the hydrogen absorption line in the spectrum of AO 0235 + 164, a distant galaxy known as a blazar, before anyone else: discovery is fun![5]

As I was coming to the end of my thesis, the issue arose of what I would do next. And also, what Riccardo would do next. And what would "we" do? Luckily for us, Arecibo was looking for extragalactic radio astronomers, offering a happy solution to our two-body problem. For both of us, being staff astronomers at Arecibo was the perfect next step. The staff was small, and we socialized a lot, ran laps around the dish, and shared Friday dinners at Tony's Pizza World. We built a wooden house by ourselves. The view out the front door of the telescope's platform and the central Puerto Rican highlands was spectacular.[6] I bought the beer and peanuts so the guys (the rest of the scientific staff) could play poker at our house while I spent those evenings at the Observatory taking advantage of the cpu cycles that they were forgoing for the game.

Cooled receivers, more capable spectrometers, "movable" feeds, and enhanced computer power kept advancing the potential of extragalactic hydrogen line science. As the "friends of the telescope," we helped visiting astronomers observe and wrote software and telescope manuals. We pushed the limits of the instrumentation and advocated further upgrades. At the Observatory with its close-knit staff, everyone shared in the excitement of discovery, and the scientific energy level was very high. We were lucky to be part of it.

During this same period, the first major surveys of galaxy redshifts were exploding concepts of how galaxies are organized into interconnected supercluster structures and how differences in galaxy environment could impact galaxy evolution.[7] We put in proposals to continue work on gas between the galaxies, to compare the gas content in isolated and cluster galaxies, and to conduct redshift surveys to map out their distances. Since remote/absentee observing did not exist at that time, our proximity to the telescope was a big advantage, and our surveys became "schedule fillers." Ironically, the airglow experiments that were a major part of the Arecibo atmospheric research program required clear weather, whereas ours did not, and, to our advantage, the weather was often cloudy, or worse.[8]

At Arecibo, lots of projects by other astronomers requested telescope time in the Virgo direction, up at night in the spring, but the fall nighttime sky dominated by the nearby Pisces-Perseus Supercluster was undersubscribed so we decided to map out its structure. Our redshift survey of Pisces-Perseus coincided with lots of studies, both observational and theoretical, of the nature and origin of the large-scale structure of the Universe. Pisces-Perseus is nearby, almost perpendicular to the line-of-sight and set apart by much less dense void regions in front and behind it. Its prominent linear ridge, which we dubbed the "Extragalactic Lizard," struck us as an interwoven filamentary structure, what is today's "Cosmic Web." That realization led to our receipt of the National Academy of Sciences 1989 Henry Draper Medal "for the first three-dimensional view of some of the remarkable large-scale filamentary structures of our visible Universe," certainly a big honor for a couple of (relatively) young astronomers.

In the late 1980s, a convergence of circumstances (the collapse of the Green Bank 300-foot telescope, the upgrade of the Arecibo telescope, and Cornell's partnership in the operation of the Palomar 200-inch telescope) turned me into an optical astronomer. Coming naturally after the redshift surveys, surveys to measure the flows of galaxies around clusters and superclusters caused by the gravitational pull of dark matter on large scales required optical imaging, better hydrogen line profiles, and sometimes optical spectroscopy. We used a pilot experiment to detect a beautifully flat rotation curve in a distant (for those days) spiral disk. Optical astronomy was also exciting stuff, but dealing with clouds and bad weather was so annoying. And the physics of the 21-cm hydrogen line is very simple. My radio astronomy heart still beat fastest.

FIGURE 16.2. Award recipients and the award presenters from the 1989 annual meeting of the National Academy of Sciences. "Neither Riccardo nor I feel very comfortable in formal situations and attending the award ceremony caused us 'dress-code anxiety.' Why is it that men don't have to worry so much about what shoes to wear? As should be obvious from this photo, I in particular felt a bit out of place among this group." *Credit*: National Academy of Sciences.

In the 1990s, a modern three-mirror system in a dome (as seen in *Contact*) was installed at Arecibo to replace the old one-pixel line-feeds (as seen in *GoldenEye*), making it possible to install a new seven-pixel radio camera, dubbed ALFA (the "Arecibo L-band Feed Array"). The advent of ALFA allowed Arecibo to map wide areas of the sky efficiently. After several years of discussions, simulations, and tests of observing strategies, we proposed to use ALFA to search for gas-bearing galaxies over about one-sixth of the sky visible from Arecibo. This program, the Arecibo Legacy Fast ALFA (ALFALFA) Survey, was awarded 4,400 hours of telescope time spread over 744 separate observing blocks scheduled over 6.5 years. It involved a team of faculty, postdocs, and students who conducted the observations remotely, wrote the software, reduced the data, and performed the science analysis. The survey strategy was driven by the aim of conducting an accurate census of gas-bearing galaxies over a cosmologically significant volume, but it enabled lots of different discoveries both by the team and by others.

As was typical of our modus operandi, Riccardo led the processing software side and grid construction, and I managed the observations, calibration and initial data processing, and archiving. With the senior team members seeing to the delivery of the data products, our graduate students focused on the science. We engaged a group of faculty and students at principally undergraduate teaching institutions, the awesome Undergraduate ALFALFA Team. Besides producing more than one hundred team publications, ALFALFA has been, as the team motto says, "more fun than human beings should be allowed to have."

Harking back to that night in Milton when I showed Saturn's rings to the police officers, sharing my thrill for astronomy has always been an important component of my professional satisfaction. I really enjoy engaging in exploration with diverse student cohorts from across Cornell, including alumni "students" of the CAU. I particularly like to tie astronomical discoveries to their historical context and to use images to introduce astronomical case studies. On a CAU tour in Italy, I've watched the Sun's image approach Cassini's meridian in the Basilica of San Petronio in Bologna at noon on the summer solstice, deciphered the intricate astronomical clock in Cremona near the violin museum, and seen the shadow on the vertical sundial in front of the Verdi museum in Busseto. Participating in Cornell's Knight Institute Writing in the Disciplines program has made me realize the importance of using the language appropriate to the audience. For many years, I have taught the writing-oriented History of the Universe class to non-scientists who like to write. Teaching this class for me is never the same one year after the next, since we follow astronomical events and discoveries as they happen. Trying out new ways to approach new topics makes teaching ever more interesting.

Over the years, I have worked with a large number of terrific students who have taught me how to be a better scientist and a better educator and have given me the treasure of being part of their lives. I would like to think that I have inspired some of them in the same way that others have inspired me. To me, interacting with students inside and outside of the classroom is the greatest reward of my academic profession. The best part is seeing so many of them happy and satisfied wherever life has taken them.

That conversation with Mort Roberts about my interest in being involved with telescope construction eventually led me to become the assistant director for Green Bank Operations. I was only thirty years old, female, and had no management experience. Six weeks after I arrived, budgetary considerations across NRAO required a reduction in the workforce. My instinct was to rely on the advice of the local management team who understood the strengths of individuals and how the Observatory could function with a much-reduced workforce. I'm sure many were dubious of my ability to manage the site, but I could not have asked for a better, more dedicated group. From them, I learned much about leadership, communication, and holistic management.

After two years, a chance arose to leave Green Bank for a faculty position at Cornell. By then I had management and operations experience, and that background seemed to earn me many invitations to participate on review and advisory committees. Although not always fun or relaxing, I did achieve a certain sense of satisfaction from being part of the practical enterprise of enabling astronomers to do research through our large facilities.

The span of my professional career maps onto the development of the Atacama Large Millimeter Array (ALMA). I have been involved in many phases of its planning and construction through my affiliations with Associated Universities, Inc. (AUI). Perhaps most critically, I served as interim President of AUI when the National Science Board approved the design and development phase of what would become ALMA. During the AUI presidency of Riccardo Giacconi, a truly inspirational administrator and Nobel Laureate scientist, I served two terms as chairman of the AUI Board of Trustees.[9] The two construction projects initiated on his watch, the Jansky Very Large Array and ALMA, have both proved tremendous scientific and technical successes. Though mostly from windowless rooms, I'd like to think I also contributed toward these great facilities.

My involvement in ALMA paralleled Cornell's pursuit of its own submillimeter telescope in the Atacama Desert. In 1994, "my" Riccardo accompanied the NRAO team on their first trip to northern Chile to explore possible sites for the ALMA submillimeter array. He came back hooked on the exceptional value of the Chajnantor region as an astronomical site and enthralled by its natural beauty and historical interest. In December 1994, we spent a wonderful vacation traveling through Chile and paid an amazing visit to the Valle de la Luna on the night of the Full Moon: simply awesome.

About ten years later, Riccardo and others led a design study for a 25-meter diameter wide-field submillimeter telescope dubbed CCAT (Cerro Chajnantor Atacama Telescope) to be located above ALMA on Cerro Chajnantor. Riccardo spent one trip exploring possible sites with Cornell alumnus Fred Young, who has provided substantial financial support for Cornell's part in the CCAT project. Although it received a strong endorsement from the 2010 Astronomy and Astrophysics Decadal Survey, federal funding for the 25-meter telescope was declined due to budget constraints, and the dream for a telescope on Cerro Chajnantor seemed dead.

However, some resources and determination remained within the CCAT partnership. Could we tackle a much scaled-down project, not simply a "mini-CCAT" but one much less costly yet still scientifically transformational? In a very short period of time, brainstorming among a small group of scientists from Cornell and our German and Canadian colleagues led to a new design of what is today the Fred Young Submillimeter Telescope (FYST), a 6-meter telescope able to scan the sky rapidly and with a precision surface that will take advantage of the exceptional site on Cerro Chajnantor for a variety of exciting and transformational survey science projects. At the time of this writing at the end of 2020, the FYST is currently being manufactured and preassembled in Germany with delivery to Cerro Chajnantor in about two years. With retirement in my future, I am thrilled by what my younger colleagues have planned for it.

On a very personal note, the demise of the 25-meter CCAT project coincided with the increasing intensity of Riccardo's battle with

Parkinson's disease. In fact, the stress and then shattering disappointment of the project not going forward exacerbated what is a horrible disease. The upside of our personal partnership is that I was able to step into his shoes.[10] My hope now is to go with Riccardo once more to the Atacama for the inauguration of our dream discovery machine.

There have been many challenges in my life and work, but overall it has been an exhilarating journey. From my first views of the sky with that three-inch telescope in 1964, I still get a thrill from eyeing a comet from my backyard, watching the Full Moon rise over the Castello di Carpineti from Sarignana, or seeing the Sun setting over the Pacific while strolling on the beach at Algorobbo. I've traveled and made amazing friends around the world. Riccardo and I have been lucky in so many respects. From Arecibo to Atacama to our Italian stone farmhouse in the hills south of Parma, there are many shared memories and feelings of satisfaction. But it has not always been easy to navigate life and career, and I wouldn't want anyone to think otherwise. I've had to be "The First/Only Woman" far too many times,[11] and we've had some demoralizing experiences with ridiculous dual-career employment issues.[12] In addition to the career challenges, over the years I have had to deal with a still-undiagnosed autoimmune condition that seemed for a while to be life threatening and is always lurking in the background. And, there was a decade when we lived three thousand kilometers apart, me in Green Bank and then Ithaca and him in Arecibo. I would not recommend such an arrangement, though it worked for us, for here we are still laughing about our first date forty-five years later.[13] And, the acknowledgment in my thesis still applies: "I'm sure I could have done it without you RG, but it wouldn't have been nearly as much fun."

To end, I emphasize again the importance to me of the support of many colleagues and friends, far too many to name here, who have enabled me to have both success and satisfaction. My message to others is: "Encourage your younger colleagues and your older ones too." Support really does matter. And to young people considering a research career in astronomy or otherwise: "Follow your dreams to wherever they lead you." And to all: "Be sure to enjoy it!"

Notes

1. One brother and one sister majored in physics, the other brother in engineering. The eldest majored in English literature and received her PhD when she was in her fifties. Both brothers had their paths interrupted by the Vietnam-era military draft, an issue I never had to face. My physicist sister found a hostile workplace in industry and left to apply her skills to raising amazing kids and forty years of community service and leadership. In 1993, she was the first woman elected to the Milton (MA) Board of Selectmen, a mere 331 years after the town's incorporation. Besides always being ahead of me, they have all always been there to cheer me on.

2. It was not her choice to give up teaching; married women did not teach the lower grades. In fact, my parents had to wait for years to get married because they both supported widowed mothers.

3. Team sports for girls in the 1960s were mostly intramural and pathetic. Field hockey was played on the football practice field, softball on the dirt diamond with only the base anchors. A basketball team had six players, three forwards and three guards. Players were only allowed to dribble for three bounces before passing. When I was in high school, liberation arrived for one of the forwards and one of the guards: these "rovers" were allowed to cross the center line. Pathetic indeed.

4. I referred to this office as the "NRAO girlie office" in the acknowledgments of my thesis, somewhat to Mort's (and probably others') chagrin.

5. My personal big "discovery," which I thought was an absorption line due to gas seen in the foreground of a more distant radio source, was too good to be true. Always skeptical, Mort suggested we confirm it three months later by looking for the expected shift in wavelength due to the Earth's changing orbital velocity around the Sun. In fact, the absorption signal went entirely missing. He concluded that the first observation was likely contaminated by transmission from a satellite, "either theirs or ours": a good lesson for this novice graduate student.

6. On December 1, 2020, this view of the Arecibo platform disappeared in a catastrophic structural failure. In my career, I have suffered the collapse of two great radio telescopes, the 300-foot at Green Bank and the 1,000-foot at Arecibo. I mourn the loss of both but am immensely grateful to have had the opportunity to use them to explore the Universe.

7. To explain my work on the impact of environment on galaxy evolution to my mother, the first-grade teacher, I referred to my research as "extragalactic sociology." She seemed to grasp that, at least vaguely.

8. It is rumored that our atmospheric colleagues considered us the "telescope vultures," circling in hope of bad weather or klystron failure. We thought we tried to be discreet, but indeed we were always eager to take up any free telescope time.

9. A photo taken of the two "Riccardo Gs" speaking to each other was circulated with the caption "When Riccardo speaks, Riccardo listens. But when Martha speaks, does either of them listen?" No comment.

10. It should be noted that his feet are substantially larger than mine.

11. Being the first/only anything adds a lot of stress in professional situations. My hard head and thick skin provided the determination needed to succeed, but, especially after four years at a women's college like Wellesley, where doubts about capability and leadership qualifications

related to gender did not exist, being the only woman in the room has often been a lonely experience. These days, I even enjoy having company during restroom breaks.

12. The most maddening experience was when my department chair asked me if I would be willing to share my tenure with Riccardo so that he could move from being head of the radio astronomy group at Arecibo to a faculty position in Ithaca. This happened just after we had received the Draper Medal and were being recruited by two other institutions. Not being sure which one of us should feel more insulted, we went looking at houses elsewhere. When my department chair heard about that, though, a full professorship for Riccardo and a promotion for me quickly materialized. Playing hardball with people's personal lives seems an all-too-common practice in academia; our profession should do better.

13. The real miracle, which will be appreciated by readers who know Riccardo as the world's fussiest eater, is that we had a second date. On the first, I served him beef stroganoff, a dish I have never made for him since. I don't remember him not eating it. I do remember him offering to do the cooking for our second date.

Chapter 17

France Anne Córdova (PhD, 1979)

The Learn'd Astronomer Discovers the Policy World

France Córdova, now the President of the Science Philanthropy Alliance, was the first woman to serve as NASA Chief Scientist and as President of Purdue University. She also served as Chancellor at UC Riverside and as Director of the NSF. At NSF, she started the Ten Big Ideas program, which included focuses on both breakthrough science and diversity and inclusion, and at UCR she started a new medical school. She was awarded NASA's Distinguished Service Medal, the Kennedy-Lemass Medal from Ireland, and the Order of Bernardo O'Higgins from Chile, was elected a Fellow of the American Academy of Arts and Sciences, the American Association for the Advancement of Science, and the Association for Women in Science, and was awarded an honorary membership in the Royal Irish Academy. Córdova Nunatak in the Pleiades range in Antarctica is named in her honor.

Richly colored paintings by Albuquerque artist Edward Gonzales illuminate my kitchen walls. They show women working maize dough to make tortillas. The images take me back to my college summer in Oaxaca, where I did a study of the sociology of bilingualism. For four months I lived in a pueblo a few kilometers off the road from Oaxaca City to the archaeological ruins of Mitla. My quarters were a room with no back door, a dirt floor, and a single wood bed. Having no bathroom, I used the corral out back. The toilet paper, stuck in a bamboo screen, was old schoolbook pages. With the women of the pueblo I carried baskets of corn in the early morning to the mill to be ground. In the cool air I wrapped myself in a black shawl called a reboso. When we returned from the mill the women set about making tortillas, as well as the stews, black bean soups, and a frothy corn milk that nourished their families. I went from adobe house to house that summer, taking my census, finding out ages and relationships, and asking questions about when, where, and to whom they spoke either Spanish or Zapotec. I attended baptisms, confirmations, and funeral parades. I learned passable Spanish and Zapotec salutations. When I returned home to finish my last year at Stanford, I had in my possession a worn pair of homemade zapatos and a handbag woven in nearby Teotitlan del Valle, where everyone was a weaver. I carried home a sack of frijoles negros, dried cakes of chocolate, and my reboso. I had comadres and compadres for life, for now I was a godmother to youths of the pueblo.

And I had in my head a book, which I promptly sat down to write, called "The Women of Santa Domingo." I submitted it to a college magazine writing contest and won a guest editorship. That experience took me the next June to New York City, the headquarters of the publishing house Condé Nast. I was sent with other guest editors to Israel, which would be featured in the August issue of the magazine. I was made the guest travel editor and wrote the story of our adventures through the country, from Bethlehem to the Golan Heights, from Jerusalem and Masada to the Negev. It was called "Shalom, We Echo Shalom."

So, you would think with this background in cultural anthropology and writing that I would have taken my Stanford English degree and made something of myself—like a writer, editor, or linguist. Yet I had a deeper wanderlust inside. I wanted to connect with a place bigger than my hometown of Los Angeles, bigger than the Earth that I loved to roam and explore through observation and writing. I wanted to connect with something wider, deeper than I could imagine—the stars and the Universe that held them. Yet there was nothing in my early background that illuminated a path to the stars—no scientists, no teachers who believed that women could become scientists. There was discouragement for even imagining oneself in graduate school. The goal of college, my parents believed, was to get a M.R.S. degree. (I did not get that degree until many years after my PhD!)

The summer that I spent in New York City writing for a magazine was also the summer that took me to Cambridge, Massachusetts, to start an education project with a few other recent college graduates. One evening I saw a public TV special about the newly discovered neutron stars. I was introduced to Jocelyn Bell, virtually of course, and her remarkable discovery of pulsars as a graduate student in England. And I was introduced to astrophysicists at MIT who explained that a neutron star was a collapsed star only as big as New York City, at the end point of a brilliant life as a large star, when it was a few times more massive than the Sun. They explained the significant gravitational pull such compact, dense stars would have. Even dropping a marshmallow onto a neutron star would liberate tons of energy! They could spin faster once collapsed (like an ice skater who whirls and then draws in her arms to whirl even faster). If they had a bright spot on them, we would see these objects appear to pulse as they rotated—like a lighthouse beam. This, the MIT astrophysicist explained, was what they believed Jocelyn Bell had seen in her data: the periodic signals of rotating neutron stars.

I was mesmerized. The next morning I hopped on a bus to visit MIT and try to find at least one of the scientists featured on the TV show. I did—and (after some talking) ended up with a lab job at the Center for Space Research! The rest of the story . . . well, it's long and complicated. But after spending the summer and early fall working at MIT I returned to California and decided to pursue my goal of becoming an astrophysicist.

Through my work and studies at MIT, I came to know about the scientists—most of them physicists—who were contributing to the new world of X-ray astronomy, founded by Riccardo Giacconi and his colleagues. (Giacconi was awarded the Nobel Prize for this work in 2002.) They employed World War II era sounding rockets to observe the heavens with instruments that could detect X-rays at high altitude, X-rays that would normally be absorbed by the Earth's atmosphere.

I pursued math and physics courses at a nearby California state university and sought a job at Caltech with Professor Gordon Garmire, one of the astrophysicists whose papers I had read at MIT. Dr. Garmire gave me a job programming for an upcoming X-ray satellite mission called HEAO-1 (High Energy Astrophysics Observatory 1), on which he was a principal investigator. He handed me a copy of Chandrasekhar's famous book *Radiative Transfer* and asked me to write a computer program to analyze X-ray data utilizing the equations. I didn't know anything about computer programming, but the Caltech graduate students were immensely helpful, and I quickly learned Fortran IV. After a year, I asked Garmire if I could audit Caltech physics and math courses, and he agreed. On the basis of my performance in the classroom that year and my work in the lab helping with building and testing rocket flight payloads, as well as computer programming, the Caltech physics faculty admitted me to graduate school. That was in the mid-1970s. I could finally start my journey through courses, exams, rocket payloads and flights, observations and analysis, writing and thesis defense! It was a glorious time for me. I just put my head down and *worked*. Actually, that's somewhat of an exaggeration, for I also played, just enough, and I spent the weekends rock-climbing with fellow graduate students in Southern California's great stony places, like Joshua Tree National Park and Taquitz Rock. I kept volumes of T. S. Eliot's *Four Quartets* nearby to give me perspective when I felt down.

When HEAO-1 was launched it afforded a deeper look at the X-ray Universe. It was an era of discovery because of the enhanced capabilities of the X-ray detectors: known and unknown objects were discovered to be emitters of X-rays. My own discovery, of soft X-ray pulsations from close binary systems called dwarf novae, came about because of the partnership I made with amateur astronomers who used backyard telescopes to look for optical brightenings from these binaries. Alerted to such events by these amateurs, I convinced Garmire to point HEAO-1 toward them when they were in outburst and, voilà, copious soft X-rays! That in itself was marvelous, but to discover that the X-ray emission was pulsed with high amplitude . . . well, that was extraordinary.

In my PhD thesis I thanked amateur astronomers, professional astronomers, friends, colleagues, and family in the acknowledgments. I couldn't have pursued this "about-face," turning to science—after graduating in English and being accepted to graduate school in anthropology—without the support of many, including my professors, who never doubted my aims. Years later I asked one of them why they accepted me into graduate school with so little background in science. He replied, "Because we thought you had a high slope." Even my parents thought I was crazy to make such a career shift, but they were proud of me and were cheering for me when I received my Caltech PhD in physics in 1979. That evening they threw me the most amazing party. Think about this: your parents and siblings dancing with your faculty . . .

If I've spent more time on my beginnings as a scientist, it's to emphasize that there are many paths to becoming a scientist. They all converge on a search for answers to mysteries, the biggest mysteries. When I was a young girl I wanted to be Nancy Drew because she was a smart (and nice) detective. She looked for clues, interpreted them, and pursued the truth until she got answers. There is a voice inside me that says, "This doesn't make sense. I need to know more. What if everyone is just making stuff up? How can I find the truth?" I am happy, beyond any words that I can find, that I found others also looking for clues, also wanting to solve mysteries. And in my lifetime, what incredible discoveries have been made—all started with hints, with clues.

Now, halfway through this narrative, dear reader, I must hurry through the rest of my life.

My first job, post–graduate school, was a permanent position at Los Alamos National Laboratory in New Mexico, managed at the time by the University of California and the Department of Energy. There I was able to keep rock-climbing, now on the cliffs above the Rio Grande, and pursue astrophysics, especially multiwavelength astrophysics. That was a new pursuit, the result of several space missions and ground-based observatories opening their facilities to users. I co-hosted the first workshop on multiwavelength astrophysics, in Taos, New Mexico, and edited the first book of articles on the subject, called *Multiwavelength Astrophysics*. I made observations with all manner of spacecraft, from optical and UV to X-ray and gamma-ray, and traveled to observatories all over the world to observe at radio, infrared, and optical wavelengths. (Little did I know at the time that "multiwavelength" would evolve into "multi-messenger" . . . but that comes later.)

I spent a decade in New Mexico, an important decade as I met my husband (rock-climbing of course) and our two children were born in Santa Fe. I often tell female graduate students to think seriously about careers at the national laboratories. I chose a permanent job, one in which I could mature as a scientist and have a family, instead of going from postdoc to postdoc (I had been offered postdocs at MIT, Harvard, and a number of other places). I was able to be a principal investigator (PI) on grants from the start and join national policy committees. I became the chair of a committee of scientists who advised the lab director. Los Alamos was a good choice for me.

Then Penn State came calling. I was asked to be head of the Department of Astronomy (whose name I changed to Astronomy & Astrophysics). I figured, if you are going to be the only woman in a department, then you should be the head of the department! Naturally I set about starting to recruit other women to join us. And I was made a tenured professor. So with my decade at the lab, I was able to skip the steps of assistant and associate professor.

At Penn State I had many grants, including a very big one from NASA. I was co-PI on an experiment that would fly on the European Space Agency's X-ray Multi-Mirror mission (XMM). Our experiment was an optical/ultraviolet telescope that would ride along with the X-ray imaging and spectroscopy telescopes, making it a truly multiwavelength mission. I had proposed this for NASA's Chandra (née AXAF) mission, but it wasn't in the "baseline" and didn't get accepted. Today, more than twenty years after their launch, XMM and Chandra still proudly surf the skies, gathering X-rays, making seminal discoveries.

I was only at Penn State for four years when the head of NASA, Administrator Dan Goldin, asked me to serve as NASA's Chief Scientist. I would be the first woman to do so. I accepted the invitation, reluctantly at first because I had no experience in policy and didn't want to be taken away from my research. But female friends (and Mom) convinced me that this would be the platform I needed to advocate for more women in science. I signed on for three years so that I could return to academia. The NASA job was a turning point for me because it introduced me to the possibility of making impacts that were not necessarily scientific but were immensely important to the world of science—its culture, its values, its processes. The NASA Administrator would say later that I brought science to NASA, because its focus up to that time was more on engineering. He wanted science to be the driver for the engineering feats that NASA was capable of doing. He wanted to explore the big scientific questions that NASA could best address, like the search for life beyond Earth. I worked closely with the Clinton administration White House, especially through the National Science and Technology Council. We made critical changes to the federal definition of research misconduct and started using the capabilities of the Internet to improve science outreach to the public.

When it came time to leave NASA, I was recruited to UC Santa Barbara by Chancellor Henry Yang. (I have been fortunate to have been mentored by remarkable people: he and Dan Goldin were two of them.) At UCSB we were on the cusp of the "interdisciplinary age." The faculty was remarkably eager to work across interdisciplinary lines; in my position as Vice Chancellor for Research, I could facilitate that spirit. I launched an initiative called Research Across Disciplines that funded faculty who took on high-risk research across multiple disciplines. These almost always led to major grant funding from the government after the initial projects were completed.

An opening for the chancellorship at UC Riverside (UCR) was the opportunity for me to lead a campus. Dr. Richard Atkinson, then President of the UC system, made the appointment, with the approval of the UC regents. He was a great mentor. (I still recall his reading to me from his book of wisdom.) So, after six years at UCSB I found myself moving my family south to Riverside. My husband and children were gracious about all the moves, although of course nervous about each change.

One of the best things about being a leader at a UC campus is that you get to know well the leadership of all the UC campuses, and the UC regents, and learn from them. I met many remarkable people in this position and we worked on initiatives to enhance all the campuses.

I continued with my space satellite project and mentoring graduate students the entire time I was in the UC system.

I asked myself when I joined UCR: What could make the biggest impact on the campus, especially regarding research? The huge Riverside-San Bernardino region had no public medical school, so I set about trying to get approval for a new medical school in the UC system. It would be the first new public medical school for California in over forty years. With persistence and much help, we got approval from the UC regents to begin on this path, and today UC Riverside has a vibrant—and diverse—medical school. It is one of the things I am most proud of initiating.

After five years at UCR, Purdue University extended me an invitation to become its eleventh president. I would have a larger platform at a prominent university known for its prowess in engineering. In 2007 I became Purdue's first female president since it was founded in 1869. I was able to accomplish record levels of student success and fund-raising while there, develop a new college, and build many new facilities. Yet it was not an easy task: the university suffered through the largest recession (2008–9) since the Great Depression. And the football team was not going to the Rose Bowl during those years.

Open one door, and another opens ahead. While I was at Purdue, the George W. Bush administration nominated me to be on the National Science Board (NSB), at that time a Senate-confirmed appointment. I became chair of the NSB's Strategy & Budget Committee. Also while at Purdue I was recruited to be a regent of the Smithsonian Institution. I later became chair of the Board of Regents, and that was a marvelous adventure—with so many wonderful people, all committed to inspiring the public through history, culture, and science. Some of the congressional regents on the board suggested to the Barack Obama White House that I become part of the administration and, before I knew it, I was nominated to become the fourteenth director of the National Science Foundation (NSF). My experience on the NSB, which co-governs NSF with the director, was an extremely helpful warm-up. Throughout my tenure as director I relished a supportive relationship with the wise members of the NSB.

FIGURE 17.1. Official portrait photo of France Córdova as NSF director. Photo by Stephen Voss. © NSF.

Every job I've had is my favorite job—until I have a new job; then that's my favorite. I really enjoyed my time at NSF. I served a complete six-year term, the first person to do so in thirty years. With a powerful, imaginative, hardworking team, we accomplished a lot—and this deserves a separate book, which I shall write one day. We identified hard problems, many of them not scientific but affecting the progress of science. We addressed gaps in our agency's support of science, and we tried to beat them down, systematically, as smartly as we could, with persistence. We originated "Ten Big Ideas for Future Investment"; these influenced the direction of NSF, increased its support from Congress, and led to renewed emphasis on the U.S. role as a global leader in science and technology. NSF set a new standard for other federal science agencies in its approach to inclusion and to sexual harassment.

Remarkably for me as an astrophysicist, during my NSF tenure gravitational waves were discovered for the first time on Earth, and

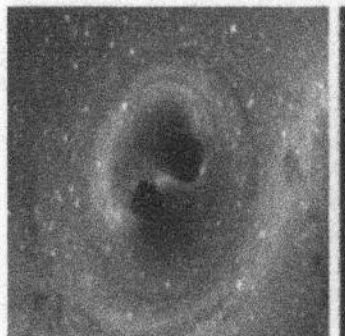
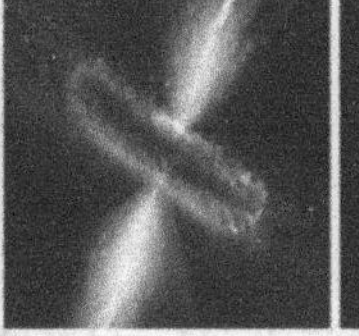
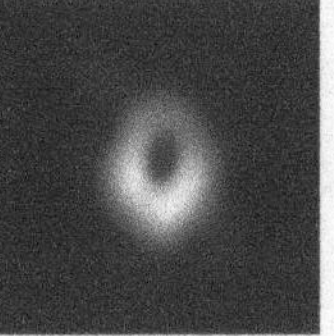

FIGURE 17.2. Breakthroughs enabled by NSF's long-term investments in basic research. Left: the detection of gravitational waves from merging collapsed stars. *Credit*: Simulating eXtreme Spacetimes (SXS) project, http://www.black-holes.org; Center: The identification of a blazar as the source of high-energy cosmic rays. *Credit*: ESA/NASA, the AVO project and Paolo Padovani; Right: Imaging a massive black hole at the center of a faraway galaxy. *Credit*: Event Horizon Telescope collaboration.

high-energy neutrinos were detected with strings of detectors deep in the ice at the South Pole. These events opened up new ways of observing the Universe. The era of multi-messenger astronomy had begun. Today we observe and decipher the cosmos not only via electromagnetic radiation but also via neutrinos and gravitational waves, very different "messengers." The world was in awe as the first image of a giant black hole at the center of a faraway galaxy was unveiled. To make all of these discoveries, international teams of astronomers had come together, using instruments distributed globally. The impact of these discoveries is profound: we now know much more about the origin of the heavy elements like gold and platinum, the origin of the highest-energy cosmic rays, and the evolution of black holes and binary stars. I was proud that NSF had supported these and many more groundbreaking discoveries and was thrilled to participate in these joyous announcements.

Why are there so many branches on my life's journey? I have long admired people who stick to one thing, a single pursuit with a great goal. Yet I have a restless spirit. I think of the poet Emily's Dickinson's words, "The Brain is wider than the sky . . ." We can imagine beyond what we see. There is always another "continent" to explore. And "I got to light out for the territory ahead of the rest" (Huck Finn).

Chapter 18

Dina Prialnik (PhD, 1980)

From Stars to Comets and Back

Dina Prialnik, the Jose Goldenberg Chair in Planetary Physics at Tel Aviv University (TAU), specializes in using numerical simulations to understand the internal structure and thermal evolution of cometary nuclei, icy satellites in the outer solar system, Mars, single stars, and cataclysmic variable stars. She has had a distinguished career as both a stellar astrophysicist—her textbook *An Introduction to the Theory of Stellar Structure and Evolution* is widely used—and a planetary scientist. She is associate editor of *Meteoritics and Planetary Sciences* and has served as Vice Rector at TAU and as Vice President of the IAU. Her asteroid is 8881 Prialnik.

I was born in Bucharest, Romania, in 1950. This was only two years after Israel's Declaration of Independence. My entire family applied for permission to emigrate to Israel, which was granted to all but my parents and me—I am an only child. Our permit arrived almost fourteen years later, so all this time we lived in uncertainty, on modest means, suffering from constant persecution. The feeling of not belonging that I experienced during that period will follow me through life in different contexts. In January 1964 we finally left Romania, arrived in Israel, and settled in Jerusalem. I didn't know a word of Hebrew at the time, but I decided to abandon Romanian in order to speed up the assimilation in my country. Thus, for a while, a couple of years or so, during which I learned Hebrew—mostly by myself, with grammar books and dictionaries—I spoke very little. Even when I had mastered Hebrew, speaking little, in any language, remained one of my characteristic traits, often to my detriment. Was I right at the age of fourteen to deliberately abandon my mother tongue? Probably.

The attraction to math and sciences began early in school; I remember being fascinated by batteries that were used in toys to light up small bulbs—electric toys were very primitive in those times and a pleasure to take apart. Although in the 1950–1960s, sciences were not very popular with girls, this interest continued and grew throughout my school years. It was my father who encouraged me to pursue this passion for math and physics, although he was a devoted humanist himself. And I should mention that his kindness, love, and support have always been a pillar I could lean on, even long after his untimely death, when he was sixty-three years old and I was twenty-seven. I kept my maiden name in academic matters in his memory.

Upon finishing high school I was looking forward to being drafted to army service; in Israel, this is compulsory, for men and women alike. This would have given me the opportunity to become completely assimilated and also the chance to decide on my university studies, whether math, or physics, or perhaps chemistry, or else architecture, to which I was also strongly attracted. But this was not to be. In late September 1968, at the recruiting army base near Jerusalem, where I arrived with all the necessary equipment and enthusiasm, I was informed that there were too many women in that year's pool and the decision was taken to discharge all single daughters. No protest on my part helped—I was naive enough to think it would. I could wait for another year and try my chance again, but instead, as though guided by an unseen hand, and without consulting anybody, I took the bus straight to the Hebrew University of Jerusalem (HUJI), right to the Department of Physics, and asked to be accepted to the mathematics-physics course of study. Registration had long ended by then and I was anxious, but they reassured me that I would be accepted—my matriculation grade in math was particularly high. Only then I announced the news to my parents, who, I must admit, were delighted to hear it.

To this day, I do not understand how I made my decision so quickly and so firmly—but I never regretted it. I should point out that 1968 was the year Princeton University reached the decision to accept women to undergraduate studies, which started a year later. There was no such discrimination at HUJI; nevertheless, the ratio of female to male students in physics was less than one in ten. The three years of undergraduate studies in math and physics were a constant fascination. I mostly enjoyed the math courses; they left a distinct imprint on my way of thinking. The supreme beauty of physics revealed itself more gradually.

Towards the end of my undergraduate studies, I had to decide on the subject to choose for the next step. The idea of pursuing an academic career had already begun taking root. The late 1960s were times of revolutionary astrophysical discoveries, such as the cosmic microwave background radiation and pulsars, so astrophysics was a natural choice. The next question was "where?" as I wished to explore new ground. The first institution I visited was the Technion—similar in scope to MIT or ETH—in the northern Israeli town of Haifa. I was granted the honor to be received by the dean of the Faculty of Physics, who was very kind and claimed he would be delighted to accept me, but for one concern. And so he said: "In our school, graduate students are asked to give tutorials in physics and I cannot imagine you standing in front of a class of forty men [there were barely any female students at the Technion] and teaching them physics." His words surprised rather than insulted me and strengthened my ambition rather than weakened it. But I left, disappointed, and turned to Tel Aviv University (TAU). Incidentally, a few years later I did stand in front of a class of forty men at the TAU Faculty of Engineering and did teach them physics—quite successfully, judging by their enthusiastic feedback.

TAU was very young at the time—one could sense the youth and energy in the general atmosphere, as well as the beginner's optimism and non-conformism. I decided on TAU while climbing the stairs of the Faculty of Exact Sciences, even before talking to any faculty member. In retrospect, it proved to be a good decision! TAU has been my academic home for half a century—and still is. My MSc thesis supervisor was Attay Kovetz, whom I first met as external examiner for the oral exam in the Stellar Structure course; he suggested two possible subjects for the thesis—stellar evolution or the equation of state for compact stars—and asked me to think about it and let him know. I immediately replied that I didn't have to think; I chose stellar evolution. This is how my decades-long research in stellar evolution started. A few years later, Attay and I got married. I should mention in passing that most female astronomers (astrophysicists) of my generation, in all parts of the world, are married to men of the same trade. We have a son, Ely, and when he was little and heard us talking about our work, he used to say: "One day I'll understand what you are talking about." Indeed, Ely is now an astrophysicist (cosmologist) himself at the Ben Gurion University in Be'er Sheva and it is my turn to try to understand what *he* is talking about.

I continued at TAU towards a PhD degree in astrophysics under the supervision of Giora Shaviv, one of the pioneers in stellar evolution simulation, with a wide range of interests in many branches of astrophysics and a great deal of contagious enthusiasm. I developed a nuclear reactions network for the stellar evolution code and applied the code to simulations of nova outbursts,[1] then a subject in its infancy, now well past maturity. I learned a lot and enjoyed every minute of it together with three other of Giora's PhD students; we worked on very different topics, but all based on heavy computing, carried out at the university's central (and only) computer, on shared time. Working late at night, when the computer was more accessible, was common; luckily, Tel Aviv was already in those times a very lively city, where one could get coffee around the clock. All four of us, Michael Shara, Oded Regev, Mario Livio, and I, pursued academic careers, stayed in touch, and collaborated in research, and when Giora turned sixty, we organized a conference to celebrate his birthday.

I was pregnant when I got my PhD diploma and shortly afterwards we travelled to California, where my son was born. I stayed home for a year and in the following one went to Stanford University for a one-year postdoctoral fellowship. I was in awe of the Stanford Physics Department's reputation and I shall never forget the colloquium talk I gave there, in front of a full auditorium, including no less than three Nobel Prize laureates—I have never felt so nervous in my entire life.

Upon our return to Israel from the United States, I began looking for a faculty position, not an easy task when decision makers were still gender-prejudiced. I started on a temporary position as instructor in the Physics Department at TAU and was very successful in teaching. Then, in an interview with the department head about future prospects, he condescendingly told me that teaching is less important than research in faculty recruiting considerations. Clearly, he was completely unaware of my research and didn't bother to check or ask about it. It was then that I had completed the study I have always been most proud of and submitted it to the *Astrophysical Journal*—single authored—where it was immediately published (it is cited to this day). It was a very ambitious endeavor to combine many different approaches to the study of nova evolution into one code and compute the evolution of a nova through a complete cycle in order to obtain a full and self-consistent picture of the phenomenon. Nobody else would dare to engage in such a task for years to come. The study took almost two years to complete—on the mainframe university computer (now ten minutes on my laptop!)—and a great deal of perseverance. I was disconcerted by that interview but did not lose hope. Instead, I accepted a temporary part-time position at HUJI that I held for a couple of years.

Then, an opportunity came up at TAU to be hired as research assistant in a project related to comets,[2] of which I knew nothing at the time. This is how I was introduced to and became acquainted with the field of planetary science, which I embraced with enthusiasm. My task was to develop a comet evolution code. This had been initiated some time before but got stuck on the simulation of phase transition fronts. Being familiar with stellar evolution codes and especially with the calculation of nuclear reaction fronts, I could implement the methods used there and overcome the obstacle. This was to be among the first full-scale comet evolution codes; the field flourished in the mid-1980s, as in 1986 comet 1P/Halley passed perihelion and was encountered by several spacecraft, which provided first-time close-up pictures and measurements. I enjoyed comet research tremendously, treading upon unknown ground; I published several papers, gave talks at many conferences, and thus established my place in the cometary research community and made many friends.

Finally, in the fall of 1989, nine years after completing my PhD, I obtained a tenure-track faculty position in Planetary Science at TAU. From then on, progress was much smoother: I became full professor of Planetary Physics in 1998, was awarded the chair in Planetary Physics in 2005, was elected Vice Rector of Tel Aviv University in 2010, and served as a vice president of the IAU between 2012 and 2018.

I did not abandon research in stellar evolution, though, but continued my two-track research during all this time. In both areas, I looked for the big picture—full-scale models, as complete as possible, and long-term evolution. Macro-structures have always fascinated me, and virtual reality created by numerical simulations incited my curiosity much more than basic physics. The old study of a full nova cycle was followed by multicycles—again a first-of-its-kind attempt—then extended parameter studies, and, more recently, combined simultaneous long-term evolution of nova binary systems, solving many puzzles on the way. In these projects my husband and I have collaborated, supervised students together, and travelled together to conferences. Attay is the best fundamental physicist I know—we make a good team.

Planetary research followed a similar trend: with students, I developed the comet evolution code to calculate long-term evolution, then a 3D-code for the entire comet nucleus—a first attempt—followed by an evolution code for minor icy planets including formation from an embryo. Compared to stellar structure, planetary research has always been livelier, involving more research students and young scientists and wider international collaborations; new discoveries have been more frequent, presenting new challenges that demanded not only improved computational tools but also input from other disciplines, such as geology or biology.

And so I have kept changing hats, alternating between the two fields, and securing my place in the two distinct scientific communities, and not too long ago I was invited to write two encyclopedia articles, one on novae and the other on comets. Glimpsed at from a distance, plots illustrating my work on comets and novae look very much the same (see

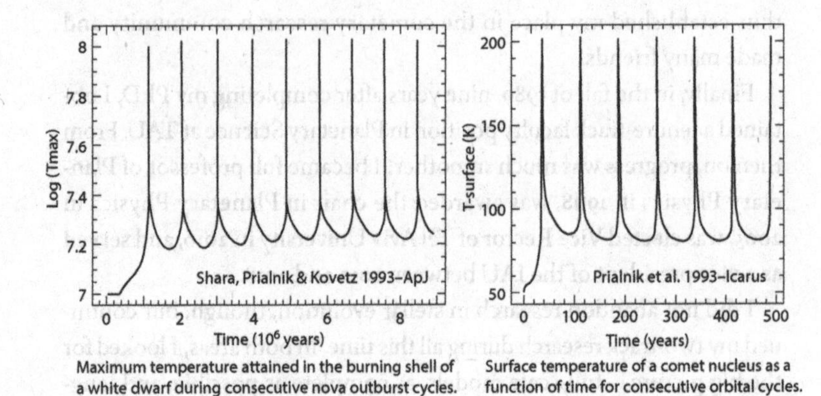

Maximum temperature attained in the burning shell of a white dwarf during consecutive nova outburst cycles.

Surface temperature of a comet nucleus as a function of time for consecutive orbital cycles.

FIGURE 18.1. These plots show an apparent similarity in the temperature cycles of novae and comets, despite being very different kinds of objects controlled by very different physical processes. Note the differences in time and temperature scales.

figure 18.1, showing periodic changes in temperature). Physically, they have nothing in common, and only by a close look that reveals the scales on the axes, one may grasp the fact that completely different worlds are involved. Nevertheless, years ago at a conference, I learned that grains observed in cometary tails bear similarity in structure and composition to those ejected in nova outbursts, and possibly originate there. This discovery made me (and probably nobody else) very happy!

In the beginning, I signed my papers "D. Prialnik"; social studies showed that papers authored by women tend to be downgraded in the reviewing process. But later at a conference, introducing myself as is often the case to somebody I knew only from reading his papers, the reaction was, "This is impossible! I don't believe it. My mind picture of you was a short fat Russian, with a beard and thick glasses!" From that day on, I have spelled my name in full.

I started teaching Stellar Structure at HUJI to undergraduate students of physics. It is a challenging course to teach due to the wealth of subjects involved. I developed the course as one builds a mathematical theory, starting from axiomatic assumptions and building up in a logical manner, and prepared extended lecture notes for the use of my students. It did not occur to me to turn the notes into a book until one day, quite out of the blue, a colleague told me, "You should write a book." Once I started, I could not stop writing, and I enjoyed it deeply, particularly digging through libraries and reading old papers. I wrote *An Introduction to the Theory of Stellar Structure and Evolution* with love and enthusiasm. Knowing very little about publishing books, doubting that the manuscript would be accepted for publication at all, and wishing to get over the expected disappointment as quickly as possible, I sent the manuscript at once to four publishing houses.

To my surprise, Oxford and Cambridge both offered to publish it (the other two expressed no interest in the subject)! Soon after it came out (by Cambridge University Press, which I finally chose), the book was put on display at the 2000 IAU General Assembly in Manchester, and I remember hiding behind a screen to watch it being leafed through and eventually bought. The sense of joy and satisfaction is hard to describe. Over the years, and with the publication of the second edition, I realized that it had become popular with students and teachers worldwide. At one time, a group of students from Spain wrote to me that they discovered my book while preparing for an exam and had a few questions. A few days later they wrote: "You may be happy to learn that with the help of your book we all got an A in the exam." I was happy indeed and moved by their letting me know!

Along with research, teaching, and mentoring, I enjoyed the academic administrative duties, and during my term as vice rector I especially enjoyed my interactions with people from other disciplines, in particular the humanities and the arts, which have always been close to my heart. Here again, as in research, I engaged in big, comprehensive projects such as setting the foundation for a multidisciplinary school of neuroscience and developing an all-embracing course of study for all undergraduate students, so that those who study humanities, arts, or social sciences get acquainted with hard-core science topics and science students get a glimpse of history, philosophy, literature, and arts.

When my term as vice rector was over, I decided after (uncharacteristic for me) long pondering, and with the support of my husband and

FIGURE 18.2. Dina Prialnik with her granddaughter Yarden, at age six in 2020: She will have an easier time if she will choose to be a scientist. We paved the way.

my son, to submit my candidacy for the position of rector. There was one other (male) candidate and a search committee had to make the choice. I failed to be selected, and I am mentioning it for two reasons: first, because the disappointment proved to be much harder to deal with than I had expected—especially the finality of it, due to my age; and secondly, because I learned a lesson that I would like to share with and pass on to younger women. My election campaign consisted of proving that I had the skills, the experience, the knowledge, and the drive to become a successful rector. This is (too) often the strategy adopted by women running for office, not only in science but in general, and it is the wrong approach. These attributes should be regarded as axiomatically true, while the right strategy—usually adopted by male candidates—is to focus on and promise to change the world, not necessarily showing how.

Although officially retired now, I keep working as usual and enjoying it as always—the privilege of artists and scientists. Looking back, I realize that the way research is carried out has changed enormously over the past half century: from individual work towards collaboration on many scales and sharing of expertise as well as data; from modest experiments to huge and costly enterprises; from seclusion to outreach. To a large extent, this was caused by the tremendous technological progress; to a lesser extent perhaps, by the fall of barriers—gender-related, geographic, and disciplinary. Keeping up with these changes has been a constant challenge and, no less, a source of excitement. Looking forward, I wish to live long enough to see my grandchildren grown-up, and also to see astronauts landing on Mars and the solution to the dark-matter puzzle.

Notes

1. Nova outbursts are recurrent eruptions that occur in close binary systems, due to explosive hydrogen burning in the surface layer of the white dwarf that has been accreted from a low-mass main-sequence companion star.

2. Comets are small, ice-rich bodies in the solar system, orbiting the Sun in highly eccentric orbits. When they approach the Sun, some ice evaporates, giving rise to the coma and tail that distinguish them among small bodies.

Chapter 19

Beatriz Barbuy (PhD, 1982)

From Stargazing the Southern Cross to Probing the Depths of the History of the Milky Way

Beatriz Barbuy, now a professor at the Instituto de Astronomia, Geofísica e Ciências Atmosféricas at the University of São Paulo, has conducted research on metal-poor stars that has deepened our knowledge about the oldest stars in the Milky Way and about the chemical changes that have occurred in stars over billions of years. She is a member of the Academia Brasileira de Ciências (Brazil), the Académie des Sciences (France), and the TWAS, has been honored as Commandeur de l'Ordre National du Mérite, and has received the Trieste Science Prize of the TWAS in 2008 and the L'Oréal-UNESCO prize for Women in Science in 2009. She has been a vice president of the IAU and President of IAU Commission 29 (Stellar Spectra) and IAU Division IV (Stars), and played a major role in establishing 2009 as the International Year of Astronomy.

I grew up in a favorable situation, as both my father and mother were university professors of philosophy, my father at Universidade de São Paulo and my mother at Pontificia Universidade Catolica de São Paulo. Their example, by spending time reading and discussing, naturally directed me toward an academic career. From when I was as young as fourteen I was reading books by philosophers (e.g., Max Scheler) in addition to romances (e.g., Victor Hugo), and doing so in different languages (mainly French, since I was studying in a French school).

Why in the end I chose a career in the hard sciences and not in the humanities was probably due to a combination of two reasons. The first was described superbly by the journalist Yudhijit Bhattacharjee, based on an interview during the 2009 General Assembly of the International Astronomical Union in Rio de Janeiro. With his British-like humor, he wrote: "She took up astronomy because she couldn't handle philosophy . . . when she tried to read Hegel as a teenager, the German thinker's ruminations on mind, spirit, and logic made Barbuy's head spin." The second reason was that I had an admiration for the sciences, including mathematics, physics, and biology. I very much wanted to engage more deeply in academic learning but struggled to choose a direction, and so I read many outreach books about different fields. When I was about sixteen years old, my brother won as a prize at his school the book *One, Two, Three . . . Infinity*, by George Gamow. After reading it my decision was firm, with no doubts: I would become an astronomer.

An earlier reason for my interest in astronomy might have been the fact that, at the age of around eight or ten, when arriving home after school at around 5:30 in the afternoon, I would climb to the highest branch of a loquat tree, which was "my branch" in the tree, whereas my two brothers had lower, stronger branches. Up there, I used to look at the sky until dinnertime. Another very important ingredient is that, although it was evident that my grandmother and her sisters stayed at home, I did not understand the prejudice against women working outside the home that existed at that time. On the contrary, my father told me that it was crucially important to have a profession and never depend on anybody but oneself.

During my undergraduate studies at the Institute of Physics (IF) of the University of São Paulo (USP), besides attending the lectures on courses in astrophysics and relativity, I attended weekly meetings under the supervision of the same professor. And before I had earned my diploma in physics from IF-USP, an astronomy group had formed at the Instituto Astronômico e Geofísico (IAG-USP), which met at a beautiful park on the other side of town. This group had only three astronomers; one of them, Sayd Jose Codina Landaberry, was assigned to be my supervisor for my master's degree work. During this period of time, I realized that I would need to go abroad to complete my education in order to start my career in astronomy.

The most important and decisive step in getting me started occurred when, in 1976, I obtained two complementary fellowships to pursue a PhD in France, one from the French Consulate and the other from the national Agency Conselho Nacional de Pesquisas (CNPq). Both were small, equivalent to about $400 each, and I was allowed to keep both. Previous to this, in the 1960s, Abrahão de Moraes, at the time the director of IAG-USP, together with Jean Delhaye, the director of the Observatory of Paris, and representative from CNRS to CNPq, had done important "political" work that was intended to boost astronomy in the country. First, as Brazilian representative in 1961 in Berkeley, Abrahão de Moraes had brought Brazil into the International Astronomical Union (IAU); second, in the 1960s they argued that training Brazilian astronomers was among the highest priorities, both at the French Consulate and at the CNPq, and succeeded in implementing this strategy. It is interesting to note that Roger Cayrel was a visitor in a mission to São Paulo in the late 1960s, and he later made important contributions to my knowledge in astronomy through collaborations we developed over the years.

It is clear that earning the PhD in France was crucial and that if I had stayed in Brazil I would have had no future in astronomy. At that time I had already expressed the intention to work on the chemical evolution of the Galaxy. The person who helped me get accepted into a degree program in France was Licio da Silva, an astronomer from Observatorio Nacional in Rio de Janeiro, to whom I am very grateful.

The group at the Observatory of Paris included several astronomers who were very important in nurturing my development, both professionally and as local family. Roger Cayrel, who initiated studies of atmospheres of low-mass old stars in France, was exceptional from two points of view. In addition to mentoring me in astronomy, he and Giusa Cayrel gave me strong support by inviting me to dinners throughout and beyond my time as a PhD student. My advisor Monique Spite immediately gave me data to work on and was very attentive to my development. François Spite guided me on articles to read.

All was not roses immediately for me, however; it took three years until I started to appreciate the city of Paris. Some high-level French astronomers treated me very kindly, but others did not. They would comment that Brazilians cared only for carnival, football, coffee, and generals (Brazil was under a dictatorship at the time).

I successfully defended my PhD in January 1982 and accepted a position at the University of São Paulo immediately afterward, starting in early February. Why did I return to Brazil? I considered staying in Europe, but I had several reasons that compelled me to return: I had signed a contract with CNPq that required that I return to my home country for twice the length of time that I had stayed abroad with the support of their fellowship; I had lost my father while earning my PhD and I wanted to be close to home to support my mother; I did not want to feel like a foreigner for the rest of my life and, as some insinuated, be accused of taking a job from a French person; and, of course, Brazil has wonderful weather.

Upon my return to Brazil, I encountered a serious problem: the absence of good computers. A law had been approved that prevented the purchase of foreign-built computers for ten years. The intent of the order was to push the manufacturing sector to create Brazilian-built computers, but this never happened. This rule delayed progress in the country in several industrial and academic aspects, and for this reason I continued to go to the Observatory of Paris for several months per year, with grants from the Observatory, in order to be able to use the computers in France. In 1983 and 1984 I also spent five months at Lick Observatory, in 1986 three months at the University of Cambridge, and in 1987 two months at European Southern Observatory (ESO) headquarters in Garching, Germany.

All these activities greatly pushed my scientific work forward. I was promoted to associate professor in 1987 and full professor in 1997; then, in 1997, I was elected to the São Paulo Academy of Sciences and in 2002 to the Brazilian Academy of Sciences.

FIGURE 19.1. Beatriz Barbuy at her induction ceremony at the Académie des Sciences in France on December 6, 2006.

My yearly visits to the Observatory of Paris went on regularly until the early 2000s. In 2006 I was elected to the Académie des Sciences in France. Considering that this Academy has only 150 foreign members, I consider this a most important honor! Also the fact of having been nominated by Roger Cayrel, who has always had my greatest admiration, made the honor even more important to me. The ceremony took place in the well-known room under the "Coupole," at the Institut de France, 23, quai de Conti, on December 6, 2006, which was, coincidentally, exactly thirty years after I had embarked for France in pursuit of my PhD.

Another important milestone for me was becoming President of the International Astronomical Union Commission 29 on Stellar Spectra (1997–2000). Subsequently, I was President of IAU Division IV on Stars (2000–2003) and then one of the six vice presidents of the IAU (2003–9). Here too, having David Lambert, an astronomer whom I greatly admire and who was, at the time, President of Division IV, put forward my name to be President of Commission 29 was very significant to me.

In 2008 I was elected to the World Academy of Sciences and in the same year was awarded the Trieste Science Prize by the TWAS, in a ceremony held at the International Center for Theoretical Physics in Trieste. In 2009 I received the L'Oréal-UNESCO prize for Women in Science. The ceremony for this award took place at UNESCO headquarters in Paris, on March 8, 2009.

These prizes have had a very important, positive impact on my career. Particularly after receiving the L'Oréal prize, I (and my work) began to receive more public recognition . I made many appearances on TV shows and was interviewed by several newspapers, among other things.

Since 2009–10, I have become even more active on committees than before, serving as a member of the Haut Comitê Scientifique de l'Observatoire de Paris (2010–15) and the Jury of the Initiatives d'Excellence within the Agence Nationale de la Recherche (ANR) (2011–24), as a member of the jury of the L'Oréal-UNESCO prize (international) (2011–18), a member of the Associated Universities for Research in Astronomy (AURA) Oversight Council for Gemini (2011–17), and a member of the Steering Committee of the Cherenkov Telescope Array (2013–20). Also, during this period I was associate editor of the *Publications of the Astronomical Society of Australia* (2013–19).

Beginning with my move to France in 1976, I have studied metal-poor stars, which was the main subject of interest of my advisor Monique Spite. I started by analyzing stellar spectra and learning about nucleosynthesis. France was very active in these fields, in particular at that time when seminars by Canadian astrophysicist Hubert Reeves used to fill the conference rooms. The scientific ambiance in Paris also was very important: seminars at the Observatoire de Paris-Meudon, the Institut d'Astrophysique, the Collège de France, and sometimes also at the CEA-Saclay included presentations by some of the most important astronomers from all over the world.

Upon my return to Brazil in 1982, I started writing proposals to obtain observing time with the instrument CES at the 1.4-meter CAT telescope at ESO in Chile. My observations led, in 1988, to my first major work, which dealt with oxygen abundances in 20 metal-poor stars.

Beginning with my PhD work and continuing in following years, one of my main activities has been working to include molecular lines in the code of synthetic spectra. This work expanded into different wavelength regions, starting with my work in the optical, then, in the PhD theses of my students Bruno Castilho, Ricardo Schiavon, and Jorge Meléndez, into the near-ultraviolet, the red, and the near-infrared, respectively; finally, in the thesis by Paula Coelho all of this was gathered into a unified line list. This work was entirely rewritten in Fortran 2003, thanks to software expert Julio Trevisan.

My work on metal-poor halo stars continued through the years, and the data for a particularly important project entitled "First Stars," headed by Roger Cayrel and Monique Spite, were obtained at ESO, with thirty-seven nights on the Very Large Telescope equipped with the UVES spectrograph. In these data, we detected a uranium line in a metal-poor star and derived the age of the star from the uranium-thorium abundance ratio. This discovery allowed us to obtain data from 60 orbits on the Hubble Space Telescope (HST) for observations of the metal-poor uranium-rich star CS 31082–001 and to produce papers XV and XVI in the First Stars paper series.

Another extremely important scientific activity, which began in 1990, is my collaboration with Sergio Ortolani (University of Padova), and Eduardo Bica (Federal University of Rio Grande do Sul) on colour-magnitude diagrams of globular clusters. Using data from ESO and from HST, we achieved a breakthrough when we found evidence of the presence of a population of old stars in the Galactic bulge, since at the time there were papers indicating that the bulge was too young to have such stars. In more recent years we have been particularly interested in the moderately metal-poor globular clusters in the bulge that constitute probably the oldest stellar population in the Galaxy. The image in figure 19.2 combines data from three near-infrared colours, J, H, and K bands, all obtained as part of the Vista Variables in the Via Lactea (VVV) survey done at ESO. Since then, a combination of photometric and spectroscopic work, using several telescopes including the New

FIGURE 19.2. The very old, metal-poor globular cluster HP1, as seen in the near-infrared colour bands J, H, and K (Barbuy et al., *A&A* 591, A53 [2016]).

Technology Telescope and the Very Large Telescope (VLT) at ESO, and HST resulted in about 150 papers from this collaboration. This work led to me receiving an invitation to be the lead author on "Chemodynamical History of the Galactic Bulge," published in the *Annual Review of Astronomy & Astrophysics.*

One of the aims of having a job as a university professor of astronomy is to educate students to obtain their PhDs in order for them to become professional astronomers. Doing this is also my most important immediate contribution to society. A good side of the Brazilian system of graduate studies is (so far) the availability of fellowships for the support of PhD students. I have been able to supervise excellent students, and most of them are now active in universities and observatories. Bruno Castilho has been the director at Laboratorio Nacional de Astrofisica, Minas Geraes, since 2013; Silvia Rossi, Jorge Meléndez, and Paula Coelho are now my colleagues at IAG-USP; Alan Alves-Brito and Marina Trevisan are professors at Universidade Federal do Rio Grande do Sul (UFRGS); Rodolfo Smiljanic has a permanent job at Nicolaus Copernicus at the Polish Academy of Sciences in Poland; Ricardo Schiavon is professor at Liverpool John Moores University; André Milone is a researcher at Instituto Nacional de Pesquisas Espaciais (INPE); and Bruno Dias is a postdoctoral fellow in Chile. Three other students directed their lives into teaching in colleges or to software development, and four of my female students abandoned astronomy. My present students, Heitor Ernandes, Stefano Souza, Raphael Oliveira, and Roberta Razera, are also very promising.

Instrumentation is another area to which I tried to contribute, given our need to develop observational astronomy in Brazil. As for international participation, I have been a member of the board of the Gemini Observatory telescopes consortium, been on the scientific advisory committee of the SOAR telescope, and attended the ESO Council meetings during the period of temporary membership of Brazil to ESO (2011–18). In the 2000s I was able to obtain the funds for construction of the SOAR Integral Field Unit spectrograph. This spectrograph, whose design and construction took more than ten years, is now in use at the SOAR telescope. It is important to acknowledge the constancy and seriousness of the Fundação de Amparo à Pesquisa do Estado de São Paulo (FAPESP), which over the years has provided funds for astronomical instrumentation. The same applies to computers—since the 1990s, when FAPESP first provided IAG with a VAX computer, we have never had a problem with getting the needed computers for our work. My present activities in instrumentation involve two instruments: (1) as I worked as PI of the CUBES spectrograph for the VLT, a group of Brazilians continues to be part of the project, which was recently relaunched and is now led by Italy; and (2) we are working to also be part of the MOSAIC spectrograph for the Extremely Large Telescope

(ELT), of which we have been a partner since 2011, and of which I am the local Brazilian PI. Work on the Brazilian side with optical fibers and robots for optical fiber positioning is being planned.

Among the activities I was involved with as a vice president of the IAU, one of the most important was to support the initiative of Franco Pacini, President of the IAU at the time, for declaring 2009 as the Year of Astronomy in celebration of the 400th anniversary of Galileo's first recorded telescopic observations. My work with the Brazilian Ministry of Foreign Affairs was a very rewarding activity, given the high level of the involved Ministry personnel and my work to gain their support. In fact, Brazilian support was important because countries like Australia, the United States, and other first world countries were against having "Year of Something" altogether. My visit to the United Nations in New York, when we essentially obtained the approval for the Year of Astronomy concept, was memorable, rewarding, and interesting. Also, as far back as 1994 we had proposed the idea of hosting the IAU General Assembly in Brazil, and this occurred in 2009 in Rio de Janeiro, in the Year of Astronomy.

I realized, over the years, that working with low-quality astronomical data is not worthwhile, which is why I have dedicated so much effort in support of efforts to fund, plan, and build high-performance telescopes and their associated instruments. In the years 2009–10, when Brazil appeared to be experiencing a period of very good economic growth and improving from all points of view, we began discussions about joining ESO as a member, which previously had seemed impossible. An agreement was signed in December 2010, making Brazil a temporary member from 2011 to early 2018. During this seven-year period, I and a few colleagues engaged in a series of activities to move this proposal to join ESO through different committees at the National Congress in Brasilia. It was extremely interesting to have discussed this proposal with dozens of congresspersons. The project was approved and published by the Congress on May 19, 2015. The subsequent economic crisis in Brazil, however, prevented this proposal from being implemented. Although Brazil did not, in the end, join ESO, our temporary membership for seven years was extremely important, and I hope my country will continue to be somewhat involved with this admirable and efficient organization. In the long term, the Brazilian astronomy community hopes that when our country is more stable we will be able to join the organization.

Through my nomination by the Académie des Sciences, my participation in the Jury of the Initiatives d'Excellence projects, coordinated by the Agence Nationale de la Recherche, extends from 2010 to 2024. This extremely important process aims at reorganizing French universities and higher-level teaching institutions.

In conclusion, I recommend that each of us make dedicated efforts to develop our interests and abilities. Firstly, constructive and progressive work, which is the job of a scientist, or any intellectual, artist, technical worker, or other highly trained professional, forms your identity. Such jobs are extremely enriching and important for personal development and satisfaction. Secondly, the life of a scientist, in particular being an astronomer and part of an international community with rather constant contacts throughout my life, has been a privilege. It is normal for an astronomer, for example, to sit on the panels of the HST or other international telescope panels, to observe in Chile and Hawai'i, and to take part in international high-level scientific and administrative committees composed of scientists from all over the world. Not only are these pleasant activities, but they raise both our scientific and personal profiles.

Chapter 20

Rosemary (Rosie) F. G. Wyse (PhD, 1983)

A Journey through Space and Time

Rosie Wyse, now the Alumni Centennial Professor of Physics and Astronomy at Johns Hopkins University, has used theory and observations to advance our understanding of the structure, dynamics, and formation history of the Milky Way and its satellites, including showing how the Milky Way's thick disk is a natural consequence of the dynamical evolution of disk galaxies. She won the Annie Jump Cannon Award from the AAUW in 1986 and the Dirk Brouwer Career Award from the Division on Dynamical Astronomy of the AAS in 2016. She is a Fellow of the American Association for the Advancement of Science, of the APS, and of the RAS, and a Legacy Fellow of the AAS.

I was born in Dundee, at that time the third-largest city in Scotland. I enjoyed a very comfortable childhood. My father was a medical doctor and my mother an art teacher, though she stayed at home while I and my sisters were young. I am the youngest of three daughters and I do sometimes wonder how my upbringing would have been different had I had a brother. In the end I was the child who took up golf, my father's passion—I was really a quite good player as a young girl but my more recent attempts are best forgotten.

I attended the local Catholic state-supported primary school, then secondary school. The Scottish education system is separate and distinct from the English system but shares the (unfortunate, in my view) characteristic of being organized by religion. Catholics were, and still are, very much in the minority in Scotland and, as was widely acknowledged, were discriminated against by the main employers in Dundee—naming your school would immediately label you. I mention this as being brought up Catholic and educated in Catholic schools did instill in me a certain expectation of having to push against society and stand up for myself. My schools were for both boys and girls, and most classes were mixed. Two teachers were particularly memorable: Mr. Hanlon from my primary school, who set me special problems to solve (the ones I remember would have been much easier with calculus!), and Sister Mary Bernadette from secondary school, who exhorted us, in her strong Irish accent, "Do not be *jellyfish*, individuals without backbone!"

I was top, or close-to-top, in all my subjects and by age sixteen was qualified to enter Scottish universities for a degree program in the discipline of my choice. At that time it was possible to stay on at school for an extra (sixth) year, during which one could specialize, taking fewer subjects at a more advanced level (equivalent to the first year of a typical Scottish university degree) and study with a more self-motivated approach. This had the added advantage for me that the sixth-year courses provided better preparation for the English university system. My interest in leaving Scotland had been stimulated by an ill-fated (for me) initiative by a Cambridge College to matriculate more Scottish students. Several Scottish schools were invited to put forward their top pupil, who, provided their paper qualifications were satisfactory, would be invited for interview and then, assuming all went well, offered a position. I was selected to interview and arrangements for my trip were proceeding, until a halt was called—the College had somehow only just realized that I was a girl, and it was an all-male College (the norm at that time). Needless to say, the top few pupils at my school were all girls, and so none of us were deemed acceptable.

I had always been interested in physics, and also in science fiction. I am lucky to have experienced, as a child, the first series of each of *Doctor Who* and *Star Trek*. Each made a lasting impression. I decided to investigate the possibility of studying physics and astrophysics. I discussed

FIGURE 20.1. Rosemary Wyse in 2016. *Credit*: Will Kirk, JHU.

this possibility with the careers master at my school, and he came up with Queen Mary College, University of London (QMC). I applied and was called for interview. I did get the impression that my interviewers did not understand my Scottish qualifications. I had obtained the highest grades in the school-leaving exams in six subjects (I added a seventh, biology, in my last year) and was taking the advanced-level physics, mathematics, and chemistry. I was bemused to receive a conditional acceptance; this I interpreted as meaning they were extremely selective, which increased my enthusiasm. I gained the required grades in the final exams and set off to London.

I moved to London in June 1974, along with my best friend from (high) school, to spend the summer working before we embarked on our undergraduate studies, myself in London and her at St Andrews. We stayed in a hostel in Bayswater, and I worked at the Biba store in Kensington High Street. This "Big Biba" store opened in 1973 and is recognized as an icon of the age—it was in a beautiful Art Deco building and quickly became established as *the* place in which to see and be seen. I took over the running of the shop in the roof garden. Biba shone brightly and, like a luminous star, was short-lived, closing in 1976. Clothes from the collection are now held in the Victoria and Albert Museum, the London museum of design. I do have a Biba black evening jacket with beautiful bronze sequined trim.

My summer of fun over, I moved out to the far eastern suburbs of Greater London to live in the QMC student housing complex. QMC is located in the east end of London, and the local area was far from gentrified in the 1970s. I soon discovered that the Scottish school curriculum had not prepared me for an English university as thoroughly as I had expected, and I took a year or so to master the material. I also quickly discovered that very many of the (majority) English students had the bare minimum of qualifications. Happily, there was a nugget of engaged physics students, and the coursework was stimulating. The majority of the astrophysics courses were taught in the (Applied) Mathematics Department; I remember in particular Ian Roxburgh's Stellar Structure & Evolution course and Michael Rowan-Robinson's lectures on cosmology. During the winter break of my first year I returned to Scotland and went to visit my friend in St Andrews. While I was there, she introduced me to a final-year physics student who is now my husband.

I did much better in my coursework in my second year, and by the end of third year I was the top physics student. Time had come to decide on where to go for my PhD. I had been very excited by the high-energy physics that we had studied but astrophysics was what really fascinated me, especially theoretical aspects. I decided against applying to any U.S. PhD program, once I discovered the process involved taking (more) exams. The Institute of Astronomy (IoA), Cambridge, was my top choice, but I was told that since I wanted to undertake theoretical work, and had not studied for my BSc at Cambridge, I had to take Part III of the Applied Mathematics Tripos first and then reapply for a PhD studentship at IoA. I received excellent advice from both Ian Roxburgh and Mike Rowan-Robinson, which was "Go to Cambridge." And I did.

Part III is a notoriously difficult year of study. To quote Wikipedia, "It is regarded as one of the hardest and most intensive mathematics courses in the world," to which I can only add that it must have gotten easier over the years, as it was said to be *the* hardest when I matriculated in 1977. I enrolled in courses given by the Department of Applied Mathematics and Theoretical Physics (DAMTP), mostly in theoretical astrophysics. The material covered was compelling, but the attitudes of many fellow students and, sadly, some lecturers were very disheartening. The "non-Cambridge graduates" were expected to fail and were told so. Largely in defensive mode, we sat together at the back of the room.

My saviors were my family, my partner, and the (lifelong) friends I made (Jean, John, and Ellen in particular). Students at the University of Cambridge have also to be admitted by a College. I had, of course, earlier discovered that I needed to identify those that would accept women, but I had little other knowledge to guide my choice. I selected Emmanuel College (swimming pool and squash courts in the grounds, housing supplied!), not realizing the full implications of the fact they had very recently voted to admit women, and only as graduate students. A mere handful of female students were admitted, and three of us were each allocated a room in the same College-owned house just off College grounds (no women were given rooms within the College gates). We soon shared the horrified response of the College cleaner on her discovery that women were in the house—after telling us we shouldn't be there, she marched over to College to announce her refusal to have anything to do with us. Her furious return soon after, with orders to carry out her duties, set the tone of our (often fractious) relationship. The uneasy acceptance of women was also manifest in the repeated challenges by the College porters to our entering College grounds; there were so few of us (my memory is just four) that it should have been possible to recognize us after a few weeks. It was again made very clear that I was an outsider.

Some of the Part III exams at the end of the academic year did come as a shock. Indeed, I felt that I had done extremely badly in one and was sufficiently stressed and upset that I decided to withdraw and not take the remaining exams. I was in the act of telephoning the DAMTP office, using the communal phone in our house, when I was interrupted by my friends. John, in particular, had experience of the exams and assured me that I had probably done better than most. I was persuaded to persevere. This was the best advice I could have received, and I have had occasion to remember his words many times over the years.

In those days, the results of Part III were posted on the notice board at Senate House—if your name appeared, you had passed, and if your name did not appear, you had failed. I steeled myself to go and get the news and was astonished to not only find my name but see it had an asterisk beside it. I had attained the equivalent of a first-class degree. This meant I would be accepted to undertake my PhD at IoA. It also meant that I became the first female Bachelor Scholar, and thus a member of the Foundation, of Emmanuel College. The High Table dinners were memorable, although "Lady and Gentlemen" rapidly lost its charm.

I started my PhD officially in October 1978, but since I was already in Cambridge Bernard Jones, who became my thesis advisor, arranged funding for me over the summer. The approved title of my thesis was "The Fate of Gas in Mergers," but in the end it was "The Formation of Disc Galaxies," rather more all-encompassing than merited by the contents! IoA at that time was small, with fewer than ten students a year and only two professors (Martin Rees and Donald Lynden-Bell), and many of the staff, such as my advisor, were on fixed-term postdoctoral appointments.

The annual summer conferences were a real pleasure and offered a wonderful opportunity to meet more senior researchers. There was always a conference photograph, and I regret deciding that I couldn't afford to purchase every one of them. The one from 1978 ("Globular Clusters") that I have displayed outside my office certainly starts conversations! Other highlights included summer schools at Les Houches (1979) and Erice (1981), where I met many of my peers and some future colleagues. I really value these experiences and prioritize sending my graduate students to summer schools, which are still most often held in Europe.

My partner moved to Cambridge and found employment in a nearby town. I published my first paper in 1981, and the last paper from my thesis research came out in 1985. These papers cover a range of topics in the general field of galaxy formation, from the evolution of density fluctuations in the early Universe to the predicted structure of the Milky Way Galaxy, a typical disc galaxy. This last topic has remained a passion of mine.

In early 1982 I had shown that the first luminous stellar component to result from the collapse of gas within a non-spherical dark halo was a thick disc, with the thin disc established later, after dynamical relaxation of the system. One day I was sitting in the canteen of the Cavendish Laboratory (where we would go for lunch) when Gerry Gilmore appeared. I had met him earlier when we were both student-participants at the summer school in Les Houches. Gerry was then visiting IoA to use the plate-measuring machine and had heard of my results. He explained that he and Neill Reid had just discovered such a component in star counts towards the South Galactic Pole. We had a lot to talk about. The relevant two papers, his and mine, were both published in the spring of 1983.

The expectation (reflected in the government funding) was that PhD research would be completed in three years. Luckily, I was awarded an Amelia Earhart Fellowship from Zonta International and could take a further year—my advisor had in the meantime moved to Paris Observatory, Meudon, which necessitated my traveling to Paris to talk to him. What a hardship! I was ready to apply for postdoctoral positions for the 1982/83 academic year and was awarded a Lindemann Fellowship from the English-Speaking Union of the Commonwealth and also the Parisot Fellowship at UC Berkeley. I arranged to take the Lindemann, which was for a year only, to Princeton University and then move to Berkeley. I was very excited, but first I had my thesis defense. The least said about that the better. Joe Silk was my external examiner, and happily for me he became a mentor and collaborator.

I flew to New York in August 1982; this was my first time in the United States. I was welcomed into the Department of Astrophysical Sciences in Princeton and also into the Gravity Group in the Physics Department. I was sitting in my office in Peyton Hall one day when, to my surprise, Gerry Gilmore walked in. He had been invited to Princeton (to the Institute for Advanced Study) by John Bahcall; the Galactic thick disc was missing from the star-count models of Bahcall & Soneira and its importance became a source of vigorous, one could say full and frank, debate. Gerry and I decided that we should work together on establishing the nature of this component and on testing whether or not it did indeed probe the early stages of galaxy evolution. Thus started a very fruitful collaboration.

I moved to Berkeley in spring 1983. I already knew several postdocs and a few faculty members there and settled in quickly. There was a strong Cambridge connection within the galaxy formation/cosmology group at Berkeley, augmented by the steady flow of French and Italian postdocs who came to work with Joe Silk. It was an incredibly stimulating environment. Adding to my happiness, my partner managed to obtain a student visa, allowing him to embark on his master's degree in electrical engineering at San Jose State, just in time for the start of the transformation of orchards into Silicon Valley.

During my time in Berkeley I held three different postdoctoral fellowships: the Lindemann, the Parisot, and lastly a UC President's fellowship. I was also a postdoctoral fellow for six months in the Academic Affairs Division of Space Telescope Science Institute, between terms as a UC President's Fellow. These were all independent fellowships, and I relished the freedom to choose my own directions of research. Indeed, UC Berkeley awarded me Principal Investigator status, which enabled me to apply for research grants in my own name, and I succeeded in being awarded a grant from the National Science Foundation to support my research about the Milky Way. Gerry Gilmore and I also were awarded a Collaborative Research grant from the Scientific Affairs Division of NATO (sadly no longer active). This funding was critical in the early years of our collaboration, supporting visits to each other's institutions; it is a real loss to the scientific community that NATO no longer plays this role—they also used to fund scientific conferences.

The time came for me to transition to a more senior position. My main choice was between returning to Cambridge (IoA) with an Advanced Fellowship or moving (back) to Baltimore to join the faculty of the Department of Physics at Johns Hopkins University (JHU). I chose JHU.

I arrived in Baltimore in January 1988. The Physics Department (now, Physics & Astronomy) had committed to an expansion across the subfields of physics, especially in astrophysics, after JHU won the competition to host the Space Telescope Science Institute. The students are outstanding, and the scientific environment is stimulating. I enjoy living in the city of Baltimore. The summer weather, however, is a challenge and the summer of 2020, with the Covid-19 lockdowns, was the first summer that I spent, in its entirety, in Baltimore.

My favorite summer retreat is the Aspen Center for Physics (ACP). I was introduced to the ACP by Joe Silk, who invited me to speak at a winter conference. I soon discovered the unparalleled environment for both quiet contemplation and robust discussions that is created during the summer workshops. I still much enjoy the fast-paced winter conferences. I organized a particularly memorable one in 1999 on the topic at the heart of my research—understanding the Milky Way Galaxy and its companions in the Local Group of galaxies within the larger picture of galaxy formation and evolution—that emphasized the power of using old stars in nearby galaxies to test predictions from cosmological simulations. This is now a vibrant area of research. And I was honored to be the first female President of the ACP from 2010 to 2013.

I have had the pleasure of advising seven PhD students at JHU, five of whom were women. Jay Gallagher and I developed a close collaboration over the decades, based on our mutual interest in the physics of star formation, and we unofficially co-advised two thesis students, one at JHU and one at the University of Wisconsin-Madison. My turn towards observational approaches, initiated by my desire to understand the Galactic thick disc in the context of my models, with Gerry Gilmore, was reinforced by my role as a founding member of the RAdial Velocity Experiment (RAVE) collaboration. This medium-sized international collaboration, using the 1.2-meter UK Schmidt telescope at the Siding

announces the

1999

ASPEN WINTER CONFERENCE

ON

ASTROPHYSICS

JANUARY 24 – 30, 1999

COSMOLOGICAL IMPLICATIONS OF THE LOCAL GROUP

The aim is to bring together researchers in the area of stellar populations with cosmologists, in order to maximize the return from advances in each of these areas, and foster interactions between the stellar population community and the cosmology community. Motivation for the meeting comes from the large observational datasets becoming available for local galaxies, that constrain star formation histories, chemical evolution etc., complementing new estimates of "cosmic" star formation. Further, N-body simulations of the formation of structure in the Universe are now achieving the resolution necessary to study the formation of "Local Groups", allowing direct comparison with the real world. There is the opportunity for synthesizing the new data and models pertaining to galaxies in the Local Group, and focussing on the cosmological implications, including such topics as the presence of, nature of and shapes of dark haloes; star formation histories; merging histories; intra-group medium; overall metal content; Local Group dynamics.

The conference is open, however due to space and format, attendance is limited. Participants are selected from applications submitted to the conference. Some support for younger participants will be available.

The Aspen Center for Physics is committed to a significant participation of women and under-represented groups in all the Center's programs.

DEADLINE FOR RECEIPT OF APPLICATIONS IS SEPTEMBER 15, 1998.

Applications may be obtained from:
http://andy.bu.edu/aspen/
or
Aspen Center for Physics
700 W. Gillespie Street
Aspen, CO 81611
(970) 925-2585 • Fax: (970) 920-1167
jane@aspenphys.org

Please send responses to:
Rosemary Wyse
The Johns Hopkins University
Bloomberg Center
Homewood Campus
Baltimore, MD
Fax: (410) 516-5096
wyse@tardis.pha.jhu.edu

SCIENTIFIC ORGANIZING COMMITTEE:

Rosemary Wyse ......Chairman, Johns Hopkins
Leo Blitz ....UC Berkeley
Russell Cannon....Anglo-Australian Observatory
Ray Carlberg ....U of Toronto
Gerry Gilmore ....U of Cambridge
Garth Illingworth ....UC Santa Cruz
Mario Mateo ....U of Michigan
David Spergel ....Princeton University

FIGURE 20.2. Poster advertising the 1999 Aspen Center for Physics conference. *Credit*: Aspen Center for Physics.

Spring Observatory in Australia, was led by Matthias Steinmetz, himself primarily a theorist, and was the first wide-area, large spectroscopic survey entirely dedicated to Galactic stars.

Forty years after I finished my undergraduate studies at QMC, I returned for the first time on the occasion of the annual meeting of the Division of Dynamical Astronomy of the American Astronomical Society (occasionally held outside the United States). I had been selected for the 2016 Dirk Brouwer Career Award and was giving my acceptance lecture at the meeting. I returned again the following year, as I had been nominated for an honorary doctorate from QMC (by then officially Queen Mary University, London), and it was conferred on me during the summer 2018 graduation ceremony for the physics undergraduates. This was the first graduation ceremony in which I actually participated, and I enjoyed it greatly. In my brief words to the new graduates, I attempted to summarize what I had learned over the years: follow your dreams and do not let anyone else define you or limit your aspirations.

Chapter 21

Bożena Czerny (PhD, 1984)

A Fortunate Sequence of Events

Bożena Czerny, now a professor at the Center for Theoretical Physics of the Polish Academy of Sciences in Warsaw, is a leading expert on the theory of AGN, in particular accretion disk theory for black holes. She has spent most of her career at the Nicolaus Copernicus Astronomical Center in Warsaw. She is a science editor for the AAS Journals Editorial Board, served as President of the Polish Astronomical Society, and was the recipient of the Ernst Mach Honorary Medal for Merit in the Physical Sciences from the Czech Academy of Sciences in 2019.

I am slowly approaching seventy, perhaps nearing the end of my long career, so this is a good moment to look back at my scientific life. This is something I almost never do. Instead, I always look ahead, toward the future. Even now. I have begun work on a forward-looking European Research Council Synergy Grant. My plans for retirement clearly have again been postponed.

Some of my colleagues planned to be scientists from their very early years. I did not. First, for a long time I did not even know that such a profession existed. Second, when I was a small girl, only a few years old, I loved fire and I loved dancing. My mother made me special long skirts that reached the ground, so when I was turning the skirt whirled in a fantastic way. And I did not like to play with other children that much, so I aspired to be either a dancer or a witch who lived with a cat deep in a forest.

I never attended a school for dance, though, so I did not follow that career path. Instead, I attended standard, very boring schools. My parents by profession were a nurse (my mother) and an administrator (my father). My mother was very devoted to her career; she loved working in hospitals, but I never learned to share her interest in interacting with sick people, and hospitals seemed to me utterly depressing. My father's education ended at the middle school stage, and he never had another chance to return to school. Immediately upon leaving school in 1920, he volunteered to participate in the war between newly created Poland and the Soviet Union. That ended his formal education. Later he worked in a factory, then had a shop, then another war came. During the 1920 war he and a friend walked two thousand kilometers from Moscow to Poland, in severe winter; thereafter, he felt that he'd done enough personal traveling, though he always indulged his broad interest in the world by reading travel books.

At school I had problems with history and geography, since I was unable to memorize all those battles and rivers joining the Vistula River from either the east or the west. My imperfect skills in these subjects annoyed me, since I was rather ambitious and always wanted to have the best grades. However, mathematics and physics were always very easy for me, and I frequently helped others do their homework. However, middle school physics did not inspire me, because it looked like a closed system of untouchable laws. The first moment when I thought that physics could be fascinating was when I read George Gamow's *Mister Tompkins in Wonderland.* This book fortunately had been translated into Polish and was broadly available. *Mister Tompkins* opened my mind to such basic questions as the real nature of time and space. At that moment I decided to study physics, since the book had not quite answered all my questions.

At Warsaw University, where I started as a university student in 1969, I was mostly disappointed with the fact that my classes still provided no explanations of the true nature of space and time; instead, we just learned numerous laws. However, I was attracted to general relativity

(GR), since GR came closest to answering one of the most basic questions physicists should ask. After attending an excellent course by Andrzej Trautman, I appreciated the theory, although I never had an impression that I fully understood it. And anyway, it seemed like a complete theory, so it was not clear to me what I could do with it.

The key inspiration that determined my future path in life came from a series of lectures given at Warsaw University in 1973 by Bohdan Paczyński, the leading Polish astrophysicist for the last part of the twentieth century. His lectures showed astrophysics to be a dynamically developing new branch of physics, with plenty of things for scientists to study and learn. Astrophysics, to me, was something completely new, like opening a window and letting in fresh air. He asked the students who were potentially interested in pursuing this area of science to contact him. I did. I told him that I liked GR. He then advised me to work with Marek Abramowicz, and indeed I then prepared my master's thesis about the theory of space-time around black holes under Marek's guidance.

However, this period still did not mark the beginning of my scientific career. In my master's thesis, I was either simply following Marek's instructions as to what to calculate or merely checking his computations, and I did not feel passionate about this equation-heavy topic. After earning my degree in 1975, I was afraid that I would have to take a job as a middle school teacher, and I did not like this idea. I like teaching people if they want to listen, but I am helpless if they do not care. Fortunately, I found a job with a publishing house in Warsaw as the editor of popular books, which seemed a perfect fit for me.

I would be doing that still if not for a fortunate sequence of events. First, I was assigned to edit a popular book on astronomy, which contained the statement that "stars evolve along the main sequence" and several other statements that were equally strange to me. I went to the Copernicus Astronomical Center (CAMK) in Warsaw, where I consulted with astronomer Joe Smak. I learned that, indeed, the idea that stars evolve "along the main sequence" from faint, cool stars to hot, bright stars was popular in the 1940s, but by the late 1970s this concept had been dismissed as incorrect. The whole book depressed me, and

I did not know what to do with it as I was not supposed to completely rewrite it. Then Paczyński presented another series of lectures at CAMK, this time on accretion disks, which I attended and on which I took detailed notes. Paczyński saw that I had done this and proposed that we write an article on his lectures, in Polish, for *Postepy Astronomii.* I was more than happy with his proposal and did the writing in my free time, but I did not think that his arrangement of the material was optimized for readers trying to understand his ideas about disks, so I rearranged the material considerably. Consequently, he withdrew as a co-author but assured me that the article was publishable. Indeed, it was published in 1978. And as a result of our discussions on this topic, he offered me a job as his secretary/assistant at CAMK. I started to work at CAMK in the autumn of 1978 and worked there for the next forty years.

At the beginning I was mostly helping Paczyński prepare his own publications, checking equations (I found an important mistake in his paper titled "A Model of Selfgravitating Disk with a Hot Corona") and helping prepare figures. But I also participated in his discussions with other co-workers, and later he suggested that I try to solve the problem of the inner boundary of the disk in the case of a geometrically thin disk around a black hole, known as a Shakura-Sunyaev disk. The problem was difficult conceptually, and the numerical code was also difficult to write. Our results were eventually published in *Acta Astronomica* in 1982. The key effect Paczyński had on me was through his style of work, and I still remember his practical advice for solving problems, which was widely different from the style of Abramowicz. For Abramowicz, all started with equations that had to be solved (at least I saw it this way), while for Paczyński, all started with a simple sketch on a piece of paper, an image. How could this work? He would then ask, what other elements do we need to solve the problem? Then he found approximate solutions, using order-of-magnitude estimates, and then, at the end, he derived and solved some equations, either analytically or numerically. Only then did I understand how new things get invented and how to do science, and I was ready to do it. I still work in this style—image, estimates, equations, solution.

A few years later, the political situation in Poland changed. Martial law was imposed in 1981, and Paczyński, who at that time was in the United States, chose to stay there. As a result, we were unable to continue to collaborate. Afterwards, I met him but once in the United States, for only two or three hours. But by this time I was ready to start my own life as a scientist, and since then I have published many papers. The most-cited one, written with Abramowicz in 1988, introduced the idea of "slim disks" into the family of accretion disks.

The key next step in my development as a scientist was my collaboration for several months with astrophysicist Martin Elvis, in the mid-1980s. When I visited the Harvard-Smithsonian Center for Astrophysics (I was accompanying my former husband on his visit), Elvis proposed that I collaborate with him, saying that he had data and I had models, so why not try to match the two. In those days, well before the development of the software package XSPEC, which is now widely used to analyze X-ray spectral data, I had to write all the code for analyzing X-ray data by myself. The project was an excellent lesson about what surprises can be hidden in the data, including anomalous numbers that indicate the existence of problems with the responses of the detector system. Since that time, I have developed various models of astrophysical phenomena, but, being fully aware of their limitations, I continuously test my models by comparing them with data.

More recently, my attention has been focused mostly on understanding the outer part of the accretion disks in active galaxies. As in all of my life, this work resulted from my meandering among people and scientific topics. Over ten years ago, my previous PhD student, Marek Nikołajuk from Bialystok University, asked me to suggest an observationally oriented research topic for his master's degree student, Krzysztof Hryniewicz. I suggested he look at the bluest and brightest quasars that had been observed as part of the Sloan Digital Sky Survey, having in mind that they should provide excellent data for comparison with accretion disk models. Krzysztof selected some quasars to study, and while one of the selected quasars at first glance looked normal, it turned out to be almost devoid of the strong spectral emission lines common to most quasars. We realized that he had identified a weak-lined quasar, these being a newly identified class of object that was quite fashionable for study in 2010. We published a paper in which we suggested an explanation based on the evolution of quasars rather than a spectral-shape-based one. That called my attention to the issue of the formation of the Broad Line Region (BLR) of quasars, which is a disk smaller than a few light years across that feeds the black hole at the center of an active galaxy. Why do the properties of the BLR correlate with the monochromatic quasar brightness at a wavelength of 5100 angstroms (green light) instead of with the total quasar luminosity or the ionizing (ultraviolet) flux? We talked with Krzysztof about it a couple of times (in the meantime, Krzysztof started work on his PhD thesis in Warsaw, since Bialystok did not offer the PhD option at that time) and then, looking at the equations (yes, occasionally equations know more than we do!), I realized that the standard size of the BLR fixed by the monochromatic flux corresponds to a fixed universal temperature for all active galactic nuclei (AGN), independent of the mass of the BLR or the accretion rate onto the central black hole. That seemed puzzling. There was only one temperature that could be considered—the temperature at which dust grains condense out of a gas (the dust sublimation temperature). Any other temperature would not be unique. Krzysztof recalculated this temperature from the sample of AGN for which the BLR sizes had been measured, and he came up with the value of about 1000 K! We were really very, very happy, feeling that we finally had an insight into the formation mechanism of broad emission lines from quasars.

In this way we introduced the Failed Radiatively Accelerated Dusty Outflow (FRADO) model. Our first paper on FRADO was published in 2011, and we have slowly progressed towards building a better model. I even became an observer (I never was before!) and joined a team to collect data from observations of three quasars with the 11-meter Southern African Large Telescope (SALT).

Our new BLR model is not yet very popular—others working in this field keep saying, "It cannot be that simple." Perhaps, but it is still a very good starting point for building a complete theory. I am continuing to make a large number of improvements to FRADO, and I have a group of fantastic young people who are helping me now.

So am I happy with what I have achieved? Is it consistent with my plans? I have a house, though it is not in the woods. It is at the outskirts of Warsaw, with a large garden, fresh air, sunshine, green grass, trees and flowers all around. I also have a fireplace and a cat, almost like a witch, I suppose. I am busy, and I am happy to have young collaborators around me. I also do some editorial work—since 2011 I have been one of the American Astronomical Society Journals editors. In this case, I do not have to accept papers that are certainly wrong. Thus my answer is: yes, definitely.

A few times, rather recently, I was asked whether it was much more difficult for a woman to be a scientist. I never thought much about this in the past. I was just trying to do what was possible. I cannot answer this question easily, since I have no exact comparison—I was never on the other side. But there are secondary symptoms that point toward some systematic differences. I recently attended a few very formal award ceremonies; the award recipients were always men, and they always started their presentation with a long list of people who helped them get the prize. This list usually included their family, who allowed them long working hours and no home duties, and supervisors—almost always men—who gave them opportunities or helped or guided them.

If I think about a list of people I should thank for help, my list would be very, very short. It would include Bohdan Paczyński and, recently, Lech Mankiewicz. Then I should add a very few other men of my age or older who clearly showed a positive attitude (most older men did not; in contrast, virtually all younger men and collaborators have been very supportive).

I received my first-ever award in 2019, the Ernst Mach Honorary Medal for Merit in the Physical Sciences from the Czech Academy of Sciences. Unfortunately, there was no ceremony because of the coronavirus. Before that, I had never received any prize, not even the smallest one. Sometimes I had the impression that men expected me to be stupid and that they were deeply disappointed when it appeared I was not. But maybe I was wrong. Anyway, it was not like that with Paczyński . He once said to me, at the very beginning, just do not be nervous and you will do fine in science. And I didn't even realize, then, how nervous I was.

FIGURE 21.1. (clockwise): Bożena Czerny, Agnieszka Janiuk, Monika Mościbrodzka, Joanna Kuraszkiewicz, Aneta Siemiginowska, and Agata Różańska.

Young women entering the field frequently pick me as their potential supervisor. Apparently, it is easier for a woman to work with a woman. I had no such opportunity when I was young, as there were no senior women for me to work with. But now I am very happy with my past female PhD students: Aneta Siemiginowska, Agata Różańska, Joanna Kuraszkiewicz, Agnieszka Janiuk, and Monika Mościbrodzka. They are doing well in science, pursuing their careers in Poland or abroad, and reaching the professor stage now. It is not that I have only had female PhD students; right now I have two men, Swayamtrupta Panda from India and Mohammad Naddaf from Iran. But girls always will have a special place in my heart.

Chapter 22

Ewine F. van Dishoeck (PhD, 1984)

Building a Worldwide Astrochemistry Community

Ewine van Dishoeck, now Professor of Molecular Astrophysics at Leiden University, innovatively combines the world of chemistry with that of physics and astronomy to study the trail of molecules from star-forming clouds to planet-forming disks. She was the winner of the Spinoza Prize from the Dutch Research Council in 2000, the Lise Meitner Göteborg Award in physics in 2014, and the Albert Einstein World Award of Science in 2015, presented the Lodewijk Woltjer prize lecture to the European Astronomical Society in 2015, and was awarded the Kavli Prize for Astrophysics in 2018, the James Craig Watson Medal by the National Academy of Sciences in 2018, the Karl Schwarzschild Medal by the Astronomische Gesellschaft (Germany) in 2019, and the Prix Janssen de la Société Astronomique de France in 2020. She is a member or foreign associate of the Royal Dutch Academy of Sciences, the National Academy of Sciences, the American Academy of Arts and Sciences, the Leopoldina Academy of Sciences (Germany), and the Norwegian Academy of Sciences. Her asteroid is 10971 van Dishoeck.

My birthplace and home for much of my life is the small university town of Leiden, the Netherlands.[1] My father, a medical professor, and my mother, an elementary school teacher, clearly envisaged an academic path for me: my birth card shows a baby wearing a stethoscope crawling toward the university with the motto "vires acquirit eundo" ("she will gather strength along the way"), from Virgil's *Aeneid*.

I was enrolled in a Montessori elementary school, which encourages self-study and collaborative projects, two traits that have served me well. Among my schoolfriends was Jette van de Hulst, daughter of the famous astronomer Hendrik van de Hulst (who predicted the 21-centimeter-line radiation of hydrogen). This led to my first encounter with the Sterrewacht (now "Old Observatory"), where van de Hulst and Jan Oort, another great Dutch astronomer, lived. Little did I know then that it would become so important to me later in my life.

My first encounter with science came in spring 1969, when my father went on a six-month work visit to San Diego where I was enrolled in junior high school. I chose classes that I had not (yet) taken at the Stedelijk Gymnasium (grammar school) in Leiden. Most notably among them was science, taught by an inspiring, female, African-American teacher; only in hindsight did I realize how unusual that must have been then.

Back in Leiden, I became fascinated with chemistry thanks to a modern teacher who was more interested in understanding how reactions happened than just bookkeeping them. Consequently, I enrolled in chemistry at Leiden University in 1973 but quickly discovered that physics was just as much fun, whereas the smelly organic chemistry labs were not. For me, theoretical quantum chemistry provided the best of both worlds, and for my MSc project I embarked on a theoretical study of the photodissociation pathways of the methane ion (CH_4^+) under the guidance of Marc van Hemert. Photodissociation, that is, the process by which a molecule falls apart under the influence of ultraviolet radiation (AB + UV photon $\rightarrow$ A + B), has been one of the main research themes throughout my career, benefiting from my deep understanding of the processes at the molecular level. I continued to take lots of physics courses, in which I usually was the only female student. I personally did not experience this as a disadvantage; in fact, it was easier for me to get noticed by the professors.

I met my future husband, astronomer Tim de Zeeuw, during my high school and university years, but not through physics or astronomy courses, as many people naively think! Both of us played the violin in the Leiden Youth Orchestra and in the "gypsy" music orchestra Csárdás. We performed many gigs at parties and used the money we earned for a three-week study trip to Hungary in 1974, which at that time was still behind the Iron Curtain. We visited the Rajko, the state school for Hungarian folk music in Budapest, and went to many restaurants with orchestras, taping their performances to transcribe later and add to our own repertoire. We also enjoyed long vacations as students (grab them while you can!), including one six-week trip driving from San Diego up to Glacier National Park and back through Yellowstone, Bryce, and the Grand Canyon. Our love for the spectacular nature and scenery in the western United States was born then, and we have since gone camping and hiking there every summer, with our forty-plus-year string broken only by the pandemic in 2020.

In 1979 I was planning to continue in quantum chemistry for my PhD research. Unfortunately, the professor had just passed away and finding a successor was going to take a long time. By chance, Tim had just taken the Interstellar Medium course from Harm Habing, which included a lecture on interstellar molecules. "Isn't that something for you?" he asked. At that time, I had not yet had a single course on astronomy. That summer, Tim and I were camping in Canada in the Mont Tremblant National Park, where the first International Astronomical Union (IAU) symposium on Interstellar Molecules took place. There I met Alexander Dalgarno, a world leader in the new field of astrochemistry, who kindly invited me to spend some months at Harvard to get started working on photodissociation processes of astrophysically relevant molecules, most notably hydroxide (OH). Dalgarno would become my PhD supervisor and lifelong mentor.

With Tim pursuing a PhD in Leiden, I did not want to stay at Harvard. But getting funding for an interdisciplinary project in the Netherlands turned out to be far from simple. Both the chemistry and astronomy streams of our Dutch Research Council (NWO) liked the astrochemistry proposal, but each looked to the other to fund it. Fortunately, Habing was able to convince the NWO to provide me with a special grant to continue my PhD work in Leiden. This was my first lesson on the difficulties of getting funding for interdisciplinary research. It also highlights the importance of mentors for opening doors and creating opportunities for younger scientists.

During regular visits to Harvard, I met John Black, then a postdoc, who patiently introduced me to the physics and chemistry of interstellar clouds. Meanwhile, Habing alerted me about a new, high-resolution spectrometer at ESO La Silla, which was well suited for observing interstellar diatomic carbon (C_2) and testing the models that Black and I had developed for how this molecule is excited. And that is how I became a real astronomer: I went to Chile in 1982 to do the observations myself, a fantastic experience.

In the same year, Thijs de Graauw, from Space Research Organisation Netherlands, invited me to a conference on planning a new space telescope, then called FIRST and ultimately named the Herschel Space Observatory, that would study the Universe at far-infrared wavelengths. I did not realize then that the first data would not appear until 2009! Thus, I learned another important lesson: the timescale for building astronomical facilities, from conception to actual operation, is typically twenty-five to thirty years.

Tim and I graduated on the same day in 1984 in a "double star" ceremony at Leiden University and subsequently started postdoctoral fellowships at Harvard (me) and Princeton (Tim). Even though we had offers for fellowships at the same place, they were not optimal, so we followed the advice of Tim's mentor, Martin Schwarzschild: take the steepest trajectory early on in one's career in order to create more options later on. We lived in two places for a number of years, but John Bahcall, at the Institute for Advanced Study, kindly gave me a visiting position so that I could work from there for a fraction of my time. In 1987, I obtained an NSF visiting professorship for women (one of their first special programs for women) and spent a year at the Department of Astrophysical Sciences of Princeton University, getting my first experience teaching. Several of my most-cited papers were written in this period with John Black, most notably on the photodissociation of carbon monoxide (CO) molecules and on the excitation of molecular hydrogen (H_2). We also defined a new type of interstellar cloud, translucent clouds, as those at the transition between diffuse interstellar clouds and dense molecular clouds. Decades later, these clouds, which are rich in H_2 but often poor in CO, have become known as "CO-dark clouds."

In 1988, Caltech encouraged me to apply for one of their new cosmochemistry positions in the Geology and Planetary Sciences (GPS) Division. The first hire was Geoffrey Blake, who, like me, was trained as a chemist but also became fascinated by interstellar molecules. I was warmly welcomed as the first female faculty member in GPS, and, even though I stayed only two years, my time there was formative in many ways. Tom Phillips introduced me to hands-on observing at the new Caltech Submillimeter Observatory and taught me much about how big science is done. Geoff Blake and I embarked on some of the first observational studies of the chemistry associated with star and planet formation, resulting in a lifelong collaboration and friendship. Our favorite object quickly became the infrared-bright source IRAS16293–2422 in the constellation Ophiuchus, which is a low-mass binary star system in formation. IRAS16293–2422 proved to be a gold mine for determining the chemical inventory of a young Solar System analog.

Although we enjoyed life and science at Caltech to the fullest, new opportunities arose in 1990 when Tim was offered the professorial chair of theoretical astronomy at Leiden vacated by van de Hulst. I was offered a senior lectureship (equivalent to associate professor with tenure), together with a big personal grant to start a group. Colleagues joked that Tim got the title, but I got the money. Moreover, de Graauw guaranteed me observing time on the Infrared Space Observatory (ISO), a mission of the European Space Agency (ESA), that would for the first time obtain infrared spectra of astronomical objects unhindered by the Earth's atmosphere, allowing observations of key molecules like water (H_2O) and carbon dioxide (CO_2). Saying yes to two faculty positions at one of the leading astronomical institutes worldwide was thus made easy. After three moves in six happy years in the United States, we returned to Leiden. In hindsight, our period abroad should not have been any shorter.

The overarching science goal during my career has been to follow the trail of molecules from star-forming clouds to planet-forming disks.

FIGURE 22.1. Ewine van Dishoeck at Leiden University. *Credit:* Leiden University.

Molecules do not emit visible light effectively; rather, they emit light best at infrared and millimeter wavelengths. And I have been fortunate to be active in the decades during which these wavelength regimes have opened up to astronomical observations, thanks to enormous progress in detector technology and huge leaps in capabilities. With the commissioning of every new facility, astronomers have been able to look either more sensitively, more sharply, or more uniquely at molecules. For example, ISO (1995–97) was particularly well suited for studying interstellar ices, silicates, and water but only for a handful of the brightest protostars and disks. The Spitzer Space Telescope (2003–9) lacked spectral resolution but made up for that limitation in raw sensitivity, allowing dozens of fainter solar-mass sources and disks around young, Sun-like stars to be observed. And the European Southern Observatory (ESO) Very Large Telescope (VLT, 1999–now) has infrared instruments with uniquely high spectral and spatial resolution that can measure the motions of hot gases in the inner disks surrounding forming stars.

At millimeter wavelengths, the 15-meter James Clerk Maxwell Telescope on Maunakea was an important platform for measuring chemical inventories on the size scale of protostellar envelopes, which are larger than protostellar disks. The wavelength window in between the infrared and the millimeter (the far-infrared) opened up with the launch of the 3.5-meter Herschel Space Observatory (2009–13), which was particularly well suited to observing gaseous water molecules. Finally, the Atacama Large Millimeter/submillimeter Array (ALMA) (2012–now) is the ultimate "astrochemistry machine" for probing the bulk of interstellar molecules. I have enjoyed co-leading major observational programs on all of these facilities and have helped push to make several of them happen. One of the lessons I learned from my research on all of these big telescopes is the impact (and fun) of large coordinated programs for moving the field forward.

These instruments now allow us to zoom in on future solar systems, at size scales of a few thousand astronomical units (au) in the 1980s (1 au is the distance from the Sun to the Earth) to scales of just a few au today. It is as if a city like Leiden, that used to be a blur with Google Earth in its infancy, is now suddenly resolved into its canals, streets, and individual houses. The "golden triangle" of combining these observations with sophisticated models and laboratory experiments in-house was a further key to success.

An example of this triangle is the study of ices. In cold and dense star-forming clouds, atoms and molecules freeze out onto the tiny dust particles, where chemical reactions can take place that normally do not happen in dilute interstellar gas. Take the formation of water itself: H and O are the first and third most abundant elements in the Universe, but they can only be brought together effectively on the surfaces of dust grains. Leiden University had set up the first laboratory astrophysics group world-wide in 1975. Upon my return to Leiden, it was clear that lab data would be essential for interpreting spectra of ices that ISO would obtain, and the lab became my scientific responsibility, a somewhat daunting prospect for a theorist. Together with two highly capable postdocs, we managed and established the first version of the Leiden ice database, which would be used world-wide for decades. The combination of the lab data and ISO spectra resulted in a much deeper understanding of ices in interstellar space than would have been possible without the dedicated lab effort. This example illustrates the need for long-term planning, looking years—if not decades—ahead, and making sure science stays in step with instrumentation along the way.

Thanks to the Netherlands Research School for Astronomy (NOVA), I was able to obtain a ten-year grant in 1998 to support the work of the lab and to build two new instrumental set-ups using ultra-high vacuum techniques borrowed from surface science. This allowed the lab, now under the direction of Harold Linnartz, to study the underlying chemical reactions on icy surfaces and their dependence on physical parameters, such as the amount of UV radiation to which they are exposed. A beautiful example of how laboratory work, modeling, and observations can be combined is how hydrogen combines with CO to form methanol (CH_3OH), which has now been shown to be a key step in making even more complex organic molecules like glycolaldehyde (a simple sugar), ethylene glycol (also known as anti-freeze), and tri-carbon species like propenal and glycerol. All of these are now being observed in interstellar and circumstellar gas observed with ALMA and have been found in comets.

One of the three instruments on board Herschel was HIFI, designed and built by a consortium under the leadership of de Graauw. In return for its huge financial investment, the Netherlands obtained a significant fraction of guaranteed observing time on Herschel. I was fortunate to lead the "Water in Star-forming regions with Herschel" (WISH) key program, a wonderful collaboration of some seventy scientists around the world that started in 2005 and lasted for more than a decade. HIFI was designed to observe the fingerprints of water vapor in interstellar space with exquisite detail and sensitivity, and revealed that water, a key ingredient for life elsewhere in the Universe, is abundantly present in regions where new stars and planets are being born. The sometimes weaker-than-expected signals imply that water becomes locked up in large icy bodies early on in the star-formation process. This also confirms one of my dictums: most new scientific information is in the weak signals!

Plans for the next generation of arrays of millimeter-wavelength telescopes started in the early 1980s, first in the United States and then in Europe and Japan. It became clear by the mid-1990s that neither continent would have the resources to put together an array powerful enough to address the top-level science goals. In 1997, scientists like myself, involved in discussions on both sides of the ocean, managed to convince the funding agencies to join forces to build ALMA; the Japanese joined a few years later. ALMA thus became the first modern, world-wide collaboration in astronomy. Making ALMA into a reality comprised a major part of my efforts in the early 2000s, first as member and chair of the ALMA Science Advisory Committee (1999–2005) and later as member of the ALMA Board (2006–12). Moreover, I was asked to step in during 2001–2 as interim European Project Scientist to help guide the project and set the specifications.

ALMA finally opened its scientific eyes in 2012, with data quality that exceeded everyone's expectation. IRAS16293–2422 was chosen as one of the first science verification targets and proved to be even more chemically rich than found in the pre-ALMA era. Jes Jørgensen led a first paper on the detection of glycolaldehyde, which received wide attention as the first "sweet" result from ALMA. A few years later, we carried out a complete, unbiased spectral survey of this source, which became the Protostellar Interferometric Line Survey.

Inspired by Spitzer observations of a set of disks around young stars that had central dust cavities, we embarked on a new program to image these disks with ALMA. Our first ALMA image was utterly surprising: instead of a full dust ring, we found a highly asymmetric, banana-shaped feature offset from the star, the first observational evidence for a gas pressure bump in the disk generated by an embedded planet pushing around tiny dust grains and confining them to locations known as dust traps.

The next era of mid-infrared astronomy will be opened up by the James Webb Space Telescope, with which I have been associated since 1997. Riding on the success of ISO, several of us saw an opportunity to persuade NASA and ESA to include a mid-infrared instrument (MIRI) on this flagship mission. As Dutch co-PI, I was heavily involved in the building of MIRI in 2000–2010 and am looking forward to the first data in 2022! MIRI will allow us to make the link from "disks to planets" by probing the chemical composition of the inner planet-forming zones of circumstellar disks.

As highlighted by my first encounter in 1980, the IAU has been important throughout my career. The IAU symposia on astrochemistry are special because they review the entire field of astrochemistry (from planets in our Solar System to galaxies in the distant Universe), provide a platform to promote young and upcoming scientists, and help scientists forge relationships around the world. I treasure my many friendships that resulted from these meetings. In 1992, I was asked to serve as secretary of the IAU Working Group on Astrochemistry and subsequently became its president, (co-)organizing four major symposia in the 1996–2011 period. One thing led to another: in 2012, I was nominated President of the new Division H on Interstellar Matter and Local Universe, and in 2015 I became President-elect of the IAU.

One of my more gratifying tasks as a faculty member has been my editorship since 2004 of *Annual Reviews of Astronomy and Astrophysics*. Closer to home, I took over as NOVA Scientific Director in 2007. NOVA—the alliance of the four Dutch university institutes with astronomy programs—spearheaded by Tim as its founding director, enables Dutch university astronomy to build up a strong optical-infrared and submillimeter instrumentation program centered on ESO. Keeping the community aligned and NOVA funded has been a significant but rewarding effort. The program is now culminating with Dutch leadership of the mid-infrared instrument METIS for the Extremely Large Telescope.

Our lives took another turn in 2007. Tim was offered his dream position, to become ESO Director General. With headquarters in Garching near Munich, this meant that we had to face once again the challenge of living in two places at a time when my aging mother needed more care. Fortunately, Reinhard Genzel kindly offered me a visiting position at the Max Planck Institute for Extraterrestrial Physics, which proved very fruitful in the Herschel and ALMA era. We enjoyed ten wonderful years

FIGURE 22.2. Ewine van Dishoeck working with students Catherine Walsh, Mihkel Kama, and Xiaohu Li. *Credit*: NOVA.

commuting between Leiden and Garching, with significant time in beautiful Chile. The Christmas periods spent there with the staff at the telescopes have left many happy memories.

The biggest joy throughout my career has been working with a very talented set of more than eighty graduate students and postdocs. Much of my success is due to their hard work and creativity. Each one of them is special, and I wish I could thank them all personally here. The fun of getting new data, finding an unexpected result, getting a complicated program to work, brainstorming together around the table in my office, and finally putting it all together and learning something new is what continues to drive me.

Throughout this memoir, I have attempted to give advice to young scientists based on my experiences. Here is a brief summary. (1) Be passionate about what you do! (2) Make sure you excel at something so that you get noticed: it is better to be an expert in one thing than average in many. (3) Dare to look elsewhere and make an (interdisciplinary) jump. (4) Seize opportunities when they arise (but also know when to say "no" so as not to overload yourself). (5) Get involved in big projects, with a visible role, and help make them happen. (6) Balance opportunities with your partner. (7) Enjoy life, including (long) vacations.

It is a tradition that Leiden University reminds every student upon graduation to enjoy the benefits of the PhD title but never to forget the obligations that it brings to science and society. Being President of the IAU (2018–21) was one of those rare opportunities that allowed me to give back to society world-wide, by promoting and stimulating the importance of outreach, development, education, diversity, and inclusion. I look forward to more years of exciting research but at the same time continuing to convey the message that we are all world citizens, living on a small and fragile planet under the same beautiful starry sky.

Notes

1. This autobiography follows closely that written for my 2018 Kavli Prize for Astrophysics.

Chapter 23

Wendy L. Freedman (PhD, 1984)

My Astronomical Journey

Wendy Freedman, now the John & Marion Sullivan University Professor of Astronomy & Astrophysics at the University of Chicago, has spent her career improving our knowledge of the expansion rate of the Universe by measuring the Hubble constant—the current expansion rate—as well as studying the stellar populations of galaxies. She is a member of the National Academy of Sciences and the American Academy of Arts and Sciences, and a Fellow of the APS and the AAS. She was awarded the Gruber Cosmology Prize in 2009 and the Dannie Heineman Prize for Astrophysics, awarded jointly by the AIP and the AAS, in 2016. Her asteroid is 107638 Wendyfreedman.

As my parents liked to tell the story, they would enter my room when I was a baby and find me sitting up in my crib, looking out at the night sky. A budding astronomer, they later joked. Actually, I had absolutely no idea what I wanted to be while I was growing up—I loved many subjects. But by the end of high school, I knew that I wanted to study science.

I have fond memories of sitting for hours on end in our basement looking through a toy microscope, a gift from my father. Cells in my blood, worms, insect wings. Mixing chemicals from a chemistry set. Looking up at the night sky while my father explained that the stars we were gazing at might not be there any longer, because we were seeing them in light that had left them long ago. Listening to him explain how the human body worked. All profoundly stirring my imagination. My mother nurtured a love of literature, a love of music, a love of art. She spent hours with me and my siblings at a public library, times when I pored over every astronomy book on the shelves, as well as classic literature.

In elementary public school I was fortunate to have teachers who nurtured my love of reading. I was even more fortunate to have friends who were fun and smart. Sadly, many of the girls I knew dropped out of math when we reached high school. Worse, I had a physics teacher in tenth grade who, while explaining a complex concept, actually said to the class, "The girls don't have to listen to this." I didn't much care for high school. But by a stroke of luck, I had a phenomenal twelfth grade physics teacher, Mr. Bowers. Bowers had a master's degree in physics from Oxford. He spoke with an imposing British accent and pushed us in a way that I had never been pushed before, forced us to think instead of solve cookie-cutter problems. Before him, subjects came easily, and I hadn't experienced the feeling of exhilaration that accompanies wrestling with a seemingly insurmountable problem and then cracking it. I fell in love with physics. Mr. Bowers also lent me his telescope, a great joy for me.

Compared to high school, university was hugely stimulating. I loved the University of Toronto. In my second year, I switched majors from biophysics to astronomy and astrophysics. The faculty were extremely supportive of me throughout and encouraged me to pursue a graduate degree, something I had not initially considered. When it came time to decide on a graduate school, the Canada-France-Hawaii Telescope (CFHT) was just being commissioned at Maunakea in Hawai'i. The opportunity to do a thesis while having access to this telescope was a tremendous draw and convinced me to stay in Canada. At 14,000 feet, the CFHT had phenomenal astronomical "seeing," that is, very little blurring or twinkling caused by the Earth's atmosphere, and therefore very high-resolution imaging capabilities. My thesis involved comparing how the formation of new stars proceeded in different galaxies. With the CFHT it was possible to distinguish and to measure the brightnesses of individual stars (with masses greater than that of our own Sun) in our nearest neighbor galaxies, and to measure hundreds or thousands of them, where previously only dozens of stars had been measured.

Observing at the CFHT hooked me as an observer. I loved being at the top of the telescope (the "prime focus cage") and looking up at the night sky while obtaining the data for my thesis, and I loved writing proposals—asking, then going to the telescope and then answering, a question. There are many joys in research and many challenges. You don't know how it will turn out; there is no answer at the back of the book. There are frustrating obstacles: at times you may hit a dead end, but if you persevere, you learn something. It usually (always) takes longer than you think. And then there are the moments when you actually see something never seen before. Or emerge from an afternoon reading about an exciting new result, marvel at the nature of the Universe, and feel grateful for the opportunity to have a life in science.

During childhood I developed the habit of waking early, and that has served me well all through my life. It is hard to imagine now when laptop computers are so ubiquitous, but while I was working on my thesis, there was a single computer (called a VAX 780) that was available to the entire physics and astronomy departments. That meant that during the normal workday, everyone was competing for cycles on this one VAX machine, and it was frustratingly slow. So I would wake up at 2:00 A.M., after the night owls had gone home, and have the VAX almost entirely to myself until others in the department rolled in around 10 A.M. Without this quirky morning habit, I would not have been able to finish my thesis in a timely manner.

Near the end of my thesis, Cepheids (a type of star) became my focus. I could never have predicted that I would spend a great deal of my career studying these (super) giant stars. Over timescales of days to months, the atmospheres of these stars pulsate in and out, causing the stars to alternately brighten and dim. It turns out that the rate of the pulsation (or period) is directly related to how bright the star is (bigger stars pulsate more slowly, or have longer periods). Suppose we have two galaxies and we want to know their relative distances. We can compare stars with the same periods. The stars in the more distant galaxy will be fainter, just as car lights in the distance appear fainter than those nearby.

FIGURE 23.1. Wendy Freedman in the control room of the Magellan Baade 6.5-meter telescope in 2017. Photo by Yuri Beletsky.

The "inverse square law of light," where the brightness of an object falls off inversely as the square of the distance from us, allows astronomers to use Cepheid variable stars to measure distances.

At that time, it turned out that astronomers measuring the distances to galaxies had not been accounting for the fact that the class of Cepheids we use for the distance scale, being young stars, are located in regions where there is gas and dust—the material from which they are formed. Light coming to us from these Cepheids is both scattered and absorbed by the dust, making them appear fainter than they actually are, as well as redder in color. If you don't account for the dimming effects of this dust, you will erroneously conclude that the galaxy in which you are measuring the Cepheid is farther away from us than it actually is. I was very fortunate to come into the field at a time when telescopes began to be equipped with a new kind of detector (called a charge-coupled device, or CCD). We use these devices now in our cell phones. CCDs replaced the earlier (less accurate) photographic cameras. With a CCD, I could measure the distance to a galaxy using different wavelengths of light, from the blue part of the spectrum to the red. The galaxy, of course, is at a single distance from us. But what I found was that the measured distance depends on the wavelength at which the measurement is made. Moreover, it did so precisely in the way that would be expected if the stars were being affected by interstellar dust. These new measurements not only illustrated that dust was a problem in the earlier measurements but also provided a way to measure the amount of dust and to correct for it. This result later turned out to be critical for measurements using the Hubble Space Telescope.

After completing my thesis in 1984, I was awarded a Carnegie Fellowship at the Observatories in Pasadena, California. This proved a dream job for me. With the unusual amount of freedom afforded to Carnegie Fellows, I was able to choose my own projects and pursue my own areas of study. As young fellows, we had generous access to the 200-inch telescope at Palomar (then the largest working telescope in the world) and to telescopes in the Andes mountains in Chile. I also had ongoing access to the CFHT. I embarked on an ambitious program to measure the distances to all of the galaxies close enough to identify and measure their Cepheids.

During this period, a rather heated debate was taking place in the astronomical community concerning the age and the size of the Universe. Resolution of this debate was one of the primary goals of the soon-to-be-launched Hubble Space Telescope (HST). Estimates of the age ranged from 10 to 20 billion years, a rather large uncertainty. Settling the issue would require an accurate measurement of a quantity known as the Hubble constant, named after astronomer Edwin Hubble, whose work in 1929 suggested that the Universe is expanding. Measuring the Hubble constant (the rate at which the Universe is expanding today) requires accurate distances to galaxies, an area in which I had developed expertise.

In the summer of 1984, a meeting on the extragalactic distance scale was held in Aspen, Colorado. We were given the news that the Director of the Space Telescope Science Institute, Riccardo Giacconi, was searching for very large programs to be carried out using HST. HST had been excitedly anticipated for decades. Giacconi was concerned that astronomers would be too conservative when awarding HST time and slice the observing pie into many small pieces. He asked a senior committee to advise him on the most important projects that would require large amounts of time and that could only be done with HST. The plan was to set aside time for "Key Projects" that would be completed within the community. One of those highly ranked programs was for the Extragalactic Distance Scale. HST was due to be launched in 1986, but the tragedy of the *Challenger* space shuttle accident resulted in a delay until 1990.

In 1985, I married astronomer Barry Madore, who to this day remains my closest collaborator. He and I have shared thirty-five incredible years of family and astronomy together. I couldn't have asked for a partner who would be more supportive of family and career. In 1987 our daughter, Rachael, was born, two months before I was appointed as the first woman to the Scientific Staff at the Carnegie Observatories. Our son, Daniel, was born in 1988. I feel fortunate to have been born at a time when I didn't have to choose between career and family. However, it's not as simple as "you can have it all," as the saying goes. It is possible, just *not all at one time*! For years when our children were young, I didn't read a newspaper, didn't see any movies, didn't go out at all, didn't read a novel. Life was work and family. And very little sleep. But this was a deeply rewarding period of my life. My family remains an unrivaled source of joy and support. My advice to young people is, if you want children have them. You can always find reasons to wait (a degree, a thesis, postdoc, tenure, etc.). There simply isn't a perfect time, and time does run out.

As in many careers, I have experienced unwelcome challenges (hostile and aggressive colleagues, unforeseen obstacles), in addition to the high points. I don't wish to minimize the painful times, but I also refuse to grant them too much weight. On balance, my enjoyment has far surpassed the low points in my career. I love what I do. Astronomy isn't a job for me, it is a passion.

Returning to the 1984 Aspen meeting, it was then that a group of us joined together to compete for the Key Project on the Extragalactic Distance Scale. The principal investigator of the proposal was Marc Aaronson, and I was the deputy. Tragically, thirty-six-year-old Marc was killed in an accident while observing in 1987. Thereafter, I and two other astronomers (Jeremy Mould and Rob Kennicutt) led the Key Project.

When HST was launched in 1990, there was a shared sense of euphoria in the astronomical community, followed by a feeling of horror as it unfolded that the primary mirror of the telescope was flawed. The science we had waited to do for so long could not be done, and newspaper headlines declared Hubble a disaster. It was a depressing time. We were lucky, however, that for the two nearest galaxies in our sample, we were able to use the unfocused telescope—we simply had to observe for longer. But for the rest, we had to wait until a space shuttle mission was launched that took up "corrective lenses," allowing HST to focus and to achieve its full potential.

In December 1993, we received the first images of one of the key targets in our program, a galaxy in a large cluster of galaxies called the Virgo Cluster. An early member of our group had concluded that it would not be possible for us to find Cepheids in the Virgo Cluster. One of the happiest moments in my research life was when, a few months later after we had collected a series of images, I did a search for stars that could potentially be Cepheids. And there they were. Not only were they definitely Cepheids, but the signal came booming through—the points on the curves I plotted were lined up like beads on a string. From that moment on, we knew that we could do the project that we had set out to do: measure the Hubble constant to an accuracy of 10 percent. Over the course of the next six years, our international team of thirty astronomers discovered Cepheids in two dozen galaxies and ultimately measured a value of the Hubble constant with a value of 72 (where previous values had ranged from 50 to 100) in the standard units in which it is always quoted. We published our final Key Project paper in 2001, and our value has stood the test of time. For the past twenty years, it has been verified by many other independent groups, to within the accuracy that we measured it.

Over the next decade, my scientific interests were focused on a program that a number of Carnegie astronomers undertook: an intensive study of a type of supernova, brilliant explosions that can be seen to vast distances. And with Barry and several collaborators, we extended our

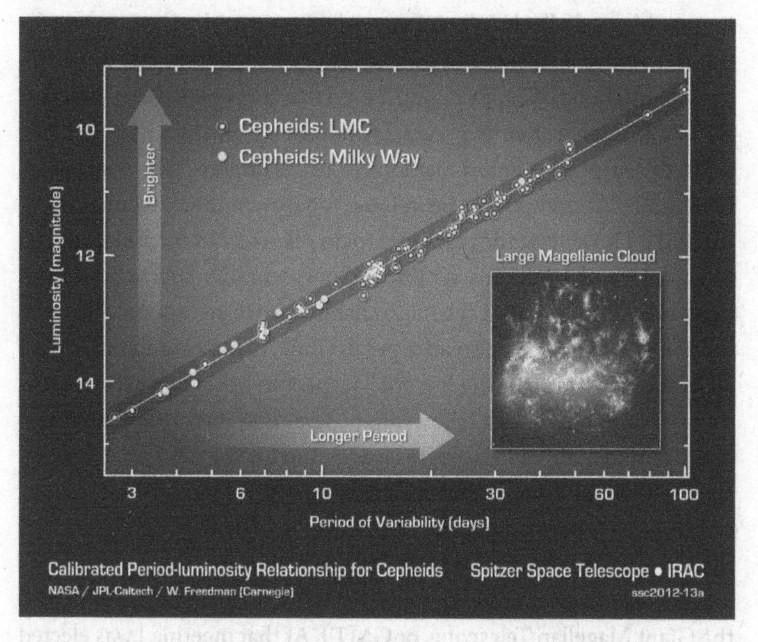

FIGURE 23.2. A plot of the periods of Cepheid variable stars in the Milky Way and the nearby Large Magellanic Cloud (LMC) versus their brightnesses, using data from the Spitzer Space Telescope. The inset shows an image of the LMC. The straight-line fit to the data reveals one of the most important relationships in astronomy, now known as the Leavitt Law: longer-period Cepheids are brighter; shorter-period Cepheids are fainter. In a distant galaxy, we obtain a measure of the apparent Leavitt law. We then need a sample of Cepheids for which the true brightness has been previously calibrated (based on geometric distance measurements). A comparison of the apparent and true Leavitt laws then yields the distance to the star or galaxy (via the inverse square law of light). *Credit*: NASA/JPL-Caltech/W. Freedman.

Cepheid studies using an infrared satellite telescope called Spitzer. But in 2003, my career took a different turn when I accepted the position of Director of the Carnegie Observatories.

When starting out in my career it never occurred to me, nor did I have any desire, to become an observatory director, let alone the leader of a giant telescope project. Twice, I had earlier declined an invitation to be considered by previous search committees for the Carnegie directorship. I was in the midst of the Key Project, and it would have been impossible to do both. When I was asked to consider the position for the third time, I decided to give it some serious thought. But I still feared I might be unhappy taking it on. I made a promise to myself that if I was miserable after two years, I would step down.

To my surprise, I found that I loved being director. The Institution, founded by Andrew Carnegie in 1904, focuses on the individual scientist and on providing the resources to do science. I immensely enjoyed interacting with both the scientific and non-scientific staff in both Pasadena and Chile. Initially I set out to enable science, to support the building of instruments on our telescopes in Chile, to increase the salaries of all of the staff (which were low with respect to other departments), to increase support for our fellowship program, to initiate a theory program, and to maintain Carnegie's leadership for the next generation of large telescopes. I succeeded in all these goals.

I began my term as director in March 2003. That spring the council that had overseen the construction of two previous 6.5-meter telescopes in Chile met to discuss building a 20-meter telescope (eventually named the Giant Magellan Telescope, or GMT). At that meeting I was elected Chair of the Board for this new telescope, a position that I held until stepping down in July 2015. It never occurred to me that I would spend twelve years of my career leading the effort to build the GMT, one of the world's largest telescopes, 80 feet (25 meters) in diameter, in the Andes mountains in Chile.

The original design of the GMT came from Roger Angel at the University of Arizona. In the early days of the project, there were three people who worked on the telescope on a daily basis in Pasadena: Steve Shectman (Project Scientist), Matt Johns (Project Manager), and me, Chair of the Board. Matt and I would regularly agonize over the looming budget shortfalls. Sometimes the horizon for funding stretched only a month or two into the future. But with time, and particularly with the support of the Carnegie Board and another generous donor by the name of George Mitchell, the project was launched, and the first of seven 8.4-meter mirrors was cast in July 2005. The National Science Foundation also provided early funding to help complete the mirror. I did a lot of traveling, putting one foot in front of the other, bringing in one partner and one donor at a time. From the original four partners (Carnegie, the University of Arizona, Harvard University, and the Smithsonian Institution), the project grew to include the Australian National University, Astronomy Australia Ltd., the Korea Astronomy and Space Science Institute, Texas A&M University, the University of Texas at Austin, the University of Chicago, the Saõ Paolo (Brazil) Research Foundation, and Arizona State University. It is an unusual consortium—composed of both private and public institutions, and international in scope. Nothing like this in previous generations of optical astronomy had been attempted before. At the start, the Carnegie Board had contributed half a million dollars in seed funding. By the time I stepped down in 2015, the partners had signed a legal agreement committing to half a billion dollars. Construction is currently scheduled to be completed by the late 2020s. Heading the GMT project was a labor of love for me, and one day I hope to use it!

By 2013, I was nearing a decade of Carnegie directorship. At that time, the GMT partnership was approaching the construction era, which seemed to me the right time to transition leadership of the project. I wished to begin a large new science project while I was still young enough to do so.

When I was offered a position at the University of Chicago, it came out of the blue, and I was not looking to move. I had enjoyed thirty immensely pleasurable years at Carnegie. After visits to Chicago with Barry and months of thought, I realized that I was excited about the opportunity to do more science and to teach students of my own, and I was ready to step down from the responsibilities of directorship and leadership of the GMT project. University life in Chicago has exceeded my initial hopes. There is a youthful energy that comes from being on a vibrant university campus. The students are bright and enthusiastic, and they keep you up to speed. Returning primarily to research science, my first love, feels like a gift.

Now I find myself in the thick of a new debate about whether or not our fundamental model of cosmology has something missing. The ability to measure an accurate value of the Hubble constant has continued to improve since the days of the Key Project, and with it comes the ability to test our models of the early Universe. We don't yet know how the controversy will be resolved. These questions will be answered as several extraordinary new telescopes go into operation in the next few years. I eagerly anticipate what this next chapter will bring.

Acknowledgments: I thank first my parents, to whom I owe so much. I would not have accomplished what I have without the love, support, and ongoing partnership with Barry. My family (up and down the family tree) enriches everything. Thanks are due to many collaborators over the years, with special thanks to Rob and Jeremy. I feel tremendous gratitude toward, and respect for, Maxine Singer and Leonard Searle, two remarkable scientific leaders and mentors. I have had the good fortune to work with many talented students and postdocs who have taught me much, and I derive so much pleasure in seeing them launched in their careers. Finally, I have great appreciation and respect for the women astronomers who came before my generation and broke down barriers: the Harvard women astronomers (including Henrietta Leavitt, who discovered the Cepheid period-luminosity relation), Cecilia Payne-Gaposchkin, Margaret Burbidge, Vera Rubin, Beatrice Tinsley, Sandy Faber, and many others.

Chapter 24

Meg Urry (PhD, 1984)

The Gentlemen and Me

Meg Urry, the Israel Munson Professor of Physics and Astronomy and Director of the Yale Center for Astronomy and Astrophysics, studies active galaxies, which host accreting supermassive black holes in their centers, and develops models for the growth of these black holes over cosmic time. She is a Fellow of the National Academy of Sciences, the American Academy of Arts and Sciences, the American Association for the Advancement of Science, the American Astronomical Society, the American Physical Society, and American Women in Science. She served as President of the American Astronomical Society (2014–16), was awarded the AAS's Annie Jump Cannon Award and George van Biesbroeck prize, co-organized the first Women in Astronomy conference, held in 1992, and was the 2015 recipient of the Edward A. Bouchet Leadership Medal from Yale for her efforts to broaden participation in STEM fields.

I grew up in an ordinary family that was also extraordinary. The second of four children, I was born in St. Louis, where my father was a chemistry professor, though we moved to West Lafayette, Indiana (Purdue), when I was two. My dad had grown up in a large Mormon family in Salt Lake City, the middle of nine kids, living in impoverished circumstances typical of the Great Depression, before leaving home at age sixteen to study at the University of Chicago.

After my mother's parents immigrated to the United States from Italy, she was born into an Italian community in New York and raised by her mother alone. She never wanted to talk about her difficult early life, though growing up we understood that she and my grandmother sometimes lacked food or housing. After excelling in school, she earned a scholarship to the University of Chicago, where she majored in zoology while working nearly full-time.

My parents met at the Museum of Science and Industry in Chicago, where both worked as exhibit presenters. Six years after getting married, they started a family. We never had much money, so we did everything on a shoestring. Vacations were car trips, with picnic lunches and nights in budget motel rooms. The kids' job was to clean the picnic dishes in the restroom or, sometimes, in a nearby stream. My mother was exempt from dishwashing. Instead, she spent the cleanup period turning over rocks in those streams, looking for new species of worms. We didn't grow up being told to make money or become famous; Mama wanted us to discover a new species of worm and name it after her.

One day on a California beach, we stumbled across a large dead sea lion. It smelled terrible, but my mother was fascinated. We visited the carcass daily to check how much progress the maggots were making, as she worried whether the waves would wash it back out to sea. Finally, she took a stick and pried out the skull. We kids were, I have to admit, completely grossed out and embarrassed by the incredulous stares from passersby, but this was normal behavior for my mom. For several days, the stinking skull hung, dripping, outside our bedroom window. Then she boiled it with Clorox for several more days until it was mostly clean. We carried that skull two thousand miles home and added it to my mom's collection.

These things didn't seem strange to us at all—not the skulls casually displayed throughout our house, not the picnic worm-hunting, nor the countless other activities that I now recognize helped turn me into a scientist. As a kid, I liked my science classes far less than math, English, or history. But thinking about how worms fit into evolution, wondering about the night sky, and witnessing sociology experiments (like the time my mother showed us kids a photograph of Black and white people

FIGURE 24.1. Urry kids on the beach in Bolinas, California, prior to acquiring our sea lion skull. Interestingly, we all became scientists. Oldest to youngest, from left to right: Lisa, a biologist; Meg; Serena, an art conservator; Tony, an engineer.

in a crowd, in order to determine the age at which we had learned to recognize race) made science part of our daily lives.

My parents were relentless proponents of education and of excelling. I spent a lot of time at the library, reading every biography I could find in an orange-bound series for kids. Most of these bios were about men—presidents and the like—but not having other options, I took them as role models anyway. Still, stumbling onto the rare biography of a famous woman was a gift. Whether it was Florence Nightingale or Elizabeth Blackwell (the first woman to get a medical degree in the United States) or, my absolute favorite, Amelia Earhart, whose biography I re-read dozens of times—women who had done something extraordinary thrilled me. I wasn't sure what my role in life would be, but I wanted to do something remarkable, like them.

I disliked my first physics class in high school. I couldn't grasp the point of what we were doing. We messed around making waves in water tanks, and our teacher, though enthusiastic, didn't clarify matters. I remember him leaping on top of a lab table, miming a golf swing, and shouting something—but I have no idea what his point was. I reacted in what I later learned was a stereotypically female way, which was to dislike physics because it lacked all context.

Far more enjoyable was chemistry because of my teacher, Miss Helen Crawley. She was a no-nonsense, authoritative woman, and a role model for me. I headed off to college thinking I should become a chemist. But my father pointed me toward physics, adding that "astrophysics is very interesting." Much later, I realized how much this advice influenced me.

In intro physics, I sat near the front of the lecture hall so I could concentrate but also to avoid noticing how very few women were in the room. The first semester of mechanics was not hard, but electricity and magnetism, in the second semester, was entirely new to me, and I had zero intuition for it. For the first time in my life, a class was difficult. I bombed the first exam and, humiliated, thought momentarily that physics might not be for me. But then I told myself thousands of people learned physics every year, I was smart, and the subject simply could not be that difficult. So I started reading the textbook and worked at understanding what was going on. Soon I saw that physics was actually kind of neat. There were simple equations, the solutions of which corresponded to the behavior of matter, in clear and predictable ways. Once I understood this, the whole subject became easy and enjoyable.

The summer after my junior year I was hired as a summer student at the National Radio Astronomy Observatory in Charlottesville, Virginia. My main job was to identify optical counterparts of newly discovered radio sources, using Palomar Observatory Sky Survey photographs. I was thrilled to be the first person to see these optical quasars (short for "quasi-stellar radio sources" but actually distant galaxies with rapidly growing supermassive black holes at their centers). I found one radio source that, unusually, was associated with not a single quasar but a close *pair* of quasars; this system was later identified as the first known gravitational lens.

I went back to college fired up to do astronomy. I applied to half a dozen graduate programs, mostly in cities with decent opera companies (as good an organizing principle as any), but was accepted to only two. Johns Hopkins proved a good place for me, and I loved Baltimore and Washington.

My first year of graduate school was terrifying and stressful and fun and ultimately encouraging. I was one of only two women in my class; the other, a lovely Korean woman, Jung-hee Kim, unfortunately left at the end of the year. I soon learned that most of the male graduate students got together on Wednesday nights to do their quantum homework. I wanted to be in their study group but wasn't invited. So the next Wednesday, I asked some of them over for dinner. After we finished eating, they looked at each other awkwardly, unsure how to excuse themselves to go do the homework. I just played dumb until finally one of them asked whether I would like to study with them that night. "Oh, sure," I said, "sounds good."

Looking back, I see that my life in physics caused a steady downward slide in confidence, from being the top of the top in high school, to being an excellent student in college but underprepared in physics, to being among men who boasted endlessly about their achievements and abilities. I felt I had wandered into the wrong locker room. One of our professors used to start class by saying, "Good morning, Gentlemen and Meg" (ignoring Jung-hee). One day I returned to the large office shared by first-year grad students to find, on my desk, a copy of *Playgirl* magazine open to the naked-man centerfold. Everyone stared at me, waiting to see how I'd react. I was shocked and embarrassed, but I closed the magazine, raised it above my head, and asked, "Who does this belong to?" No one answered, and the moment passed but it felt like another indication I wasn't part of the team.

I did my PhD research on blazars (an unusual kind of quasar), working with my advisor, Richard Mushotzky, in the X-ray group at the NASA Goddard Space Flight Center. I showed that the radio-through-X-ray spectral energy distributions of blazars could be understood if they have relativistic jets pointing toward us. I also developed a novel framework incorporating relativistic beaming that proved radio galaxies were the misaligned parent population.

After graduating in 1984, I became a postdoc at the MIT Center for Space Research (CSR) with Claude Canizares. My office on the sixth floor had a nice view. A few days after my arrival, a well-known astrophysics professor walked by my open door, stared at me, stared at my name on the door, stared again at me, then marched in with his hand outstretched, saying his name and asking, "Who are you? Why have I never heard of you?" That seemed a little rude, but I just laughed and said, "I have no idea why you haven't heard of me." Then he challenged me to explain my work. Not the warmest welcome ever. I felt my worthiness was on trial, with him as judge.

Not surprisingly, this professor and I didn't have a lot of conversations about science. He was friendly enough but had an outsized personality and tended to be abrupt and judgmental. He was also very into art. One evening I had dinner at his house, where I met his then wife, a physicist and a lovely person. In a discussion that touched on feminism, he announced there had never been any good women artists. I mentioned a few names from my art history class, which he immediately dismissed.

Having never taken women's studies, I was underequipped for this debate. Years later, I stumbled across Linda Nochlin's famous essay "Why Have There Been No Great Women Artists?" in which she describes how most feminists, when they hear this question, "swallow the bait . . . that is, [they] dig up examples of worthy or insufficiently appreciated women artists throughout history." Exactly what I did. "But in actuality," Nochlin explains, "as we all know, things as they are and as they have been, in the arts as in a hundred other areas, are stultifying, oppressive, and discouraging to all those, women among them, who did not have the good fortune to be born white, preferably middle class and above all, male. The fault lies not in our stars, our hormones, our menstrual cycles, or our empty internal spaces, but in our institutions and our education."

I emailed a copy of the essay to the MIT professor art-lover. "Just as I said," he replied promptly, "there are no great women artists."

As in art, women in STEM had few role models and few or no champions. Expectations of our abilities were low ("Who are you? Why have I never heard of you?"), and as I was to see in the years that followed, colleagues reacted differently to women and men. One man told me,

"You know, when I am at a meeting and a woman is speaking, I can't even hear what she says for the first five minutes, I'm so busy checking her out." Then he asked, "Don't you have the same problem?" I thought about that. "Do you mean, checking out men? Because *all* the speakers are men. If that were a problem for me, I'd never hear a word." Still, I used to have a specific problem when listening to the rare woman speaker. I was tense the whole time, afraid for her failure and what it would mean for all women in the profession. I'd heard plenty of lousy talks by men, but people saw those as lapses or exceptions, not as representative of whether any other man could give a good talk. It felt, on the other hand, as if a bad talk by a woman would tarnish us all. When a woman gave an outstanding talk, I was relieved, grateful, and, if I'm honest, a little surprised. I had grown up absorbing from my surroundings the norm of expecting very little of women (codified by Virginia Valian in terms of "gender schemas").

A formative moment happened at our weekly CSR volleyball game. As we batted the ball around, we ended up talking about job searches. A new assistant professor, whom I liked and admired, told me that, as a woman, I would have no trouble getting a faculty job—because of affirmative action, I had a built-in advantage over men like him. This stung, since no one was running after me with job offers—in fact, quite the opposite. I challenged his assertion. That's when he told me this story, which I disguise here with pseudonyms. A faculty search at Prestigious State University was created for Joe Astronomer, a postdoc at said university. But the dean insisted on going around the faculty, elevating some woman who wasn't even on the short list, and ultimately insisting she get the job, while poor Joe had to find a faculty job elsewhere. All because of affirmative action, this underqualified woman got the coveted faculty job.

This felt like a knife to the heart. No one wants to believe they have any kind of unfair advantage in a competition. But something struck me as off about this story, so I pushed back. "Who was the woman?" I asked. He didn't know. Hmmm. "What was her field?" He didn't know. "Wait a minute," I said. "You don't know who the woman is, you don't know her work, but you're sure she was not qualified and Joe was far better?" He wasn't chastened at all; he simply said, "Everyone knows what happened."

About a year later, I ran into a senior faculty member from Prestigious State University, and I mentioned this story. "Oh, that wasn't what happened at all," he said, explaining that he was on the search committee, the job was never earmarked for Joe Astronomer, the woman (he named her) was definitely on the short list, and after she blew everyone away with her job talk, she was the clear choice. It's kind of a double whammy: even as women are discriminated *against* in job searches, the script says we have an advantage, so we get to feel even worse when results don't go our way.

Having failed at my first attempts to get a faculty job, I took a second postdoc, at the Space Telescope Science Institute, back in Baltimore, and married my boyfriend, who was a scientist at Goddard.

STScI was a great place to be a postdoc. Scientifically, I flourished, further developing my work on blazar unification. With Paolo Padovani, I wrote five key papers and a review, "The Unification of Radio Loud AGN," that is one of the most highly cited papers in astronomy. After a productive few years, I was definitely more salable. Three excellent universities offered me associate professor positions, with generous start-up packages and salaries. In contrast, STScI offered the equivalent of an assistant professor position, with no start-up package and a substantially lower salary.

I gathered my courage and made an appointment to negotiate with the STScI director, Riccardo Giacconi. He was a giant in X-ray astronomy, and I admired his genius. He was also straightforward though not known for his people skills. Our discussion did not go well. He declined to change the offer in any way and advised me to go elsewhere, adding, "Maybe you aren't as good as you think you are."

Many such cutting remarks are burned into my memory. An MIT professor saying, "We'd never hire someone like you" (for the MIT faculty job for which I'd applied). A Harvard colleague, after I described my latest research, asking, "What's interesting about that?" A JHU professor telling me, "Meg, every year there is a superstar faculty candidate, and you just aren't it." I never got used to such put-downs.

In the end, I took the STScI job. I wanted to have kids, and my husband loved his dream job at Goddard. I knew I could force us to move, but then I would have to carry the burden of his disappointment, as well as trying to start a research group, raise money, teach, and have babies. So I took the easy way out. It felt bad, choosing a place that so clearly didn't care if I stayed. A few months after starting the new job, I got a letter from the American Astronomical Society (AAS) announcing I'd won their Annie Jump Cannon Award for early career women in astronomy. After all the discouragement and disparagement from my employer, it was a much-needed boost.

Gender discrimination at STScI was a problem. Talented young women were treated badly, passed over for leadership roles, criticized, underestimated. Few women were on the faculty track. When I arrived as a postdoc, Neta Bahcall, head of the Guest Observer Program, was the only woman of 60 ladder faculty. (For a brand-new institution, at a time when women were getting roughly 15 percent of the PhDs in astronomy, this was not good.) Neta is the reason HST grants are generous enough to fund the actual work (an order of magnitude larger than contemporaneous NASA grants). Yet she had a reputation as a difficult boss. I saw lots of difficult male bosses at STScI, but they were called strong leaders. All the women scientists at STScI were underpaid. We knew this because we women all got "equity raises," which meant we were falling off the bottom of the pay scale for people of similar vintage. Male peers were not only paid more, they were raised up in other ways as well.

One day after I started my tenure-track position, the STScI science staff met to discuss new hires. We were shown a long list of applicants, with first names indicated only by initials. Every candidate being discussed seemed to be a man, so I finally asked, "Are any of these applicants women?" You would have thought I had committed a crime. Everyone started yelling at me. "We don't care about gender! We only care about excellence!" I was suddenly exhausted at the thought of having this argument, as the only woman in the room, so I just shut up.

Fifteen minutes later, back in my office, I got a phone call from Giacconi, who had been at the same meeting. "What just happened down there?" he asked. I explained it was hard to be the only one asking why we weren't hiring more women. That was when Riccardo began pushing for change. At the next presentation from the hiring committee, *he* asked about women on the list. Now people fell all over themselves to point with enthusiasm to female candidates. We hired three women and two men that year, and the women were all highly successful. Leadership from the top can definitely make change happen.

I kept speaking out about equity. Goetz Oertel, then President of AURA, suggested we hold a meeting, which became the historic 1992 "Women at Work: A Meeting on the Status of Women in Astronomy." Anne Kinney and I set up the organizing committee, which included Bob Brown, Laura Danly, Doug Duncan, and Ethan Schreier (then deputy director at STScI). At our first meeting, I thought we would talk about which speakers to invite, but Bob Brown wanted to talk about whether there *was* a gender problem in astronomy. To me the answer was blatantly obvious, with so few women astronomers on staff, but starting with data seemed like a good idea, so we invited the AIP and NSF to present the numbers. We added some outsiders to the organizing committee, including France Córdova, then at Penn State, who invited the historian Londa Schiebinger to talk about the overlooked contributions of women in science. Sheila Tobias gave a fiery feminist lecture that was received surprisingly well by male colleagues.[1] We held breakout sessions that ultimately led to the Baltimore Charter for Women in Astronomy,[2] a sort of Magna Carta for improving the situation in our profession.

Perhaps the most life-changing aspect of the conference was simply seeing 150 women astronomers in one room. The 1992 conference didn't solve problems overnight, but I think it started a wave of change in our field.

In early 2000, I ran into a colleague at an AAS meeting who said that Yale was hiring an astrophysicist. Would I be interested? The next week, the Department of Physics chair invited me to visit. I had a wonderful time, meeting nice people, getting a feel for the university, starting to imagine myself there. Within weeks I had a tenured offer.

Women in Astronomy Photo List

1. Thomas Hamilton
2. Anouk Shambrook
3. Judy Fleischman
4. Anne Gonnella
5. Susan Hoban
6. Andrew Wilson
7. Angela Olinto
8. Pete Reppert
9. Cindy Blaha
10. Janna J. Levin
11. Sylvanie Wallington
12. Olivia Lupie
13. Elizabeth Roettger
14. Jennifer Christensen
15. Susan Lamb
16. Mario Livio
17. Eric Wyckoff
18. Cathy Mansperger
19. Elizabeth Griffin
20. Sally Oey
21. Cherilynn Morrow
22. Jane Holmquist
23. Debbie Elmegreen
24. Lisa Sherbert
25. Douglas Duncan
26. Merri Sue Carter
27. Nancy Oliversen
28. Bruce Elmegreen
29. Sheila Tobias
30. Jaylee Mead
31. Reva Williams
32. Deoborah C. Fort
33. Andrea Schweitzer
34. France Cordova
35. Robin Lerner
36. Nancyjane Bailey
37. Ulysses J. Sofia
38. Jenny Wurster
39. Hashima Hasan
40. Sherri Godlin
41. Mike Meakes
42. Joy Nichols-Bohlin
43. Patty Trovinger
44. Andrea Tuffil
45. Sidney Wolff
46. Deepa Iyengar
47. Noreen Grice
48. Barbara Becker
49. Emily Xanthopoulos
50. Mercedes T. Richards
51. Stefanie Wachter
52. Liese van Zee
53. Suchitra Balachandran
54. Janet Levine
55. Riccardo Giacconi
56. Neta Bahcall
57. Lynne Billard
58. Margaret Burbidge
59. Pat Parker
60. Susan Stolovy
61. Judith Irwin
62. Eline Tolstoy
63. Diane Fowlkes
64. Svetlaina Hubrig
65. Charlene Anne Heisler
66. Peter Boyce
67. Ethan Schreier
68. Julie Lutz
69. Denise Kitson
70. Mimi Bredeson
71. Judith Perry
72. Goetz Oertel
73. Debra Schwartz
74. Isabel Hawkins
75. Bianca Mancinelli
76. Elise Albert
77. Sethanne Howard
78. Vera Izvekova
79. Alycia Weinberger
80. Karen Lezon
81. Kellie McNaron-Brown
82. Duccio Macchetto
83. Regina Schulte-Ladbeck
84. Debra Shepard
85. Roberta M. Humphreys
86. Londa Schiebinger
87. Vera Rubin
88. Lea A. Shanley
89. Luisa Rebull
90. Adrianne Slyz
91. Joanna Rankin
92. Svetlana Suleymanova
93. Prudence Foster
94. Melissa McGrath
95. Emily Sterner
96. Joanne Eisberg
97. Rachel Webster
98. Paul Vanden Bout
99. Elizabeth Bonar
100. Liana Johnson
101. Katherine Wright
102. Lance Wobus
103. Patrizia Caraveo
104. Eileen D. Friel
105. Fred Chromey
106. Stephen Levine
107. Martina B. Arndt
108. Samantha Osmer
109. Nancy Chanover
110. Emily Mason
111. Daryl Weinstein
112. Nathalie Martinbeau
113. Sue Madden
114. Michael Eracleous
115. Laura Kay
116. Adair Lane
117. Jennifer Johnson
118. Linda French
119. Lori K. Herold
120. Robert Hanisch
121. Mira Franke
122. Claudia Robinson
123. Jacqueline Fischer
124. Laura Ruocco
125. Lisa Buckley
126. Nancy Houk
127. Tania Smirnova
128. Anne P. Cowley
129. Cindy Taylor
130. Lisa Wells
131. James Lowenthal
132. Susan Neff
133. James Wright
134. Anne Kinney
135. Meg Urry
136. Laura Danly
137. Ira Kostiuk
138. Marla Moore
139. Grace Chen
140. Jim Etchison
141. Dave Soderblom
142. Robin Shelton
143. Omar Lopez-Cruz
144. Inger Jorgensen
145. Linda Grant
146. Todd Hurt
147. Claude Canizares
148. Daniel Golombek
149. Kp Kuntz
150. Rodger Doxsey
151. Krista Lawrance
152. Anuradha Koratkar
153. Vicki Balzano
154. Dorothy Fraquelli
155. Lauretta Nagel
156. Debora Katz-Stone
157. Nino Panagia
158. Helen Hart
159. Shireen Gonzaga
160. Fabienne Van De Rydt
161. Geoffrey Clayton
162. Alex Storrs
163. Debbie Kooleck
164. Carmelle Robert
165. Martha Anderson
166. Barbara Whitney
167. Karen Meech
168. James Hesser
169. Frances Verter
170. Diane Gilmore
171. Maitrayee Sahi-Sharma
172. Penny Sackett
173. Heidi Hammel
174. Windsor Morgan
175. Diane Alexander
176. Ellie Lang
177. Helmut Jenkner
178. Kathryn Mead
179. Derek Busazi
180. Doris Daou
181. Anne Gilden
182. Karen S. Bjorkman
183. Peter Stockman
184. Linda Skidmore

FIGURE 24.2. Attendees of the 1992 Women in Astronomy conference. Seeing so many women astronomers together was transformative and helped catalyze the movement toward equity in astronomy. Names of those pictured are available at "workshop group photo list" at https://www.stsci.edu/contents/events/stsci/1992/september/women-at-work-a-meeting-on-the-status-of-women-in-astronomy. *Credit*: STScI and NASA.

(Later, I learned I was the first woman ever appointed to a tenured position in the physics department.) I did need a long time to consider the offer, as my family was not enthusiastic about moving. But then I realized: I have two daughters, and if I turn down this job, they might learn to turn down choices they really want. To New Haven we went.

Being at Yale has been wonderful. I love teaching. I love helping undergraduates understand introductory physics, appreciate the wonders of the Universe, and get comfortable in the laboratory. I love working with graduate students, who are endlessly talented. Thanks to them, my group has done groundbreaking work on the co-evolution of supermassive black holes and galaxies over the past twelve billion years. I also love being part of a community of scholars across all fields.

Sometimes I think about how differently my career might have unfolded had I been a man—or rather, had my profession been equally open to women. Certainly, the path would have been much easier; I might have had champions, smoother collaborations, more recognition of my science. But I also got a lot of satisfaction advocating for equity and widening the path for those coming along behind.

We don't get a do-over in our lives. I wish I had known more when I was younger. I wish I had been more enlightened (every few years I cringe at my obtuseness of just a few years earlier). But I also feel incredibly lucky to have landed where I am and to be able to contribute to the greater good. I am delighted to see multitudes of brilliant young astronomers entering the field, including nearly as many women as men, all generating important new knowledge. No doubt we still have a long way to go, especially along other dimensions of exclusion, but change continues to make our profession healthier than ever.

Notes

1. Proceedings of the 1992 Women in Astronomy Conference are available at https://www.stsci.edu/contents/events/stsci/1992/september/women-at-work-a-meeting-on-the-status-of-women-in-astronomy?fromDate=01/01/1992&timeframe=past&toDate=01/01/1993.

2. See the Baltimore Charter at https://www.stsci.edu/stsci/meetings/WiA/BaltoCharter.html.

Chapter 25

Cathie Clarke (PhD, 1987)

An Astronomer (Not a Pirate!) of Penzance

Cathie Clarke, now Professor of Theoretical Astrophysics at the Institute of Astronomy at Cambridge, studies astrophysical fluid dynamics, including accretion, protoplanetary discs, and star formation, and developed the widely used theory for the photoevaporation of disks around forming, low-mass stars. She was the first woman appointed to the staff of the Institute of Astronomy, Cambridge. She was awarded the Eddington Medal by the RAS in 2017, becoming the first woman to receive this award since it was instituted in 1953, and is the co-author of the textbook *Principles of Astrophysical Fluid Dynamics.*

I was born in Penzance, the westernmost town in England, only ten miles from where the English county of Cornwall terminates in the Atlantic Ocean at Land's End. This experience of growing up within sight and sound of the sea and my sense of living in a remote locality was very important to me. As I grew up, I spent much time exploring the wild landscape of the West Penwith peninsula on foot and bicycle and developed a passion for the natural world in all its aspects—astronomy was part of this, though, at this stage, probably less of an enthusiasm for me than the study of the rocks and butterflies I encountered on my rambles. We lived a rather isolated family life, and my sister and I entertained ourselves by creating imaginary worlds and writing stories. My parents

FIGURE 25.1. Cathie Clarke in 2016.

encouraged us to be interested in everything ("Only boring people are bored" was my mother's motto): I inherited my interest in science and natural history from my fiercely rationalist mother and my love of music (especially Bach) and learning foreign languages from my gentle, cultured father. (My parents' personalities and interests inverted traditional gender roles, and, perhaps in consequence, I am slightly discomforted by people who do conform to such stereotypes.) Another deep-seated interest of mine was mathematics. From the point at which I learned to count, I visualized numbers on what I now know to be a logarithmic spiral; later, I would always be pondering the formulae that were presented in school mathematics, trying to prove them from first principles.

School was a great disappointment. I found most of the teachers preferred to teach to the median ability in the class and deplored outliers in either direction; I was very bored in class and hounded by my classmates for being "different." Perhaps, with hindsight, this was a useful preparation for being an outlier, that is, a rare woman in the astronomy world, since I have never expected to "fit in" with a dominant group.

Aged eight, I decided I wanted to go to Cambridge, because my mother (one of the first generation of women to receive a full degree from the University of Cambridge in the 1950s) told me stories of the ancient Colleges and how choirs would sing madrigals on lighted punts floating down the River Cam. However, my school was not geared up to send pupils to Cambridge and gave me no help in preparing for the very tough entrance examinations. I am, however, very grateful to my headmistress, who wrote strong recommendation letters for me and to the admissions team at Cambridge's Clare College, which, unusually for the 1970s, decided to take the gamble of admitting me, despite my poor preparation.

I was overjoyed but found the first year of Cambridge a great shock, puncturing the high levels of self-assurance I had developed as a "big fish in a small pond." I momentarily wondered whether I should lower my sights but, having a determined streak, decided to try to regain my old mastery of mathematics and physics. I did not realize that I was succeeding in this until I gained a First at the end of my second year, a result that was all the sweeter for being totally unexpected.

I didn't have clear ambitions at the end of my degree. The things I was good at were naturally preparing me for careers in technology, but I had a deep-seated dislike of technology, which I saw as being in opposition to the natural world. On the other hand, I had long ago decided that the subjects that take the natural world as their starting point are mainly descriptive, involving large quantities of information to be classified, and I found this to be much less intellectually appealing than the "first principles" nature of physics. With hindsight I think that the appeal of astronomy was that it catered to both these aspects, relating firmly to the natural world and yet comprehensible through the simplicity and elegance of physical reasoning. So I set out to do a DPhil in Oxford in theoretical astrophysics.

Oxford was not a success academically (the department was small at that time and my supervisor became ill during my first term and was unable to supervise me so that I relied on a kindly offer of remote supervision from Jim Pringle in Cambridge). I found Jim to be rather fierce, but I admired the insight that allowed him to look at problems in simple ways—I have always felt that things should become simple when you really understand them, and, although I now realize that this is not necessarily true, it's been a guiding instinct to me to try to achieve that simplicity. Nevertheless, I enjoyed Oxford immensely in non-astronomical ways, and it was a welcome interlude in which I could grow up, between the pressures of undergraduate life in Cambridge and postdoctoral life in the United States.

I was studying accretion discs, the spiraling structures of gas that feed material onto massive objects. As often happens in astronomy, a particular type of physical structure can be replicated on vastly different spatial scales. Accretion discs, for example, form when gas in galaxies drains onto a central black hole whose mass is millions of times that of the Sun. But accretion discs also form around tiny white dwarfs within binary star systems, where a companion star feeds them with gas torn off by the strong tides induced by the white dwarf. In both cases, the accretion disc is heated by the energy released by the inspiralling material, and consequently these discs shine at a variety of wavelengths. In my thesis I used computer simulations to predict how the light from these discs should vary with time in order to compare with observations.

By the end of my DPhil, I knew plenty about the contents of my thesis but little about the wider astronomical world. It was only when I went to the University of California, Santa Cruz (attracted by its redwood forests and Pacific coastline) that I became caught up in the currents of the global astronomical community and gained a broader perspective. I extended my accretion disc studies in new directions, considering discs on the scales of entire galaxies as well as protoplanetary discs around young stars (these being the sites of planet formation). I was much inspired by my mentor Doug Lin, who was a veritable fountain of possible ideas for me to work on. Always genial, Doug was never in the least offended if I seemed sceptical of a particular line of enquiry—"Okay, I give you another . . . ,"—confident that we would alight on some topic which we could productively advance. It's a great temptation for theorists to become overattached to their own theories, which can shut down debate among their junior colleagues, and so I found Doug's disinterested enthusiasm very refreshing.

After two happy and productive years in Santa Cruz I was nevertheless glad to return to England, taking up a postdoctoral position at the Institute of Astronomy, Cambridge, one of the largest astronomical institutes in Europe, and one whose faculty was responsible for some of the most influential work in theoretical astrophysics in the latter half of the twentieth century. The IoA was an excellent place in which to diversify my investigations (theoretical work is often opportunistic rather than programmatic, and so theorists thrive in interactive environments where there is a wide range of topics being studied). I initially examined how black holes "feed" in the centres of galaxies, investigating if stars can be trapped in the surrounding gas disc and ultimately swallowed by the hole. I also started working on star formation, having been immersed in its controversies during my visits to the Center for Star Formation at NASA Ames in California. This topic was being transformed by new observations at infrared and radio wavelengths, which could penetrate the obscuring dust of star-forming clouds; one of the surprising new discoveries was that even though stars like the Sun live today in relatively uncrowded environments, most stars in fact start their lives in clusters. I went on to conduct many investigations of how these clusters dissolve and how this process of dissolution affects the production of binary and triple stars.

During these years I acquired a large group, attracted by the prospect of making their mark in a relatively young field and in pioneering hydrodynamical simulations of star formation. I also started supervising undergraduate students, that is, conducting the small-group intensive teaching in which most of the "real learning" in Cambridge actually happens. To teach fundamental physics to Cambridge students demands a lot of the supervisor: thought processes to keep one a step ahead of the brightest students, an alertness to those who are not really understanding but are hoping that this is going unnoticed, and the honesty, when required, to say "I don't know . . . but I will think about it by next supervision." I felt that the Cambridge supervision system had provided me with a superb training when I'd been an undergraduate and I wanted to give back some of these benefits to succeeding generations.[1]

In 1995 I became a lecturer in Queen Mary and Westfield College, London, but within a year I was appointed to a lectureship back in the IoA. These were eventful years on many fronts. I met my future husband, Christopher Knight, a theologian with a PhD in astrophysics. Chris had three daughters by a previous marriage, and I loved the holidays we spent together on the Norfolk coast. I had been equivocal about having children up until then; however, the pleasure of their company convinced me that parenthood would not be a condemnation to dull adult respectability, but in fact could open up some of the magic of my own remembered childhood. My son, Rupert, was born in 1997, and his companionship has been a huge joy to me at every stage (yes, including the teenage years!). When Rupert was young, life became very hectic, even though Chris, who was taking an extended sabbatical at the time and working mainly from home, could take on primary childcare. I benefitted greatly from this relatively unusual arrangement, and I am very grateful to Chris for this. Nevertheless, on odd occasions when Chris undertook paid work outside the home, the practical arrangements required to ensure continuity of childcare could be complex and bizarre (I remember meeting Chris on remote railway stations and passing the baby between us as we went our separate ways).

Early in the 2000s I produced my most highly cited paper in which Bo Reipurth and I proposed a new mechanism for forming brown dwarfs, objects too low in mass to ignite hydrogen but which faintly glow with the energy released from their gravitational contraction. We proposed that these are, more often than not, a by-product of the formation of an unstable multiple star system—in our theory, more massive stars tend to remain bound to each other while brown dwarfs are ejected from the system by slingshot gravitational interactions. One of the theory's predictions is that one does not expect to find significant numbers of very small gas clouds, which would be required if brown dwarfs instead formed in isolation; to date this prediction appears to be borne out, although even now the search for isolated brown dwarf precursors continues.

I also started an important new direction, studying the dispersal of protoplanetary discs by *photoevaporation*. Photoevaporation results when the surface of a disc around a young star is heated by high-energy (ultraviolet or X-ray) radiation so that gas can escape the gravitational attraction of the central star and form a *wind*. I realized that the importance of photoevaporation as a disc-dispersal mechanism—already recognized in massive stars, following the seminal work of Dave Hollenbach and collaborators—had been significantly underestimated in the case of low-mass stars such as the Sun. The crucial insight here was that even the relatively weak wind in the case of a low-mass star would eventually come to dominate the accretion flow in the disc. I showed that photoevaporation would then drive a rapid phase of inside-out disc clearing. Although I have continued to explore a variety of other topics (such as the formation of wide binary stars and spiral structure in galaxies), my research since 2001 has been dominated by developing the theory of photoevaporation, work that benefitted immensely from the inputs of many excellent students and postdocs. We have recently shown that *external* photoevaporation (driven by ultraviolet radiation from neighbouring massive stars) is the dominant *environmental* effect controlling the lifetimes of protoplanetary discs, thus indirectly affecting their capacity to form planets.

In 2014 I was awarded a five-year European Research Council Advanced Grant, timed to coincide with the period during which ALMA (the Atacama Large Millimetre Array) was first taking spectacular high-resolution images of protoplanetary discs. I found it very rewarding to assemble a large group of scientists with the diverse skill set necessary for the modelling and interpretation of ALMA data, while also continuing to work on theoretical models for disc evolution. An observational highlight was our obtaining high-resolution ALMA data on the first disc-bearing star hosting a hot Jupiter (i.e., a massive planet in close orbit around the parent star). We wanted to know if the empirical association between hot Jupiters and outer planets, observed in mature exoplanetary systems, extended to this very young (million-year-old) star: if it did, we hoped to see evidence of disturbances in the disc's dust emission. The first glimpse of the ALMA image—showing a series of prominent rings and gaps in the dust emission, as would be expected in the presence of multiple planets—was one of the most exciting

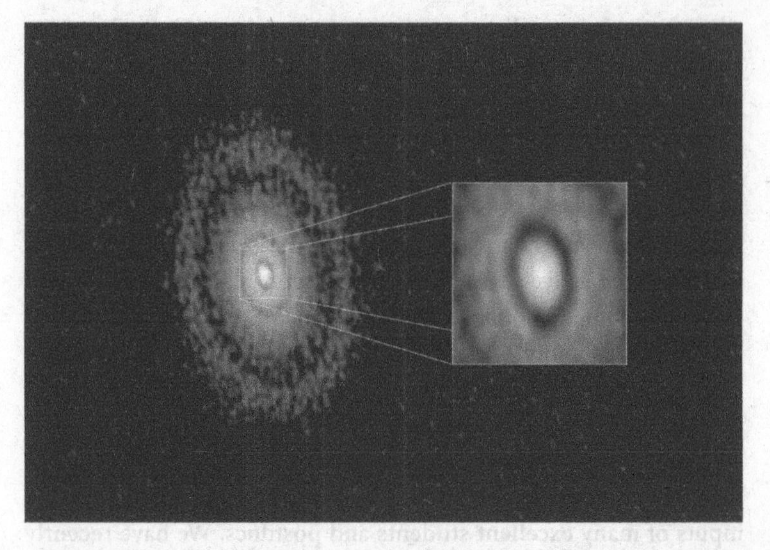

FIGURE 25.2. Image of the dust rings surrounding the young star CI Tau acquired using the Atacama Large Millimetre Array (Clarke et al., *Astrophysical Journal Letters* 866 [2018]: L6). Each of the dark rings is believed to be cleared of dust by the presence of unseen planets of around the mass of Saturn. The inset shows a zoom-in on the innermost ring located at around ten times the Earth-Sun distance.

moments of my career: several of us literally jumped in the air when we saw it!

However, for me the most satisfying achievement has been a theoretical advance when, in 2016, I developed a mathematical model describing photoevaporative disc winds. Astronomical theory has become sufficiently complicated that most problems have to be solved on a computer, raising questions about the accuracy and generality of the solutions. It is relatively hard these days to find problems that can be cast in terms of pen and paper theory. I was therefore pleased to be able to derive the equations governing a disc wind, analogous to the classic "Parker" wind for spherical outflow derived in 1958, but now accounting self-consistently for the more complicated streamline topology and flow structure in disc geometry. In 2017 I was awarded the Eddington Medal of the Royal Astronomical Society for my works on photoevaporation.

When I reflect on being female in an overwhelmingly male-dominated world (I was the first and, for many years, only female academic staff member at the IoA), some things stand out. I want women to have certain freedoms—to be treated with complete respect and seriousness in their pursuit of science but also to have the freedom, as men have, to pursue other paths. I always think it's sad when female students feel they have to justify giving up physics or when female graduate students are deeply apologetic about leaving the field. To me these decisions are not necessarily bad—I just want to try to ensure that the decisions are right for the individuals and not based on macho hype about "what it takes to be a scientist." I also want women to be able to be unapologetic about their family lives—I love it when, these days, my young male colleagues leave meetings to pick up their children from school without embarrassment. I always felt the need to downplay this aspect of my life in order to portray my "seriousness." The fact that family life is no longer seen as an exclusively female issue is surely a great advance. I also feel that a sort of smirking, casual sexism in academics' conversations has become rarer over the last decade or so, and this can only be a good thing in promoting an environment in which women feel comfortable. (When I started as a graduate student in an all-male department, the sport was to see how I would react to the soft-porn posters placed above my desk—hopefully unimaginable in the current climate.) What is much harder to gauge, however, are the behaviours behind closed doors. I have been fortunate that I have never experienced any serious harassment, but such cases still clearly exist and at a level that is hard to quantify.

When I was at school, some of my male teachers refused to believe that girls could do maths and denied us assistance in class, openly sharing with the boys their hostility to girls in their classes! Such blatant sexism—laughable to me at the time because I could so clearly refute it—is certainly not something that I have experienced subsequently, and as an undergraduate I felt my questions were treated with equal seriousness to those from male students. Over the longer haul of a career in science, it is, however, harder to gauge the subtle assumptions—gender-based or otherwise—that determine how one is perceived in the field. Perhaps trying to figure this out is an unhealthy preoccupation.

Overall, I believe that my life as a scientist would have been slightly more comfortable had I belonged to the dominant gender group, but this has not greatly detracted from my enjoyment of the enterprise. Being a woman definitely gives me the opportunity to do something useful in the field—just by standing up in lecture rooms, by being part of the academic fabric, I am making a tacit point (though, since I never consciously modelled myself on anyone else, I don't much like the notion of being "a role model"). If I can help convey to people the excitement of the field and that astronomers are a diverse collection of human specimens, I will be well satisfied.

Notes

1. https://www.mythic-beasts.com/blog/2015/10/13/professor-cathie-clarke-ada-lovelave-day/.

Chapter 26

Saeko S. Hayashi (PhD, 1987)

From Six Meters to Thirty Meters, Ever Expanding Horizons

Saeko Hayashi, one of the few women astronomers in Japan at the time she completed her PhD, studies protoplanetary disks and the formation of planetary systems. She is Associate Professor at the National Astronomical Observatory of Japan and also the Graduate University for Advanced Studies and now works on the Thirty Meter Telescope project. As one of the founding members of the Subaru Telescope team, she worked to commission, tune, and maintain the telescope's complex optics mechanism. She was Vice President of the Astronomical Society of Japan (2017–19), was a Steering Committee member (2015–21) of Division C (Education, Outreach and Heritage) of the International Astronomical Union, and is passionate about sharing the excitement of astronomy with the public. Her asteroid is 6250 Saekohayashi.

After several months of reading the book *Star Wars* over and over again, I was standing in a long line in front of a movie theater anxiously waiting for the doors to open. That theater in the middle of Tokyo had a big screen, bigger than that of any other theater in the region. It stretched from the floor to the ceiling and curved in such a way that if one sits in the center of the front row it feels like truly being inside the movie. That

FIGURE 26.1. Portrait of Saeko Hayashi in 2017 at the National Astronomical Observatory of Japan's headquarters, in front of a model of the mirror she is currently working on. The hexagonal shape, which is 1.44 meters across, is a segment of the primary mirror of the TMT. Each one of these is close to her height. Once together, 492 of these mirrors will make an enormous telescope for all humanity to use to explore other worlds beyond our Solar System.

is where I was, all excited, and all the more during the show and afterward. The most memorable scene to me was the one that showed two suns setting over a desert planet. Professors at the university would scoff at that, or at least frown at the concept of planets outside of our solar system, mostly because it was impossible in 1977 to detect them.

The year was 1981 and I was inside a dome next to a telescope, sitting on a heated pad on a chair, manually guiding the telescope with a joystick. After an hour, I pointed the telescope to another target. At the end of the nightshift, I turned off the power to the photomultiplier and the telescope drive and closed the dome shutter. The weather was not great. Yet, I experienced the sheer joy of being able to learn something new about the Universe. The telescope was a 36-inch reflector, built in 1960 by Nippon Kogaku Kogyo Kabushikigaisha (now Nikon). The photomultiplier was most likely made by Hamamatsu Terebi Kabushikigaisha (now Hamamatsu Photonics).

By a combination of luck and determination, I was allowed to begin my studies at the university level. In 1977, I competed with other students, many from the elite schools in big cities, by taking a standard entrance examination. At that time, the exam was given at the campus in Tokyo. There I was, a girl from a very rural area in northern Japan. The exam rooms for the STEM-oriented students were filled with boys only. All the bathrooms had "boys' bathroom" signs. One of them had an additional paper sign affixed below the permanent sign: "girls can use during the entrance exam." So what did I do? I had no choice, and so I entered that door. When I saw the bathroom scene in the movie *Hidden Figures*, I recalled that experience.

Japan has a Ministry-funded, that is taxpayer-funded, national university system. The tuition was affordable, equivalent to about $300 per year at that time. Textbooks were not expensive either. Only a third of the high school graduates from my generation went to four-year colleges, and the percentage was much lower for girls. In 1946, after World War II, the University of Tokyo allowed girls to take the entrance examination for the first time. Because of the notion that that university was for the most elite, that change was almost a scandal. At the time I passed the exam, about 10 percent of the students were girls. The ratio was only 1 percent for STEM majors, though. Today, girls still make up only about 20 percent of the STEM majors.

So there I was, in winter wearing a heavy coat inside the classrooms, like the rest of the students, since the heating system did not reach beyond a few feet. We also had no air conditioning during the summers in most of the classrooms. The library was our sanctuary when the temperature was high. We did not feel that was poor treatment. Everybody was delighted to be at the prestigious university and worked hard. It was a different era then, as Japan was still developing and being industrialized.

After two years of general education, I went on to major in astronomy. This was a bit of good fortune for me, as only seven students per year were allowed this opportunity. That subject was (and is) popular among nerdy students who are fascinated by nature. I think I was the fourth or fifth girl ever to enroll in the astronomy program at the University of Tokyo.

I had many opportunities to get hands-on experiences making optical observations, though the climate in Japan is not suitable for observing faint objects, and Japan had no large-aperture telescopes at that time. Was that a hindrance for students? No, not at all. Any chance to get data was extremely valuable, because we had chances to collect and learn how to properly calibrate our data.

My advisor in my first-year graduate course strongly advised me to do data reduction using the chart recorder outputs for variable stars that filled up big shelves. The data were recorded on paper that was already yellowed.

Instead of following his suggestion to go into an already established research area, I was inclined to pursue a brand-new topic, star and planet formation, which most professors in the Astronomy Department regarded as a crazy and wild idea. They did not think the study of star and planet formation was real science.

One of the institutions attached to the university was the Tokyo Astronomical Observatory (TAO). The radio group at TAO was engaged in the construction of a large facility near Nobeyama, in a radio-quiet area in the middle of mountains in Japan. At TAO headquarters in Mitaka City, in the outskirts of Tokyo, was a six-meter radio telescope. A band of young researchers and students, left alone by the more senior staff who were on-site to oversee the construction project, had the freedom to use it as a training ground. Fortunately for me, a new receiver system was being developed and commissioned there. Test observations went like this: with a printout from the very "experienced" computer, the drill sergeant, namely a young researcher from a nearby university, called out for the students at the frequency synthesizer to turn the knobs to the proper frequency. When done properly, the spectrum analyzer showed faint, flickering signals.

By using data from this radio telescope, I was able to write my master's thesis. The spectra from the test observations of the new receiver system were drawn on tracing paper with Rotring pens; otherwise, the computer output would not have been legible.

The commissioning of the new radio telescopes at Nobeyama was picking up rapidly. The engineering and scientific tests summoned researchers and students from outside of the TAO because that observatory was heavily dominated by optical astronomers. People with training in physics, chemistry, information science, communication technology, and electrical engineering joined the project. It was unheard of in academia in Japan that experts from those different disciplines would work together. Radio astronomy elsewhere had a similar history—it started with engineering experiments, and the sensitive measurements of the signals from the sky evolved into astronomical research.

It was quite often the case in the early operations phase that graduate students were at the helm, commanding and steering the telescope. When the time arrived for proposal writing, senior researchers planned low-risk studies of H I (atomic hydrogen) or CO. Meanwhile, students, who relied on the spec sheet of the 45-meter telescope rather than on their advisors' recommendations, proposed more challenging observations of CS, HCO+, and HCN molecules, which the senior scientists did not consider feasible.

Eager students ran ahead of the system as it was getting readied. As soon as one set of observations finished, the signal was recorded on a large reel-to-reel tape. One student took the tape and ran to the main building where the computer system for data reduction was located. Early mornings (or deep into the night), one would see young researchers working on the computer output using rulers and pocket calculators to convert the "distance between the axis and the plot" to a physically meaningful number. Meal and sleep schedules followed the times of day when observational targets were high above the horizon. In winter, the Taurus and Orion team got the evening shift, and when the galactic plane set, the external galaxy group showed up in the telescope control room.

Winter is very cold at Nobeyama. The temperature can go down to well below −20° C (below −4° F). Yet the Observatory was filled with enthusiasm all the time.

Nobeyama is, by its nature, far from cities and convenient modes of transportation. Yet, astronomers arrived there from around the world. Students lived in the Observatory dorm alongside those visiting astronomers. During the observations, over meals, and in the computer terminal room (but not at each other's desks!), we found ways to interact. There was not enough staff to cover the round-the-clock operations of the telescope, so graduate students were heavily involved in the operations by being at the telescope control console. Those visiting astronomers, often famous, were patient enough to try to understand my explanations, made with my limited English-language skills, when they needed help.

These initial observations were taking place while the tuning of the 45-meter telescope was still in progress. One of the major items I worked on was improving the beam pattern, which required adjusting each of the 600 panels that make up the primary reflector. The panel adjustment system had sufficiently high resolution. On the other hand, measuring the surface accuracy over such a large area was a huge technical challenge. Dr. Norio Kaifu took up that challenge and led almost all the campaigns to improve the surface accuracy of the main reflector of the 45-meter.

Radio holography, a special technique for measuring the surface shape of the dish antenna so that astronomers can remove irregularities and adjust the dish shape to be closer to the ideal one, became popular at radio observatories at that time, replacing the mechanical scale, or theodolites, such as the one manufactured by Sokkia. For the 45-meter telescope, we needed to install a small reference dish on top of its subreflector during this radio holography campaign. First, a radio source was placed on top of a nearby mountain, but the distance to this source was not large enough. After consulting with industry and government institutions, we received permission to use the signal from a communications satellite that was almost but not quite geostationary. Its trajectory was actually a figure eight. Since that satellite was not a classified one, we were able to obtain the positional information for it with sufficient accuracy to complete our work.

In the beginning, I did not like climbing to great heights. Once above the main reflector, however, the ground is not even visible, so it became manageable for me to climb up the stay to reach the sub-reflector and install the small antenna plus other attachments for the holography measurements and to make frequent adjustments during the holography campaign. The real danger occurred when I reached the driving mechanism of the panel adjustment system, located under the main reflector panels. In order to minimize the weight of the overall structure, the thin shade panels below the reflector panels were not designed to take the load of humans. Yet they gave me a false sense of security when I walked around on their narrow frames.

Professor Haruo Tanaka was the marshal of this holography campaign. In 1951, he had measured the temperature of the sky to be 0 to 5 K, in a single direction which was straight up. This value is very close to the one measured by Penzias and Wilson a decade later, and this measured signal turned out to come from the cosmic microwave background. His experimental results were not well known at the time because he published his results in a Japanese journal. His persistent approach in making precise measurements was a great lesson for me. At the end of the campaign, the 45-meter telescope became truly efficient at making millimeter-wavelength observations.

My observational results regarding the structures of molecular clouds in the vicinity of forming stars and my experiences with the engineering aspects at Nobeyama were the two major stepping-stones for my doctorate, as I became the first female to receive a PhD in astronomy from the University of Tokyo. The University itself had only 110 years of history at that time, but the Astronomy Department has its origin in the previous feudal systems and the imperial systems that date all the way back to the seventh century.

Was there a job for me inside Japan after getting a degree? No. Positions for astronomers in academia were almost nonexistent outside of TAO, and TAO had only one opening per several years.

Luckily a new facility was coming on line in Hawai'i, and I got a job there. I thought it would be great to be able to work with a telescope designed to work in the relatively new wavelength range of the submillimeter. I landed on the Big Island of Hawai'i one day before the dedication ceremony of the 15-meter James Clerk Maxwell Telescope (JCMT), which is situated near the top of Maunakea, at 4,200 meters (or 14,000 feet) above sea level. I quickly discovered that the telescope was still being commissioned, and there was a lot left to do. In this exciting phase of my career, I joined the radio holography team led by Dr. Richard Hills from Cambridge, UK. He did not sleep much, and, according to some of my colleagues, Dr. Hills was also like that years later when he was the project scientist of ALMA in Chile.

During one rigorous campaign, I was working with Dr. Hills and an electronics engineer from the Netherlands. JCMT was run by the UK-Canada-NL Joint Astronomy Centre, so many staff are from those countries. Daytime work was to adjust panels, with help from other engineers and the day crew. Sometimes visual confirmation of the movement of a panel was necessary. To witness the motion, I placed myself up at a certain place I should not identify! When the daytime portion of the work ended, we would come down to the mid-level facility, at only 3,000 meters (10,000 feet) above sea level, to eat supper in a hurry and grab some sandwiches for our night lunch. We then would drive back up to the summit again in order to make measurements during the night. One time, we forgot to lock one of the wheels to properly engage 4WD before the steep ascent, and so the vehicle we were in wiggled a lot on the narrow, windy road that had no guardrail. We quickly remedied the situation; otherwise I would not be writing this.

After what we thought was the final adjustment of the panels, we were eager to get confirmation that our work was completed correctly, and so we started our measurements early. We ended up getting a weird pattern on one of the sections, located at the top right corner near the west wall of the enclosure. The heat from the wall was distorting the panels, and our highly precise measurements showed that effect, and so we made one more adjustment. It was truly rewarding to see the improvements we made in that campaign, which enabled us to configure the JCMT to explore the heavens in the submillimeter wavelength windows.

Toward the end of the three years of my initial contract to work at JCMT, multiple paths opened up for me—stay with JCMT, move to a neighbor telescope, return to Nobeyama, or apply for a position just advertised to work on a new telescope in a different wavelength regime. During these three years things had changed significantly in Japan. In 1988 TAO became a national observatory and started to build the Japan National Large Telescope (JNLT), an 8.3-meter telescope for optical and infrared observations. At that time, the largest optical telescope in Japan was only 1.88 meters in diameter. From 1.88 to 8.3 meters, the telescope mirror is not just 4.4 times larger, as the area of the mirror is 19 times larger.

My decision was to boldly move to where the discoveries might transform our view of the Universe, and so I took a job helping design and build the JNLT. At telescope builders or related topics conferences, I overheard not once but many times experts from abroad express doubts that we would succeed, saying, "Japanese astronomers won't be able to build such a big and sophisticated telescope and the instruments for it."

My primary responsibility was to oversee the fabrication of the telescope optics. I also conducted experiments to demonstrate the effectiveness of using CO_2 to clean the mirrors and to prototype the mirror-coating chamber. During the time when I was flying back and forth between Japan and the United States, where the fabrication of the mirrors was taking place, the intellectual shock wave of the first detection of an exoplanet hit in 1995. As a result, I wanted to make sure that the telescope I was helping to build would be capable of directly imaging such exoplanets in the infrared.

JNLT, later blessed by Japanese citizens with the nickname Subaru, has a coating chamber that uses a conventional method—heat up filaments and vaporize the metal pre-annealed in the filaments. The material is either aluminum (for the primary, two secondary, and one tertiary mirror) or silver (for one secondary and one tertiary mirror). There is always competition between the observers who work at either optical or infrared wavelengths, since they can observe only during the night, unlike the radio astronomers who can observe during the daytime. Ground-based telescopes are affected by the lunar phase. Half the month, centered on New Moon, is dark, and the other half, centered on Full Moon, is bright because scattered light from the Moon affects the background level of the night sky significantly. As a result, the telescope is mostly used during the dark period by optical observers and during the bright period by infrared observers. Subaru Telescope's secondary mirror is switched from the aluminum-coated one for the optical observations to the silver-coated one for infrared.

Another door opened for me in 2017. I packed up my residence of almost twenty years in Hawai'i to join the Thirty Meter Telescope (TMT). Japan is one of five countries partnering to build this telescope. Its Japanese project name translates to "Next Generation Extremely Large Optical-Infrared Telescope."

Again, my primary responsibility is with the big mirror. TMT's primary mirror consists of 492 hexagonally shaped mirrors called segments. Each segment is 1.44 meters, from corner to corner. The segments will be constantly exchanged with spares for recoating their surfaces, and because of the variety of the mirror shapes there will be 82 spares. This means almost 600 circular blanks of super low thermal expansion material (Ohara's Clearceram-z) had to be ground by Okamoto Optics to the right shapes. Polishing requires a very long time, and four TMT partners (the United States, Japan, India, and China) share that responsibility.

As the production of the mirror blanks was making good progress in Japan, I moved to California in 2019 to work directly with my counterparts there. I have been very fortunate to be able to work with excellent colleagues in California. I was pleasantly surprised to find that highly trained engineers here welcomed me onto their team. That acceptance is not something I ever experienced in my school days, nor while working in this field in Japan. Here, my colleagues are working together to create something way beyond what a single person could make, something that can bring knowledge and perhaps wisdom for survival and sustainability to all.

FIGURE 26.2. Telescopes Saeko Hayashi has worked on. Left: telescopes for millimeter and submillimeter wavelengths observation. Top is the 45-meter radio telescope in Nobeyama, Japan; bottom is the 15-meter JCMT on Maunakea, Hawai'i. Right: mirrors (to be) for optical-infrared observations. Top is the 8.3-meter primary mirror of the Subaru Telescope. Bottom is one of the 1.5-meter meniscus blanks for the primary mirror of the TMT. Hayashi inspected this piece in a factory in Japan before shipment to a processing factory in the United States. *Credit*: Canon.

At Friday at 7 A.M. local time, an online meeting starts between U.S.-based and India-based scientists. It is late in the evening in India, so even the vendors there connect from off-site. The technical discussion takes place. There is a mutual understanding that we are building something incredible even if we have to manage moving materials across country borders and physical boundaries or deal with budgetary restrictions. Throughout my years in this profession, I have met engineers and technicians who are proud and excited to participate in this kind of endeavor. They do not or may not get much recognition for their efforts, but nevertheless they have the passion to transform an impossible idea into a working machine.

From the terrace of the apartment unit in which I live, I can see the top of some facilities of Mount Wilson Observatory. Here I am, looking up at the Observatory where the expansion of the Universe was discovered. Often in a graduation ceremony, I hear words like "the sky is the limit" or "the possibilities are enormous." I would say, instead, that the sky does not have a limit since the Universe is expanding. For the next generation of humanity, I would like to say, there is a vast amount of space to explore, in terms of knowledge.

I imagine attending a conference two decades in the future. Speakers report on their most recent discoveries of plate tectonics in the planets in a nearby star system. Others discuss the aurorae in the primordial planetary objects and refer to the classical papers on streamers first resolved in ALMA observations made at submillimeter wavelengths. The data collected by the network of 8-meter class telescopes on the ground and in space, followed up by 30-meter telescopes, are rapidly filling up data storage units, as well as the front pages of the online news services. Observatories on Mars are just about to start regular operations. A few of them are equipped with Gamma Ray Burst and Gravitational Wave detection facilities. Detailed analyses of soil samples from asteroids and volatiles from comets brought back to laboratories on Earth provide templates for the study of exoplanets.

Is this just a daydreamer's dream? I am sure we can get there if we are determined to do so. This carbon-based, wayfinding species, like the ancient mariners who used the stars to guide them on the deep-ocean voyages that led them to Hawai'i, will continue to strive and thrive as we look to the stars to learn about deep space.

Chapter 27

Gražina Tautvaišienė (PhD, 1988)

The Unfading Joy of Being an Astronomer

Gražina Tautvaišienė, now Director of the Institute of Theoretical Physics and Astronomy at Vilnius University, uses spectroscopy to measure the abundances of the elements in stars in order to understand the chemical evolution of the Milky Way Galaxy and to understand the parent stars of exoplanets. She was among the pioneers in modeling the production of chemical elements in the Milky Way and the Large and Small Magellanic Clouds. She has played a major role in increasing the activity of Lithuanian astronomers in the activities of the IAU, in which she has held numerous positions, including as President of the IAU Commission "Local Universe" and as IAU National Outreach Coordinator, and she served as Vice President of the International Union of Pure and Applied Physics and Chair of the IUPAP Commission on Astrophysics. She is the recipient of the National Science Prize of Lithuania. Her asteroid is 135561 Tautvaisiene.

Since primary school, I have always liked to try to understand things rather than learn by heart. This had a large influence on my choice of a professional path, and astronomy, a field with perhaps the largest amount of secrets to be unveiled and understood, has become my lifelong passion.

Many circumstances have shaped my life. I was born in Lithuania during the Soviet times. My mother and father had quite different political backgrounds. My mother had suffered greatly from Stalin's repressions on a personal level. In fighting against the Soviet occupation, one of her brothers was killed, and two brothers, a sister, and her mother were exiled to Siberia. Her father had remained in Lithuania but had needed to hide from officials. After a very hard and lonely childhood, my mother married at just twenty years old in order to have a more stable life. My father came from an opposite political environment: from people who had tried to adjust to the Soviet regime. His father had been chairman of a Soviet collective farm but had been killed by those who were fighting against the Soviets. Despite these differences, we never had political conversations in our family, although the relatives of my parents never became friends, and I was advised not to become a Young Pioneer and not to wear a corresponding red tie. It was not easy to resist pressure from teachers to become a Pioneer, but they didn't seem to take it out on me, and I succeeded in being a very good student in all subjects, especially in physics.

Apart from studying, I liked to read books very much from an early age. When my parents told me to go to sleep and switched the light off in my room, I often used to put a book on the floor near the shaft of light shining through from another room and continue to read. Once, probably when I was in sixth or seventh grade, I was awarded a prize for being the person who had read the greatest number of books at school. I liked to read novels about nature and about the lives of people in different countries, especially biographies of famous people. The biography of Marie Curie impressed me most of all. The working language at my school was Lithuanian, and we had many books written in my native language or translated from Russian or other languages. I also liked to listen to the radio—not the official channel but foreign programs with very different music. The sound quality was horribly noisy since the USSR blocked receivers everywhere, and I had to sit very close to the radio set to avoid disturbing my parents with the loud, noisy sound.

At the end of eighth grade, after participating successfully in a physics Olympiad for schoolchildren, I was invited to join a good high school with specialist classes for physics and mathematics. There, I started to think about which field of science I wanted to pursue as a career. Taking part in a meeting of the Lithuanian Astronomical Union proved pivotal in making the decision. I met very friendly professional and amateur astronomers, and I was elected to the board of the union. Following my success in a Young Astronomers quiz, I was asked to represent Lithuania at the Young Astronomers Meeting of the Soviet Union. In order to prepare for this, I started to read professional books, took a telescope from the school, and set about investigating the sky. All our neighbors came round and asked me to show them the stars, and this inspired me to set up an amateur astronomy club in my native town, Kaunas. Looking back at my journey to becoming an astronomer, right from the start I aimed to do everything as professionally as possible. I understood at once that astronomy is not just a romantic hobby of stargazing, but a serious and interesting field of research. This helped a lot, and I've never been disappointed.

Vilnius University was my only option for studying astronomy. My friends, who were University students, warned that I would need to become a member of the communist youth organization to be accepted by the University. Thus, at the end of my high school studies, I realized that I would have to submit a request to join. The officials in charge were very suspicious about why I suddenly decided to join the organization, but they finally accepted my application.

In the program of physics, there were very few girls. Initially, I struggled to receive top marks, but when I received the highest mark in programming, all other lecturers started to treat me more or less as a normal student. From the very beginning, I started to work at the astronomical observatory and organized an astronomy club for students. Instead of going to the summer work camps for students in Soviet collective farms, I was allowed to travel to Uzbekistan and to work on building the Maidanak Observatory of Lithuania. I enjoyed my studies at the Vilnius University very much!

I knew that my marks at the University would have to be as good as possible, since students would be ranked, and the top students would have priority in picking the best places from a list of work positions provided to the University. When I graduated in 1982, only one position in astronomy was available. I was almost ready to be locked up in jail,

FIGURE 27.1. Gražina Tautvaišienė in 2014.

rather than agree to take up any other job for the mandatory two years. Luckily, I was accepted into the Institute of Physics of the Lithuanian Academy of Sciences, which had a Department of Astrophysics. Later this department, together with several departments of theoretical physics, evolved to become the Institute of Theoretical Physics and Astronomy, of which I became the deputy director in 1998 and the director in 2003.

Life in the Soviet Union was very closed, both in terms of doing research and even more so in having no possibilities for traveling abroad. Only one professor from Lithuania, Vytautas Straižys, was sometimes accepted in the delegation representing the USSR at the International Astronomical Union General Assemblies. However, I was quite lucky even during this period. At the time, the world's largest optical telescope of 6 meters was at the Special Astrophysical Observatory of the USSR Academy of Sciences, and I was given permission to observe on it. My PhD supervisor, Professor Straižys, was a photometrist but was curious to observe the spectra of metal-deficient stars and to improve their photometric classification, as there were relatively few high-resolution spectral studies of these objects. Thus, in 1983 I was given not just a topic for my PhD thesis work but also the task of kick-starting spectroscopic research in Lithuania. Fortunately, astronomers were writing papers with very detailed descriptions of analysis methods, and our institute was the first in Lithuania to have a supercomputer. I was extremely grateful to the American astronomer Robert Kurucz for giving us his models of stellar atmospheres and a code for computations. The work was not easy, but I succeeded in overcoming all difficulties.

Everything changed when Lithuania gained back its independence in March 1990. My collaborators at the 6-meter telescope told me that our work was terminated and that I should look for telescopes in Western countries—the new friends of Lithuania. Fortunately, scientists from Sweden soon saw our difficulty, as well as the possibilities for collaboration. The Nordic-Baltic Astronomy Meeting was due to take place at the Uppsala Astronomical Observatory in June 1990. A delegation of ten Lithuanian astronomers, including myself, went abroad for the first time. It is difficult to say what made a larger impression: the scientific presentations or the surrounding environment and the variety of the car models. To my great surprise, there were no women astronomers at the Uppsala Observatory at the time. When I asked why, the answer was that "it was very difficult to find the first one to employ." Well, I thought, maybe gender problems are not so bad in Lithuania after all.

During the next meeting, held in Lithuania, I was invited to apply for observing time on the Nordic Optical Telescope. My first application was rejected with a comment that I seemed to want to solve every single problem in the field (this was how I had been trained to write when applying for time on the 6-meter telescope). However, I rewrote my application to be more realistic, and soon this telescope became like a second home to me.

While observing metal-deficient giant stars on the 6-meter telescope, I had noticed that another type of star might be even more interesting and lacked investigation. The horizontal-branch stars, which we see in stellar globular clusters and as red-clump stars in open clusters, burn hydrogen in a shell and should be present in the Galactic field as well. Thus, I looked at possibilities of photometric systems to identify the red horizontal-branch stars. Soon, I published investigations of high-resolution spectra for 10 such stars, obtained on the 6-meter telescope, with abundances of up to 22 chemical elements determined. At the time, this constituted the largest single high-resolution abundance study of the red horizontal-branch stars located in the Galactic field.

My interest in studying the stellar chemical composition alterations caused by the evolution of stars was growing. With limited options to access foreign telescopes with high-resolution spectrographs, I started an intensive search for collaborators. This was not an easy task, as I had no funds for travel, even to conferences. In order to go to a conference, I had to propose an interesting presentation and ask the organizers for financial support. A desire to go to international conferences was my main driving force to work hard—sometimes incredibly hard. However, I managed to participate in the European Southern Observatory (ESO) Workshop on High Resolution Spectroscopy in 1992, as well as several other conferences and, most importantly, the General Assemblies of the International Astronomical Union.

In 1991, I had the option of going for a longer visit to the Nordic Institute for Theoretical Physics (NORDITA) in Copenhagen. However, I was pregnant with my son, Dainius, and had to postpone this trip. In 1993, I finally packed two large suitcases filled with dry food and set off for Denmark to meet the famous scientist Bernard E. J. Pagel, who proposed a collaboration on modeling the chemical evolution of the Milky Way. Bernard presented his previous publications to me, I collected a reliable observational data set, and we started a really enjoyable research collaboration, playing with stellar yields, chemical element production time delays, and other parameters. Influenced by my Vilnius colleagues, who were physicists and theoreticians in atomic and nuclear physics, I managed to convince Bernard that the model should not necessarily be the same for all elements. While certain elements are all produced through the same thermonuclear process, atoms are different and their production may also contain differences. This idea freed our hands in modeling the evolution of the abundances of 18 chemical elements in our Galaxy. It was fairly easy to model the abundance evolution of light elements; however, modeling the heavy neutron-capture elements was a more challenging task. I remember a point when I thought we would not succeed. I showed Bernard the results of my latest attempt and told him I was going to give up. However, he told me that the models already were very good.

Inspired by the success in modeling the chemical evolution of the Milky Way, we turned our attention to the neighboring Large and Small Magellanic Clouds galaxies. For this reason, instead of the usual visits of two months, in 1997 I was invited to work at NORDITA for seven months. Initially, we tried to understand why the available studies of the Magellanic Clouds showed lowered abundances of oxygen. There were models explaining this through depletion of oxygen by Galactic winds. However, I suggested that there was evidence that the lowered oxygen abundances in supergiants could be caused by inaccuracies in the analysis of oxygen lines. When we factored in the influence of Galactic winds on the evolution of abundances of all the investigated chemical elements, we succeeded in completing another influential study. My collaboration with Bernard Pagel lasted for more than ten years until his unexpected death. This great and sincere person had the most significant impact on my scientific life.

Over these ten years, my son, Dainius, was growing up with his mother disappearing from time to time for quite long periods. I felt quite guilty about that; however, the year that I stopped my annual trips to NORDITA, he unexpectedly reminded me that I should be going there. This greatly reduced my worries, and I am thankful to my husband and mother for taking care of family life while I was absent.

Quite soon, the American astronomer George Wallerstein invited me to collaborate on the analysis of the chemical composition of stars belonging to the in-fallen dwarf galaxy Sagittarius and the irregular dwarf galaxy IC 1613. It was really amazing to see spectra of supergiants in such a distant galaxy as IC 1613, which is approaching the Earth with a speed of 234 kilometers per second from a distance of 2.4 million light years.

It was clear that progress in understanding Galactic evolution depended on how well we understand the evolution of stars. I decided to investigate evidence of mixing in stellar atmospheres by observing evolved stars in clusters and in the Galactic field. Together with Bengt Edvardsson (Sweden), we targeted alterations of carbon and nitrogen

FIGURE 27.2. Gražina Tautvaišienė and Bernard E. J. Pagel near the sculpture of Fred Hoyle in the garden of the Institute of Astronomy, University of Cambridge, in 2002.

abundances as well as carbon isotope ratios, since these are among the most sensitive indicators of material mixing in stars. In addition to collaboration with quite a few respected foreign colleagues (all their names can be found in our joint papers), I started to nurture PhD students and build a research group at my home institute. Together with my talented young collaborators, we investigated how mixing processes in stars depend on stellar mass, metallicity, magnetic activity, and other parameters, and studied the chemical composition of stars belonging to the thick disk or newly found kinematic stellar groups.

Another exciting experience beyond my wildest dreams has been the Gaia-ESO Spectroscopic Survey, initiated by Sofia Randich (Italy) and Gerard Gilmore (UK). With more than 300 observing nights on the 8.2-meter Very Large Telescope, thousands of high-resolution spectra, and about 300 collaborators, this survey, for astronomers, is the equivalent of a sailor exploring a new ocean. Since 2011, ten years full of new discoveries have flown by, with still a lot remaining to be done. Information on the chemical composition of stars has to be combined with the kinematic and asteroseismic data flowing in from space missions. Thus new, exciting investigations await us.

New possibilities are always followed by the need for new qualifications and courses, especially for students and young researchers. In 1998, I started to organize Nordic-Baltic training courses at our Moletai Astronomical Observatory. Here I must mention Professor Jan-Erik Solheim from Norway, who was the main instigator of this activity in Lithuania and contributed tremendously. In 2006, I proposed and supervised a European Structural Funds project implemented by the Lithuanian Academy of Sciences (Promotion of Lithuanian Scientists in Usage of Large Scientific Infrastructures of European Union), which aimed to disseminate new knowledge not just among astronomers but also physicists, mathematicians, philologists, sociologists, and others. After such a large and bureaucratic project, leading other new ventures has been a real pleasure. Since 2016, we have organized research courses on ground-based observations for space missions and on science communication through the European Commission–funded Europlanet 2020 Research Infrastructure (RI) and Europlanet 2024 RI projects. It is a great joy to see the happy faces of young people at our summer schools and even more exciting to meet them after ten years saying that they are thankful for the great time in Lithuania.

Lithuania has been an independent country now for thirty years and all the world is more or less open to Lithuanian scientists for scientific collaboration. However, due to quite low local salaries, our scientific teams still have to rely on local people. My scientific group is perhaps a very rare exception—currently we have three PhD students from abroad and four scientists from Aarhus University. We are extremely grateful to Professor Hans Kjeldsen, who, together with members of his group, has joined our large project "Stellar and Exoplanet Investigations in the Context of the TESS and JWST Space Missions," funded through the European Social Fund.

I have been very fortunate to meet many supportive people who have shaped my life in a positive way. Only a few dear people have been mentioned in this short story. I hope I have had a positive influence on many people too, especially my son, Dainius. He has grown up and become a very nice and intellectual man. Educated at the University of Edinburgh, he came back to Lithuania and is now working at an international biotechnology company. In 2015, he initiated a team of students from Vilnius University for the iGEM (International Genetically Engineered Machine) Olympiads. Since then he has continued to advise students, and the Lithuanian team has already been recognized as the best in the world twice. Recently, he told me that quality not quantity was most important for him in communicating with parents.

During one of my first conferences, a woman told me that if I am not afraid to smile, everything will be fine in my scientific career. My advice to the readers of this story is to smile more often, to find nice people to collaborate with, and to look for the most exciting research opportunities.

Chapter 28

Carole Mundell (PhD, 1995)

Inspired by a Maths Dress

Carole Mundell, now the Hiroko Sherwin Chair of Extragalactic Astronomy and founding Head of Astrophysics at the University of Bath, studies cosmic black holes and gamma ray bursts. In 2018 she was appointed as the Foreign and Commonwealth Office's first female Chief Scientific Adviser and in her current role as Chief International Science Envoy in the Foreign, Commonwealth and Development Office advises on the impact science can have beyond the confines of the lab on social, economic, and political objectives. She also serves on the Scientific Advisory Group for Emergencies and has been involved in the UK's Covid-19 response planning. In 2016 she was named FDM Everywoman in Technology "Woman of the Year" and is a vocal advocate for diversity in science and for ending discrimination and sexual harassment within academia. She is a Fellow of the Institute of Physics and in 2021 was elected to the presidency of the UK Science Council.

I am a mother, an educator, a diplomat, a scientist, and a communicator of science. I feel privileged that my career has given me the opportunity to push the frontiers of knowledge and inspire and connect with people all over the world. The Covid-19 pandemic, although globally devastating, brought into sharp relief the power of science to connect individuals around the world, to bring hope, empathy, and understanding and, ultimately, life-saving protection. I have been humbled and inspired by the openness and appetite I encountered for urgent exchange of scientific insights and ideas—whatever the political backdrop—and by the warmth of strangers grateful for new knowledge or willing to generously share their expertise in return. This is the best of science diplomacy.

My earliest—possibly formative—memory of science was a dress. My maths dress! An old family photo shows me at a birthday party—of which I remember little other than the sense of pride in giving my new dress an outing and being fascinated by the mysterious symbols scattered across the fabric in between pink representations of some kind of fruit. Enlarging the photo recently I was surprised to discover that the dress held more than simple "fruit" arithmetic—instead, the symbols are integrals, sigmas, brackets—code that neither I nor my parents could have recognized or understood at the time but which, later in my career, became the key to unlocking the world of physics and the universe of astrophysics. Maths confidence forms early!

A career in science was in no way guaranteed for me. Only one generation ago, British society regarded careers for young women as temporary and unimportant, mere stop gaps until marriage. My mother moved from the private to the public sector to be allowed to continue to work after marriage but was required to resign her job when she found she was expecting me. She vowed that I would have an education, be independent, and be able to support myself and any children I might have. She was a quiet revolutionary. When I was three years old, my Uncle Stuart visited. He was a kind, well-read man, a skilled labourer who had experienced a world war, poverty, and mass unemployment. He commented that there was no hope of a university education for me, even if I was clever enough to attend. I was a girl in a working-class family; it was prudent to manage expectations. I am fortunate my parents were keen to prove him wrong. A good comprehensive education opened doors for me, including into the male-dominated field of physics and the diverse, fast-moving world of government, that my Uncle Stuart could never have imagined. But these opportunities do not happen without sustained effort from supportive families and inspiring teachers, structural changes, and political will to remove barriers to opportunity.

FIGURE 28.1. Carole Mundell, age five, in 1975, in her maths dress. © Carole Mundell.

My parents never wavered in their support of my ambitions. The year I was born, men landed on the Moon. My mother watched late into the night, entranced by the ghostly images of the first human setting foot on another heavenly body. She remains convinced that this prenatal exposure marked the beginning of my career in astrophysics.

My father was my first role model in science. He ultimately achieved the rank of Senior Chief Medical Laboratory Scientific Officer in a local hospital, after joining the National Health Service Pathology labs, aged eighteen, and studying for years in the evenings, after long workdays.

He taught me about blood groups and coagulation cascades and was my go-to resource for all things biomedical—although my squeamishness precluded a career for me in medicine. He had a special talent for typing rare leukaemias, and, I expect, he would have become a research scientist had he had a more affluent family background or the opportunity of a university education.

Instead, he wanted me and my brother to have that opportunity. His thirst for knowledge and lifelong learning reached across science, politics, and sport. My mum, similarly smart, with no formal education beyond school, has always been good at listening to me struggle through a homework problem or think through a research question. They both prized justice and equity and instilled these values in us.

As a teenager, although ballroom dancing and jazz piano were my passions maths and physics were my forte. My father brought home old copies of *New Scientist* magazine, which whetted my appetite for stories about science, science policy, and the wider impact of science. I was hooked and begged for an annual subscription as my Christmas gift. Every Thursday morning, from then on, I would leap out of bed early to wait for this cherished magazine to drop on the doormat. Perhaps this was the germ of my interest in international science policy and geopolitics.

Shortly before I began my undergraduate studies at Glasgow University, the first detection of neutrinos from Supernova 1987A was announced. These exotic and elusive particles fascinated me and heralded my unexpected detour into astronomy. In my first undergraduate year I studied astronomy—after my request to study music was turned down—alongside maths and physics. Unlike some of my future astronomy colleagues, I had not been a keen amateur astronomer as a child, nor an avid watcher of famous TV programmes such as Patrick Moore's *The Sky at Night*. Instead, I loved solving equations and initially found the rough approximations used in astronomy classes unsatisfying compared with the precision of physics.

The male bias I encountered at university was a subtle and bewildering barrier. In school, I had been one of only three girls out of forty students in physics, but in my final year, all of my maths and physics teachers were women, which had a positive effect on me. I did not doubt I could become a physicist. At university, I recall being taught by only one woman lecturer in four years, which was a shock to the system. You cannot become what you cannot see. The clear subconscious message: physics is not for women. I was fortunate to discover an inspiring female role model when Professor Dame Jocelyn Bell Burnell visited and presented a thrilling lecture on pulsars to the student astronomy society. For me, her talk brought to life the elegance of the theory of stellar structure and evolution. More importantly, I discovered that women do astrophysics as a career!

My undergraduate studies led to an attractive offer of employment in the nuclear industry and also an offer, after being interviewed by a panel of all-male professors, of a PhD place at Jodrell Bank Observatory at the University of Manchester. I was torn. A very good salary, thriving social life, and employment seemed exciting and grown-up. But the company was happy to defer my entry in order for me to take three years off to do a PhD. It seemed the ideal choice. The excitement of real engineering, technology, and hands-on data science was too good an opportunity to miss. Interestingly, though Jodrell Bank was male-dominated, it was also an incredibly supportive environment. And every day, driving to the Observatory around the Cheshire countryside, seeing the giant Lovell Radio Telescope towering above the trees, was uplifting and inspirational. "Wow, I work here!" I'd tell myself. Sir Bernard Lovell himself, long retired, often came to the Observatory and was always keen to hear about our research. He also encouraged the musicians amongst us to use the beautiful Bechstein concert grand piano in the public planetarium before opening hours to "keep it in tune."

I grew up like any inquisitive child, always asking "Why?" and "How?" I wanted to understand, to know how things worked, why they were the way they were, and how to change things. As a graduate student, I had the opportunity to explore some of the most extreme places in the Universe, to use my skills in physics and technology to solve big questions. That lit a spark for me.

The intellectually stimulating environment at the Observatory and the standing offer to move to industry after my PhD enabled me to focus on the science, without the pressure to publish that many young academics may feel today.

My PhD research was pioneering but unfashionable. It involved mapping the distribution and kinematics of atomic and ionized gas on a wide range of size scales in a class of nearby spiral galaxies called Seyferts, which were thought to contain supermassive black holes that were accreting material and shining brightly. During this time, I was somewhat of a misfit; the community that studied galaxy evolution was rather separate from those studying high-energy radiation from accreting supermassive black holes. I found it good training for future life to have to make my case to each community and persuade people that the two should be considered together!

Throughout my career, I have had the opportunity to pioneer new technologies and analysis techniques, sometimes innovating on a shoestring budget. My PhD was the start of this and involved two particularly memorable episodes for me. The first was being part of an experiment with Russian scientists and military officers in which the Russian GLONASS navigation satellites—which were causing interference in protected astronomy radio bands—switched off their radio transmissions one by one over a twenty-four-hour period so that we could demonstrate the interference and use the Lovell Telescope to observe in a part of the spectrum that had previously been inaccessible. Looking back, I appreciate what a remarkable example of bilateral cooperation and science diplomacy this experiment represented. The second was performing observations with the MERLIN interferometer that led to discovery of the first observational evidence for a small rotating disc of atomic hydrogen close to a supermassive black hole at the heart of a local galaxy.

After my time at Jodrell Bank, I accepted a postdoctoral position at the University of Maryland. This was a formative time for me as a scientist, and I made lifelong friendships and joined international collaborations. The U.S. science system is energetic and fast-paced; the UM environment was stimulating and friendly. And although taking this position meant managing a long-distance relationship and wedding planning with my future husband back in the UK, it quickly set my career path as an internationally active, independent research scientist.

In 1999, I won a UK Royal Society University Research Fellowship that provided me with, ultimately, a decade of research funding. In choosing where to hold my fellowship, I realized that I liked creating and doing new things, so I opted to join the astrophysics program at Liverpool John Moores University (LJMU), becoming their first Royal Society research fellow. At that time, LJMU was building the world's largest autonomous robotic telescope—the 2-meter optical Liverpool Telescope, which still operates today on the mountaintop on the Canary Island of La Palma. But as the focus of my fellowship was surveying the atomic hydrogen properties of active and inactive galaxies, the Liverpool Telescope was not useful. Around this time, however, the first optical afterglow of a Gamma Ray Burst had been discovered, and I commented one day to the head of research, "I presume you are going to use this telescope to search for optical afterglows of Gamma Ray Bursts?" He replied, "It's ideal for that, but we have nobody to lead the science team. Do you want to lead that project?" I knew the learning curve would be steep but the scientific opportunity huge, and so I founded the Liverpool Gamma Ray Burst (GRB) group for rapid follow-up observations. GRBs result from the merger of two neutron stars or the catastrophic collapse of a massive star to a single black hole. They produce intense but short-lived flashes of high-energy gamma rays that are gone in seconds; their occurrence is entirely unpredictable and can appear at any location on the sky and at any time of the day or night—hence the need for autonomous robotic telescopes and fast follow-up. Their dying embers produce light across the electromagnetic spectrum, and the physics of the progenitors, explosion dynamics, and shocks are encoded in the emitted light. Our fast-response optical telescope opened up a new window, finding faint optical counterparts just minutes after the blasts.

At this time, I'd recently given birth to my first son. Ironically, he would sleep through the night while I was awake with my team, capturing the birth cries of new black holes, disseminating our discoveries in real time to over a thousand astronomers around the world, and negotiating data sharing and publication leadership. Two clones of the Liverpool Telescope (the Faulkes Telescopes) were built in Hawai'i and

Australia, where we also ran our GRB software, giving us almost 24/7 sky coverage. The unique, groundbreaking physics niche we carved out in Liverpool with these telescopes was to measure the magnetic fields of GRB afterglows by characterizing how the objects' brightnesses changed with time and to devise a series of new robotic, optical polarimeters (named RINGO) which we built based on a novel design developed by my undergraduate lecturer Dave Clarke. From our work, I think we now understand that GRBs have magnetized jets.

In 2015, I was recruited to a new Chair in Astrophysics at the University of Bath to found and lead their first astrophysics research group and undergraduate degree program. Until that point, Bath was the only major UK university Physics Department with neither an astrophysics nor particle physics research program.

In 2016 I was surprised to be named FDM Everywoman in Technology "Woman of the Year." This award comes from the global tech sector—quite different from my academic community. I was a finalist in the innovator category for my GRB research and my support for women in STEM. After the announcement that I had not won the category award, I sat back to finish my drink and enjoy the rest of the event. The awards ceremony was a breathtaking evening—six hundred incredible women leaders, transforming the technology sector. And then they announced that I was the overall winner! An unforgettable, humbling moment and a long walk to the stage.

In 2016, I also became Head of the Department of Physics at Bath and began to focus on leading and transforming a traditional, white-male-dominated department. My astrophysics group grew to have a 50–50 male-female split, with PhD students, postdocs, and staff from up to twelve different countries, but my group was not representative of the department as a whole, which otherwise had only two other women faculty, out of about 45 total members. This was not dissimilar to the Astrophysics Institute in LJMU, where previously, for many years, I had been the only woman faculty member.

I focused on mentoring, career progression, and recruitment processes, tackling behaviours that are disempowering to women and minorities or that even constitute raw sexism. I worked with senior women

FIGURE 28.2. Carole Mundell in the Control Room beneath the Lovell Radio Telescope, Jodrell Bank Observatory, 2016. *Credit*: University of Bath.

across the university, the wider research community, learned societies, and funding agencies to improve processes and tackle sexual harassment and its enabling factors. Making these changes is slow but important work for the entire international research community that continues in many groups and organisations.

In 2018, I was approached to consider applying for the role of Chief Scientific Adviser (CSA) at the Foreign and Commonwealth Office (FCO). I had held senior science leadership and governance advisory roles in an academic capacity in the UK and internationally, but I had never considered a position in government. As a junior academic, I had worked with Sir David King, Government Chief Scientific Adviser to the Prime Minister of the day, but I was unaware that there existed a network of eminent Chief Scientific Advisers across Whitehall. After some persuasion by the lead recruiter, I submitted my application and went through the selection process. I was aware that, again, the traditional stereotype of a Chief Scientific Adviser was an older white male scientist. I was particularly proud to be offered the position and become the first woman CSA in the FCO.

So began a whole new career experience. I split my time 60:40 between the FCO and my research at Bath. I found the two roles complemented one another. My credibility as a research-active academic was important in my FCO role in international science diplomacy, overseeing the work of a global network of Science and Innovation Officers in our embassies and policy development on emerging technologies. In turn, my experience of how government works, particularly through major historical milestones, such as the UK exiting the European Union and the SARS-CoV-2 global pandemic, has given me a deep appreciation for the breadth and pace of work done by our committed civil servants, the importance of our universities and the wider science ecosystem for public good, and the influence of the UK in the world.

My work in the Scientific and Advisory Group for Emergencies has been both rewarding and humbling during the pandemic. Seeing so many dedicated scientists and health professionals reorient their work to deal with the Covid-19 crisis has highlighted the importance of breadth and depth—deep subject matter experts coming together, able to communicate and work across disciplinary and national boundaries to tackle a major crisis, as the forefront of knowledge quickly evolves.

One thing we have learned as a result of our experiences with Covid-19 is that many people have jobs whose productivity has been boosted and creativity unleashed by working in highly distributed ways, enabled by new technologies and good digital infrastructure—things I have benefitted from throughout my career, whether chasing black holes through the night or providing international science advice to colleagues around the world, in different time zones.

Working part-time in both roles has also given me firsthand insight into the frustrations of women who work part-time in one or more roles, the inequality of opportunity, pay, and recognition that women still face in the workplace at all levels, and the disproportionate impact of the pandemic.

Unlike in my mother's youth, parenting is no longer exclusively an issue for women. Men and women have families, aspirations, and meet challenges at different stages in their lives.

In my experience, conferences and meetings have been transformed by necessity and, in turn, have driven technological improvements. A surprising new degree of inclusivity has been created online, for example, reaching far more attendees from a wide geographic spread, levelling up traditionally marginalised voices. As we emerge from this pandemic, I hope we will afford people a greater amount of autonomy in how and where they work and how they balance their jobs and home lives; I believe this will harness broader talents and create more equitable societies and resilient economies.

In addition, it will give us the opportunity to reduce our carbon budgets by more consciously assessing when an in-person meeting will add value, or when a creatively designed virtual meeting will be equally good or better—without losing the benefits of making new friendships and initiating new collaborations. We must harness the best of technology, be inclusive across and beyond gender and race, and create much-needed diverse pathways into science and science-related careers if we are to stand any chance of tackling the immense global challenges our planet faces today.

The world is unimaginably different from when I first wore my maths dress, but many of the same prejudices remain stubbornly present in the world; it is not beyond our power—if we choose—to eliminate carbon from our energy production, to clean our air and protect our water, to rebuild our natural habitats and reverse biodiversity collapse, and to tackle the growing global threat of anti-microbial resistance, whilst marvelling at the wonder of the Universe and treating one another with respect and kindness. I hope we choose this path.

Chapter 29

Gabriela (Gaby) González (PhD, 1995)

Gravitational Love

Gaby González, now the Boyd Professor of Physics and Astronomy at the Louisiana State University, has been at the forefront of the search for gravitational waves and served as the spokesperson of the LIGO Scientific Collaboration for six years, including in 2016 when the LIGO team announced their first firm gravitational-wave discovery. She is a member of the National Academy of Sciences and the American Academy of Arts and Sciences, shared the National Academy of Sciences Award for Scientific Discovery in 2017, and was honored with the Bruno Rossi Prize from the AAS in 2017 and the Edward Bouchet Award from the APS in 2007. She is the editor-in-chief of the journal *Classical and Quantum Gravity*.

I was brought up in a working-class family in Córdoba, Argentina. Our first house was on a dirt street. My dad worked in an aircraft factory while going to college (in the tuition-free Argentinian National University) and graduated as an accountant; my mother was a high school math teacher. None of their parents had finished high school; perhaps that's why my parents always emphasized education for my brother and me. As a kid I was very, very curious. In high school, I thought physics could explain everything, since plants, people, and planets are all made of atoms. I wanted to study physics in college, at the Córdoba National University, to learn those explanations. I didn't imagine myself as a scientist. I had never met one, and I imagined myself teaching physics, since I liked teaching very much (I worked as a tutor while in high school). In college I met real physicists and astronomers and learned that they worked to answer questions that didn't have answers yet; they even asked new questions. Not everything was already in books! I loved that, and then wanted to become a scientist myself. Plus, all the scientists seemed to be teachers too.

I met my future husband, Jorge Pullin, in college. He was pursuing a PhD in theoretical general relativity on the subject of gravitational waves, the same topic I was working on for my senior thesis. We got married in 1988, proving Einstein wrong: he once said "gravitation cannot be held responsible for people falling in love," but in our case, it was! In 1989 Jorge was offered a postdoctoral fellowship at Syracuse University in the United States and I went with him: I had always lived in the city in which I was born, and now I was moving to a different country, with a different language and a different culture—a big shock! I managed difficult times then and still do now with other activities: knitting and crocheting, reading fiction, and driving fast convertible cars.

I started my PhD in theoretical physics, but then I met Professor Peter Saulson, who was working on what appeared to be a long-shot idea: detecting gravitational waves. After falling in love with such a field, I decided to pursue a PhD in a topic related to the noise limits on the sensitivity of the LIGO detectors (more on this later), with Peter Saulson as my advisor. Jorge and I became friends with him and his wife, and we still are.

In Einstein's general theory of relativity, gravity is not a force, as it is in Newton's theory. Instead, Einstein's theory says that masses and space-time affect each other. The Earth does not go around the Sun because a force acts on it; instead, it does so because the Sun "curves" space-time, and in the curved space-time, the natural trajectory for Earth (and for light!) is not a straight line anymore. Imagine marbles on a firm mattress: they will travel in straight lines. But if a heavy mass is on the mattress, the surface (the "space-time") will be curved, and marbles will not travel in straight lines anymore. When masses are accelerated, for example when two objects orbit around each other, their

FIGURE 29.1. Gaby González and Jorge Pullin camping at Lake Wallenpaupack in the Pocono Mountains, Pennsylvania.

motions produce "ripples" in space-time that travel away from the system at the speed of light: those ripples are gravitational waves.

Unfortunately, according to Einstein's theory, gravitational waves are exceedingly difficult to detect because gravity is very weak. Only cataclysmic astrophysical processes, which involve massive stars moving at very large speeds, even if far away, can produce measurable gravitational waves that pass through Earth. These are phenomena like the explosion of a supernova, or two black holes that inspiral around each other with speeds close to the speed of light in their final moments before merging. As gravitational waves travel, they induce changes in the distances between objects, especially if those objects are free of other forces: this is the effect we use to detect them.

The effect of gravitational waves changing the distance between two objects is proportional to the distance being measured, so large detectors are needed: the LIGO detectors are four kilometers in length on each of the two perpendicular arms, and there are two LIGO detectors, one in Hanford, Washington, and another in Livingston, Louisiana.

Viewed from an airplane, the detectors look like giant L-shaped instruments. Heavy mirrors are suspended from wires, and we measure the distances between them. The mirrors are not completely free, but they are almost free to move in the horizontal direction. When a gravitational wave goes through the detectors, it changes the distances between the mirrors. The changes in distances are measured with laser beams to determine two perpendicular distances in each detector using the interference pattern of the laser beams: that's why the detectors are called "interferometers." The changes in distances are minuscule, smaller than a proton, so we need to minimize the motions due to anything else. My PhD thesis, finished in 1995, studied how the motion of atoms in the fibers of the pendulums would affect the measurements in the interferometers, something that was not completely understood at the time.

The first detection, made in 2015, was from the collision of two black holes, each with about thirty times the mass of the Sun, moving at half the speed of light a billion light years away, but the distance change measured by LIGO was tiny, much less than a hundredth of the diameter of a proton! Detecting such minute displacements requires quite advanced technologies. The pendulums are mounted in complex suspensions that isolate them from the natural motions of the Earth. Each mirror has a reflecting surface and laser beams are bounced between them, traveling the 4-kilometer distance back and forth a few hundred times. The beams and the mirrors are enclosed in a vacuum system to prevent air currents (or just molecules!) from affecting the propagation of the laser light.

Of course, LIGO took a long time to be built. The ideas for the detector began to be developed in the 1970s; the LIGO (Laser Interferometer Gravitational-Wave Observatory) project was funded by the National Science Foundation at the end of the 1980s, when a conceptual design was clear. Construction started in 1994, and, although the sensitivity of LIGO at that time was very limited, the first measurements were taken in 2001. But that was just the beginning: like a precious musical instrument, such a complex instrument takes many years to fine-tune. In fact, it was initially known that with the technologies installed in 2001 it was very unlikely that gravitational waves would be detected. The idea in this early phase of LIGO was to learn about the instrument while in parallel developing more advanced technologies.

After finishing my PhD in 1995, I went to work in Rai Weiss's LIGO group at MIT; in the 1970s, Rai had started the concept of a kilometer-long interferometer in the United States. Along with Peter Saulson, Rai Weiss has have been one of my lifelong mentors. At MIT, I worked on a prototype interferometer to test one of the technologies to be used and on the design for suspensions used in the initial LIGO detectors. In 1997, I accepted a faculty position at the Pennsylvania State University where my husband, Jorge, was also on the faculty. Previously, we had spent six years (!) living apart: after he finished his postdoc at Syracuse, he spent two years as a postdoc at the University of Utah and then started as a professor at Penn State in 1993, all while I was finishing my PhD in Syracuse. We kept meeting as often as we could, but sometimes we were apart for months at a time. Later, when we were separated by only the ten hours of driving distance between Penn State and MIT, we purchased a camping trailer that we parked midway between the two universities in a beautiful campground and met there on weekends: we were very creative! We are now past our thirtieth wedding anniversary.

While at Penn State, I traveled to the LIGO Observatories to help with the commissioning of the new instruments and also prototyped a double suspension system in my lab. In 2001 we moved to the Louisiana State University. This offered a tremendous opportunity since the LIGO Livingston Observatory is only thirty miles outside of Baton Rouge, where LSU is based. My husband set up a gravity theory group that among other things would use supercomputers to calculate what the gravitational waves from a binary black hole collision would look like. LSU has had a commitment to gravitational waves going back to the 1960s, when they hired Bill Hamilton, who constructed a gravitational-wave detector with a different technology from LIGO, a "bar" detector. He operated the best such detector in the world, but the sensitivity was difficult to improve, and LIGO had much more promise. When I arrived at LSU, there was already an assistant professor working on the suspensions of LIGO, Joe Giaime, who is now Observatory Head of the LIGO Livingston Observatory (LLO). I continued working on the commissioning of the instrument at LLO, and when we started taking data I became a hunter for noises in the instrument that needed to be eliminated—this is still the main research topic in my group.

LIGO is operated like most observatories by a permanent staff. The "LIGO Laboratory" has about 150 people based at Caltech, MIT, Hanford, and Livingston. But to develop the technology used in the instruments and to analyze and interpret the data taken with the detectors, a much larger group of people has been needed: this is the international LIGO Scientific Collaboration (LSC), created in 1997—I was one of the LSC members from the beginning. This collaboration has grown over the years and now counts more than 1,300 scientists spread across more than 100 institutions around the world. The first spokesperson (scientific leader of the collaboration) was Rai Weiss. The first elected spokesperson was Peter Saulson, my PhD advisor; David Reitze, who currently directs the LIGO Laboratory, followed him. I was elected spokesperson in 2011 and then re-elected twice (each term is two years) before stepping down in 2017.

The LIGO detectors took data in 2001–10 with increasing sensitivity, reaching the promised low noise level, but made no detections. This was not a surprise; it was known from the beginning that still better technologies needed to be installed, but first, the operation of such large instruments had to be demonstrated, and it was. Also, in 2007, we joined forces with the European Virgo collaboration, which operates a 3-kilometer-long detector in Italy. In 2008 funds were approved for Advanced LIGO, a detector that would be ten times more sensitive than Initial LIGO. In 2010 the new technologies that had been developed over the previous decade began to be installed: better lasers, suspensions, seismic isolation, mirrors, and so forth. Only the vacuum system was kept the same. The installation was finished in 2014, and in 2015 the detectors were able to be in "operation" (with control systems producing data), but the sensitivity was not yet ten times better than before: achieving that goal would take many more years. In 2015, however, the detectors were a factor of three better than before, so the volume of the Universe around Earth from which events could be detected was almost thirty times larger (about three cubed) than before. This was far from a

FIGURE 29.2. Gaby González at the LIGO facility in Louisiana.

big enough volume of space to guarantee detections, but we had decided we would take data for a few months to at least gather experience with the new instruments.

Being the collaboration leader since 2011, I wanted to get everything ready for a potential detection, even if it was unlikely—we couldn't be caught unprepared. Many people didn't see the urgency; this required a paradigm shift. How would we know we had a signal? How would we proceed with reviewing the evidence? What was needed to make an announcement? These efforts proved worthwhile when we were utterly surprised on September 14, 2015, before we started taking data continuously, just a couple of weeks after the instrument had come into a stable state. After some late-night tests at LLO were done, the Livingston detector took data for a couple of hours after 9 UTC, before the daytime noise proved too overwhelming; the sister detector at Hanford had been operating from just before 8 UTC for many hours. Whenever the two detectors were operating, data analysis algorithms were used to test their efficiency. And at 9:50 UTC, those algorithms produced results that indicated the presence in both detectors of a strong gravitational-wave signal corresponding to the collision of two large black holes. The signal was so strong that it could be seen with simple visualizations of the data, without any sophisticated signal analysis. But instruments are noisy and have many transients of unknown origin: maybe this "signal" was just noise? But we hadn't yet learned about the instrumental noise; it was a new instrument! We needed more data to have statistical confidence that the apparent signal was a detection of a gravitational wave; that process would take at least a couple of months, and the analysis to resolve the details of the source even longer.

Rumors started to fly across the astrophysics community that a detection had occurred, but most scientists understood and respected the need for gathering enough data and building confidence that this detection was a real signal and did not press us for answers. However, David Reitze and I had to field questions from specialized journalists and had a hard time convincing them that until we were confident about results, we could only have candidates, not discoveries, and we would not talk about candidates that might go away, as often happens in sensitive, cutting-edge physics experiments. Eventually the data taken in the next month showed that a candidate like the one observed would happen by chance only once in 200,000 years. We had indeed detected a gravitational wave! Another detection took place on December 25 (U.S. time). This further convinced us that what we had was real and not some kind of noise in the system.

After writing an article and having it approved by independent peer reviewers, we made the announcement of the first detection at the National Press Club in Washington, D.C., on February 11, 2016. A pleasing coincidence was that UNESCO had decided that every February 11, starting in 2016, would be celebrated as the International Day of Women and Girls in Science. Among the five people participating in the announcement on the stage, there were two women, France Córdova, an astronomer and director of the National Science Foundation, and myself. The other scientists were David Reitze and two of the founding fathers of the field, Rai Weiss and Kip Thorne (both would later share the Nobel Prize in Physics for the discovery with Barry Barish, a former director of the LIGO Laboratory). Of course, there were other women in the audience, including Virginia Trimble (author of another chapter in this book), invited as the widow of the first person to build a detector of gravitational waves, Professor Joseph Weber. (The technology he pioneered was not as sensitive as interferometers today.) We took a picture of several of the LIGO women present that day celebrating both the discovery and the Day of Women and Girls in Science.

The announcement appeared on the front cover of newspapers around the world. President Obama tweeted about it. The president of Argentina called me the next day. That night a group of LIGO scientists and friends went to a bar to celebrate and we hung a banner displaying the gravitational wave (a banner I had also worn earlier). When we were about to leave, another group of people had sat below it. When we asked their permission to retrieve it, this random group of people from Washington, D.C., said, "You are those physicists from Louisiana on the front page of the *New York Times*!"

Since the announcement, I have been invited countless times to address a variety of different groups of people, ranging from elementary kids to business people—and I love doing it. In all cases I have seen an enormous excitement about the field of gravitational waves and about science in general. People have endless questions about Einstein's theory, black holes, and other topics (including "What is it like to be a woman in science?"). Sometimes in our modern world one can be somewhat cynical about people being too preoccupied with material things. But there clearly is a hunger for understanding the Universe that cuts across social and age differences. At Stephen Hawking's seventy-fifth birthday celebration banquet I sat among several donors, mostly retired investment bankers. These people were incredibly wealthy and yet all they wanted to talk about was astronomy. One of them was taking a cruise around the world, during which he wanted to tackle a book on elementary astronomy. Another one told me: you scientists look so much happier than us in the finance industry!

I have received many awards since the discovery, some as part of the LIGO team, and I always emphasize that any award given to me represents recognition of the work of hundreds of people, even if my name is the one attached to the award. The award I am most proud of is the 2017 National Academy of Sciences Award for Scientific Discovery, given to David Reitze, Peter Saulson, and myself, as former elected spokespersons of the LIGO Scientific Collaboration: this award recognized that the discovery was a product of a team working for many years.

After 2015, gravitational wave discoveries kept being made, and not just of mergers of black holes but also of neutron stars. Many more discoveries are yet to come, including, I hope, surprising signals from unknown sources. Since finishing my spokesperson term in 2017, I have been a normal contributing scientist to the LIGO collaboration, although I continue to give many (real and virtual) talks around the world in English and Spanish, I also help in many scientific committees and enjoy teaching at LSU.

While studying physics, I met more than one professor who said loudly to groups of people that "women can't do physics." Once, when I was attending an award ceremony for my husband, a well-known scientist told me I should go home and give Jorge kids (we don't have children). I realize now how wrong they were, and I think about the number of women and the talent that were lost to science because of people like them. However, when I heard those words, I felt mostly sad for those people, and also very eager to prove them wrong—and I did.

If my career leads me to a reflection, it is that it takes all sorts of people to do science. When people, particularly young kids, think about scientists, they tend to imagine people like Einstein, with incredible logical and mathematical skills, "geniuses." Some scientists have those skills, but most scientists are not geniuses—I am not. A lot of people are needed to make progress in science, as are people with many different skills. Skills in computing, simulations, analysis, testing new instruments, creating new ways to put technologies together, understanding and reducing the noise in sensitive instruments, and so on. And different people have different ideas, so diversity is essential—geniality is not. Some skills not often associated with science are also important, like giving clear presentations in order to accurately communicate new scientific results to colleagues and to excite young people about scientific discovery so that they might become scientists, addressing agencies in order to receive funding, and organizing teams of people to work together. And no one person can have all the needed skills in science: we need many people, all kinds of people.

If I look back on my career, and the strengths that made me successful, technical knowledge was definitely involved. But I know a lot of people who are better than me in many aspects of scientific work. The desire to work with a diverse group of people, which sometimes meant I had to swallow my pride and keep cool in tense situations and not antagonize people who had not behaved well, definitely played a key role in my performance as spokesperson of the collaboration during a time of dramatic discovery. This is my advice to young people considering a scientific career: you do not have to be the one who works out the problems faster than anyone else in your physics or astronomy class; you do not need to be a genius. All you need is the ability to ask questions and be passionate about what you do, and you will eventually find the role in which you can contribute and even excel. That was certainly my story. And it continues to be.

Chapter 30

Vicky Kalogera (PhD, 1997)

Not Taking "No" for an Answer: Learning How to Persist and Persevere with a Smile

Vicky Kalogera, the Daniel I. Linzer Distinguished University Professor and Professor of Physics and Astronomy and the Director of the Center for Interdisciplinary Exploration and Research in Astrophysics at Northwestern University, studies how white dwarfs, neutron stars, and black holes interact in binary systems, how such systems are born, how they evolve, and how they end their lives. She was awarded a Guggenheim Fellowship in 2021, the Dannie Heineman Prize, awarded jointly by the AIP and the AAS, in 2018, the Hans Bethe Prize by the APS in 2016, the Maria Goeppert-Mayer Award by the APS in 2008, and the Annie Jump Cannon Award by the AAS and AAUW in 2002. She is a member of the National Academy of Sciences and the American Academy of Arts and Sciences, a Fellow of the American Association for the Advancement of Science and the APS, and a Legacy Fellow of the AAS.

In many ways my career path has been rather linear and traditional: a student who liked math and science, an undergraduate physics major, a PhD student in astronomy, a few years of postdoctoral research, a faculty position, earning tenure, balancing a heavy but exciting load of research, mentoring, teaching, administrative leadership, and service duties. *In many other ways my path has been quite improbable and unexpected.*

I was born and grew up in the small city of Serres, Greece. Throughout my school years I was drawn to math, and in high school also to physics and computer programming. I was inspired by reading biographies of Marie Curie and Albert Einstein, articles in *Scientific American*, and science fiction literature. As a junior in high school I heard and read obsessively about this amazing supernova explosion that changed astronomy, SN 1987A, and I watched documentaries about it that highlighted top researchers. As they told their stories of discovery and described the emotional excitement associated with their work, I remember being fascinated. I chose to major in physics as an undergraduate, and I truly fell in love with astronomy and physics during my first year in college. Somewhere between the end of high school and college I was struck with the idea of pursuing a graduate degree and spending my life doing research. I finished my undergraduate studies in four years and then applied to astronomy PhD programs in the United States. I chose one of my offers, for a variety of reasons, and had an encouraging experience as an international graduate student. Then, after three transformative and yet highly self-doubting years as a postdoc, I still felt strongly that becoming a professor was my top choice for my career path, and in 2001 from two junior-faculty offers I chose to become an assistant professor at Northwestern University. Now, twenty years later, it is interesting to look back and think about how it all developed.

In research, I was attracted to two things: (1) anything related to dead stars, also called compact objects (white dwarfs, neutron stars, and black holes); and (2) questions of origin—"How do things form?" "Why do they exist at all?" "Where do they come from?"

My love affair with compact objects goes back to my undergraduate years: despite the breadth of the astronomy and astrophysics course curriculum at the University of Thessaloniki, I remember spending most of my "extra" time reading everything I could find about neutron stars, probably because of Professor Jiannis Seiradakis's mentorship. As a sophomore I attended a NATO Advanced Study Institute in Greece focused on compact objects. *This turned out to be a pivotal decision for me!* I absorbed everything like a sponge and with awe. So many concepts I was trying to understand by reading on my own were being explained in mini lectures by all these people—Gordon Baym, Matthew Bailes, Ed van den Heuvel, Michiel van der Klis, Fred Lamb, Andrew Lyne, Jan van Paradijs, Chris Pethick, Mal Ruderman, Ron Webbink, and others—whose papers I had been attempting to read and understand for months. (Did you notice? No women, yet at the time their absence did not draw my attention.) Many of them became future mentors of mine (van Paradijs, Baym, and my PhD advisor, Webbink), and at this meeting I met students from different countries who continue now to be my peers and collaborators. I feel extraordinarily fortunate to have developed the beginnings of my science family so early in my studies.

Attending this conference was not a straightforward step. My undergraduate mentor prudently advised me against it at the time: "It is too early! You have not taken enough courses yet; the talks will be too advanced for you to follow; you will be lost; you will get discouraged and it will be harmful instead of beneficial"—this advice was completely reasonable, but I ignored all of it. My savings from a software engineering job I had held the summer before university studies (driven by both my family's financial hardships at the time and my love of computer coding, which I had pursued as a high school student as an extracurricular activity) provided the means for me to attend the conference without official research support. My undergraduate advisor, the kind, caring person that he was, was not upset; he was just worried. Attending this event was life-changing: I did not mind not understanding a lot of the material; instead I was grateful to have the opportunity to listen and learn. I was not hesitant about asking questions at coffee breaks and meals, and I met a lot of my science idols; I met non-Greek undergraduate and graduate students and got a glimpse of their academic lives and aspirations; and meeting them led me to my next step. Thanks to my new European student friends, I discovered the existence of a European Union scholarship program for undergraduates, and I insisted on squeezing in a Saturday meeting with the Greek professor who ran the program for Greece, even though the application deadline had passed and I was getting close to missing my boat home. By that time I was starting to understand how my own actions could have a positive impact on what I am exposed to along my journey as a scientist.

A lot that followed stemmed from that conference experience: I did start my research experience on a scholarship at the University of Amsterdam (working with Jan van Paradijs and Ralph Wijers), studying binary star systems with white dwarfs (the kind of compact remnant our Sun will leave behind after nuclear burning stops in its core) that are accreting material from their stellar companions. I continued to do research, next on a project with Seiradakis at the Max Planck Institute for Radio Astronomy in Bonn to calculate the population of Galactic radio pulsars (neutron stars, the compact remnants of stars tens of times more massive than our Sun, acting as radio pulse beacons).

When the time came to apply to graduate school, I included all the PhD programs in the United States I had heard about at that first conference. I ended up at the University of Illinois at Urbana-Champaign, home of three of the professors I had heard lecture at the NATO conference. For my PhD, I chose to try to understand how low-mass X-ray bright binaries form. Each of these binary star systems has a neutron star that formed when a massive star died, and the neutron star is growing in mass by accreting material from its low-mass companion. We still do not have a complete understanding of their formation, but back then knowledge that my students now consider to be unshakable truth was under intense debate. I worked directly with my PhD advisor (Webbink), but he was rather hands-off most of the time (except for a couple of times when I remember getting really stuck). I learned a lot from his physics explanations, but I also learned from his way of approaching research problems and strategy as well as from his relations with colleagues. I also learned from him to not expect much in terms of explicit praise but instead to read "between the lines."

Overall I have an incredibly fond memory of my PhD experience: I felt fortunate pursuing what I loved, in a supportive, top research environment, and being able to take care of myself financially. Yet, I now realize that I was unaware of the big picture in terms of career issues.

For one thing, I was much more hopeful and optimistic than job statistics and other realities would have justified at the time, but these facts did not register with me back then. In retrospect, I find this highly naive and unexpected of me: having come from a family of very low means (as a teenager I experienced my parents' small, always struggling business go bankrupt), I should have been more worried about securing my professional future.

Another realization is that, for all these years (until the end of graduate school), I was completely oblivious of the challenges, explicit and implicit, faced by females (and of course other marginalized groups) in STEM. I grew up in an unapologetically sexist society in the 1970s and 1980s, and I was used to brushing things off (e.g., from interruptions during discussions to outright insults). I could see, of course, the reality of women's underrepresentation in the sciences, but nothing ever fazed me when it came to seeking my place in this world. My father's confidence in my academic abilities and his encouragement of education (despite his overall traditional, sexist, full-of-limitations upbringing when it came to everything else) were crucial advantages for me. I had a sense that the status quo of women being rare in professional positions around me was the past and that things would change as my peers became professionals (improvements did occur, but it is still a work in slow progress). I had no sense of the effects of unconscious bias, cultural expectations, leaky pipelines, peer pressure, and the barrage of subtle, unspoken messages to girls and women when it comes to careers in math and science, even when they do *demonstrate success*. Although my being oblivious to all of this is embarrassing to admit now, I think in some ways it shielded me from the negative impact of all these effects. By the time I became aware of them, I was rather well committed to staying in the academic STEM pipeline.

Last, as a finishing graduate student I was completely unaware of the intricacies of the postdoctoral job market. I had happily and gratefully accepted a very early offer to work with a professor in Europe. When the European funding fell through, I was in a panic. No regular positions for working on the astrophysics of compact objects were advertised; instead, almost everyone at the time seemed to be working on some aspect of cosmology. I was convinced I had zero chances and that my career would end right then, before it had even started. Thankfully I was proven wrong and I was fortunate enough to choose the Harvard-Smithsonian Center for Astrophysics, among other possibilities, as my destination for postdoctoral work. My experience was exactly the opposite and eye-opening. Pretty quickly, for the first time, I became aware of what the term "job market" means.

My years as a postdoctoral fellow at the CfA were transformative in yet another way. I was exposed to the full breadth of research being done by the astrophysics community, having moved from a relatively small astronomy department in the Midwest to an institution with hundreds of researchers, a constant flow of visiting scholars, and multiple lectures and seminars I could attend every day. I felt like I was drinking from a firehose. I would be lying if I did not admit that it took quite a bit of adjusting. Being placed in the Theory Division helped me, as I had the feeling of being in a "small" (by comparison) department.

For the first time, however, I experienced the negative feeling of being a minority. Ironically, it was not so much my being one of just a couple of female postdocs in my everyday environment; instead, it was the fact that in the late 1990s my field of study, stars and compact objects, was not popular. Rarely did anyone seem to be interested in my research and related topics. It did not take me long to realize that the way to function in this environment was to knock on doors, literally, so I did!

I figured out how to open up new lines of research for myself through well-chosen collaborations. I had the wonderful opportunity to engage with astronomers who used X-ray telescopes to study X-ray binaries; working with them, I could focus my simulations and theoretical work on observations, either by interpreting data in-hand or planning future observations. I discovered that I enjoyed working on the scientific motivation for proposals seeking time using telescopes to make new observations. This kind of work in X-ray astrophysics became a big part of my life for more than a decade.

During those years at CfA, I made another major career decision, although at the time it was not clear how enormously impactful it would be: I joined the LIGO Scientific Collaboration (LSC), as the "first card-carrying astronomer," as Rai Weiss (Nobel Prize in Physics, 2017) liked to say. This step was not intentional on my part. My work on X-ray binaries and compact objects had led me to studies of how binaries with two neutron stars form. A two-minute visit to my office by Ramesh Narayan, who asked me to do a quick calculation of double-neutron-star formation rates (i.e., how often do they form in a galaxy like the Milky Way) he needed for a committee report, led to a major paper I led, written together with him, Joe Taylor, and David Spergel. That led to an invitation to an Aspen Center for Physics workshop to discuss birth rates of binary pulsars, which led to another invitation, this time to a conference in the French Alps to make a presentation on double-neutron-star birth rates. It turned out that was a conference of almost exclusively gravitational-wave researchers—the formation rates I was calculating were important predictors of the likelihood of detecting gravitational waves!—but how could I say no to a conference in the French Alps?

My presentation led to a question from someone in the audience: "Can you quantify the expected distribution of spin orientations in binary systems with one black-hole and one neutron star?" I said, "Yes, I can" (though I was not sure I could). I then worked furiously and was able to post a paper with my results on the astrophysics archive just a few weeks later, which led to yet another invitation, this one from two LSC senior members asking me to join the Collaboration. So, first a clear message: if you want to get into all kinds of scientific adventures, when you are a junior researcher and people ask you questions about potential projects and calculations, it is OK to say yes first and then figure out how to do it later. It requires sacrificing what you are working on at that moment (for a little while), but it opens up new paths that could change your work experience.

Up until then, I was enjoying working on new topics, writing papers, giving talks, meeting new people, and going to new conference destinations. However, when I received the invitation to join the LSC I stopped in my tracks. In the late 1990s, the astronomy community did not think very highly of the potential for detecting gravitational waves, the LSC did not include any astronomers, and every one of my peers with whom I talked thought I would be committing professional suicide if I joined.

Almost all of the senior people I asked also had nothing positive to say about joining LSC: "Why would you start something that is a dead end, a physics experiment with no chance of ever impacting astronomy?" they said. Nevertheless I was intrigued; I had just entered a completely new world that appeared both scary and fascinating! After I explained all the pluses and minuses of this opportunity to my friend Deepto Chakrabarty, he suggested I ask Rai Weiss for his opinion. I did not think Rai would respond (at that time, I had not yet interacted with him), but he did, and he suggested that we meet to discuss my concerns. I remember being terrified going to see him. He asked me a lot of questions, and at the end he said, "I think you should join, but you should think really hard about what you, as an astronomer, bring to the LSC." I left with a clear understanding in my own mind: I would join because doing so seemed irresistible, but my LIGO work wouldn't come at the expense of the rest of my astrophysics work. This decision actually turned out to be the right one, because in the couple of years that followed it became clear that most people who interviewed me for faculty positions were not thrilled that I was involved, even part-time, in LSC.

The next chapter of my professional life started twenty years ago and yet it does not seem that far back in time. Being on the faculty at Northwestern University has been extremely fulfilling. I found unique satisfaction, motivation, and intellectual freedom as a professor. I was immensely fortunate that my senior astronomy colleagues, especially Ron Taam and Dave Meyer, were supportive and encouraging at every step. They cleared my way so I could focus on what mattered most to me as a young faculty member. Building a group of students and postdocs and pursuing a wide range of projects in multi-messenger compact object astrophysics was hugely energizing, and every year felt better than the previous one. Eventually building something more impactful than my group, I founded and direct the Center for Interdisciplinary Exploration and Research in Astrophysics at Northwestern. Doing this, with the encouragement and support of many of my colleagues, administrators, our staff, and all the students and postdocs over the years, has been my honor and my joy.

From a research perspective, my gravitational-wave work expanded a lot more than initially planned, and obtaining tenure provided the security for me to focus on it, even though LIGO detections did not seem likely back then. In addition to source modeling and predictions, I built a subgroup in gravitational-wave data analysis, and we led a lot of the work on extracting astrophysical source properties from the signals (methods, tests, computational improvements), on merger rates of binary systems with neutron stars and black holes, and developing the astrophysical interpretations of individual sources and populations. I engaged in detector characterization and through that expanded into interdisciplinary collaborations with computer scientists in machine learning, citizen science, human-computer interactions, and computational education research. In parallel, work on X-ray astrophysics and radio pulsars remains a staple in my group, while I expanded into work on short-duration gamma-ray bursts and their host galaxies. None of this would have been possible without my students and postdocs who keep me on my feet. As a professor, I discovered that I love mentorship the most, brainstorming with other people, formulating questions, coming up with research plans, hitting walls, and figuring out how to break through them or change directions and eventually get to the answer or discover a new question. This process is never-ending and remains highly motivating; it is a gift to be able to share it with young people.

There is no question that the first gravitational-wave detection (GW150914) of a signal from the coalescence of two black holes and the multi-messenger detection of the double-neutron-star coalescence (GW170817) coupled with a burst of gamma-rays and the detection of optical emission from the production of heavy elements (like gold and platinum) were, so far, the most impactful events in my career. The experience of working on these analyses (and others that came in between and since then) has been humbling and precious beyond what words can express. I could never have imagined the transformations these discoveries have brought about in astrophysics, in how they were received by both the scientific world and the public. When I joined the LSC, I looked at it as an exciting intellectual adventure, even though the prospect of ever detecting actual gravitational waves from an astronomical source seemed something unreal and my involvement something I almost had to hide; this memory still makes me chuckle.

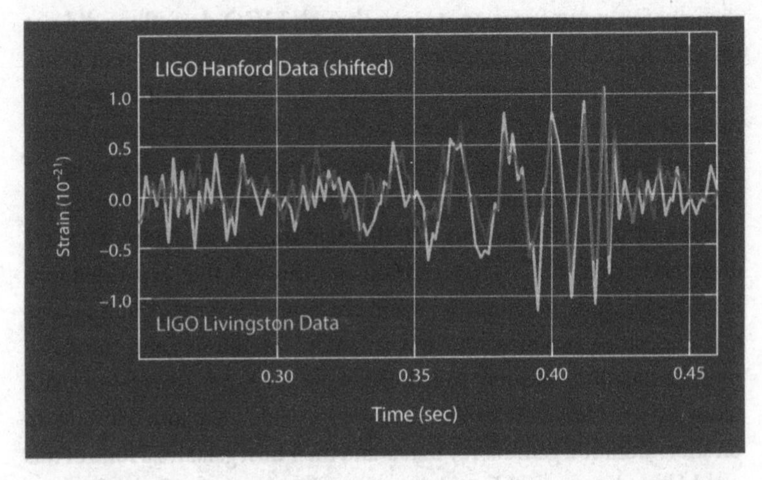

FIGURE 30.1. The very first gravitational-wave signal from GW150914 generated by the coalescence of two "heavy" (about thirty times the mass of the Sun) black holes detected by the Laser Interferometer Gravitational-Wave Observatory. *Credit*: LIGO.

In many ways my path has been quite improbable and unexpected. I was raised lovingly by two poorly educated parents. For my father, despite graduating at the top of his class from high school, college was out of the question, as it was up to him to feed his parental family in post–civil war Greece; my mother received an education only through elementary school, as her father decided this was enough schooling for a girl. Yet my father valued and encouraged my education. I am the first one in my extended family (among siblings and cousins) with a college education. My immediate family has faced severe financial challenges since my teenage years, as well as mental health challenges (my mother). Our social circle was very small, and until I entered university with one exception I had never traveled more than two hundred miles from my hometown. The only exception was a trip to Athens for a math competition, but at the time I didn't have enough time or money even to visit the Acropolis.

When I look back now, in hindsight, I see a pattern of—at the time not so major—incidents when I made choices against prudent advice, without having a clear plan in my mind, only because things felt exciting and intriguing. Finding ways out was an unconscious need (although I

FIGURE 30.2. Vicky Kalogera and her research group of postdocs and graduate students. Back row from left to right: Pablo Marchant (postdoc), Chris Pankow (postdoc), Chase Kimball, Mike Zevin, Kyle Kremer. Front row: Niharika Sravan, Vicky Kalogera, Eve Chase, Fani Dosopoulou. *Credit*: Braden Cronin.

did not know this at the time); doors closed or barely opened seemed to me to be challenges for me to push them wide open and explore. This state of pushing against obstacles effectively (or sometimes just going around them) gave me a sense of control over my life; I did not let circumstances not of my choosing determine my path. Education was my ticket, and I did not blink an eye even if I was the only girl in the room.

My father's upbringing, encouraging my love of academics, and specifically math and science, while in every other aspect of life he imposed traditional gender-driven blockades, helped me figure out how to be comfortable with not taking "no" for an answer. I learned to not take "no" as a personal attack; instead, I learned how to move ahead down the paths that felt important to me despite what people said or thought. I saw the answer "no" as motivation for me to prove otherwise to them and myself. Truth be told I have not always been proven right, but when I have been it's definitely been worth it!

Chapter 31

Priyamvada Natarajan (PhD, 1999)

Adventures Mapping the Dark Universe

Priya Natarajan, now Professor of Astronomy and Physics and Director of the Franke Program in Science and Humanities at Yale University, is an expert on gravitational lensing, dark matter, dark energy, and the formation of supermassive black holes. She also holds an honorary professorship at the University of Delhi. She is a member of NASA's LISA Science Team and of the National Astronomy and Astrophysics Advisory Committee to NASA, NSF, and DOE. She is a Fellow of the APS and was a Guggenheim Fellow. Her book *Mapping the Heavens* won the Gustav Ranis International Book Prize and was a finalist for *Physics World*'s Best Science Books of the Year in 2017.

I was always attracted to a life of the mind. I grew up in a wonderful warm home filled with love, books, art, music, and people. I was born in Coimbatore, Tamil Nadu, in southern India, where my maternal grandmother lived. My father is an educationist, who fundamentally transformed and opened up access to higher education in India. He was at the forefront of research in educational testing and assessment and is an internationally recognized expert. My mother is a sociologist and geographer, intellectually formidable in her own right, and she is acclaimed for her work on Argentina, Brazil, and Chile, the subject of her first book.

I spent most of my childhood in New Delhi, in northern India. Culturally, South and North India were starkly different, and it took us a while after moving from Chennai, in southern India, to settle down and feel comfortable. I often felt like an outsider. This is a feeling that would come to permeate my entire life—the tension between yearning to be an insider and belong while feeling like an outsider. The eldest of three children, I have two younger brothers. Their families and my parents all still live in and around New Delhi. I had an idyllic childhood, which I spent running around and playing outside. My parents were avid gardeners and were very proud of their flowers, especially their vintage roses. We had fruit trees and a lone kapok tree, around which I remember seeing coiled green tree snakes that terrified me. As kids, the three of us would chase butterflies, catch them, speak to them, and then set them free.

My first encounters with science hark back to vivid memories of "scientific experiments" I conducted in our garden while still in Chennai; I must have been younger than five years old. The extensive gardens at our beautiful house in Adyar in Chennai had two-foot-high cement water storage tanks dotted around the perimeter. I remember jumping into these tanks with various toys of mine to test and see which ones floated and which ones sank.

We lived a simple but comfortable life. It was clear to me at a young age, from both my own understanding of the world around me and my parents' reminders, that we kids had won the birth lottery. Growing up privileged, as the children of middle-class academic parents in India, a country rife with inequities, we were brought up to feel deep gratitude for our fortuitous circumstances. Our house was filled with visitors, and on any given evening we never knew who might join us at the dinner table—writers, artists, and other academics were frequent guests. My mother, the incredibly gracious host, would rustle up food at a moment's notice. My father traveled a lot internationally when I was growing up, and he always brought back for us the latest toys, gadgets, and books from every trip. And these included a Sinclair ZX Spectrum, an 8-bit vintage-1982 home computer, and eventually a slightly more powerful Commodore-64 computer when I was in high school.

My parents were busy and let us all be in terms of our academic pursuits in school; there was no pressure to perform. All they wanted us to do was to explore and find our passions. They were both extremely indulgent of any and every interest we expressed. We were a close-knit family and traveled a lot. Every long vacation we were back in Coimbatore, at the foothills of the Nilgiri mountains in South India, at my grandmother's home. I have very fond memories of spending every summer vacation there with my gaggle of cousins.

I enjoyed many aspects of school and was interested in every subject, but science and mathematics exerted the strongest pull. I was apt to be bored, and so early on I developed a reading habit—devouring fiction, poetry, and whatever else I could lay my hands on. I was fascinated with maps and atlases of Earth and sky, an obsession that I carry to this day.

I had many important mentors in my early life, from my middle school English teacher, a preppy American named Dr. Carter, who encouraged me write; to Dr. Nirupama Raghavan, Director of Nehru Planetarium in Delhi, who gave me my first taste of research. This foray into research hooked me and changed the course of my life. For my first research project, I put my Commodore-64 to its first scientific use and computed star maps for the Delhi sky. The program I wrote was used to generate a monthly star map, which was published in the national newspaper. Stoked by the desire to do research as an undergraduate, which was not an option then in the Indian educational system, I realized that I would have to go to the United States for college. It was extremely unusual at that time for a young middle-class girl from India to go to the United States for undergraduate study. I was fortunate to be admitted with a full scholarship to several top universities. The Undergraduate Research Opportunities Program at MIT was a key attraction for me. I arrived at MIT as a young undergraduate—it was an exciting and heady time for me, filled with intellectual exploration, immersed in learning about all manner of subjects, engaged in research, and deeply involved in campus activities and activism.

I raced through the curriculum and requirements, took challenging graduate courses, and worked on research that would culminate in my undergraduate thesis, done under the supervision of Professor

Alan Guth. I also had been taking courses in history, anthropology, philosophy, linguistics, and literature. Through them, I discovered that my intellectual interests transcended disciplinary boundaries, and at the end of my undergraduate studies I felt deeply torn. I was on a fast-track to a career as a theoretical physicist, yet that path felt oddly self-indulgent.

The Carroll Wilson Award, which I won for an appropriate technology project to redesign the hand-pump for rural India, was deeply satisfying to the young activist in me. However, I decided that my own desire to engage and change the world was best accomplished by making a difference intellectually, with ideas. I decided to forge a new way forward. It took a lot of courage. I deferred graduate study in physics and instead entered the MIT Program in Science, Technology & Society (STS) to pursue a doctoral degree. The plan was to pursue two intellectual interests: physics and the history and philosophy of science in graduate school. I started on this track, deeply immersing myself in a humanities PhD, learning to read, write, and think about ideas and arguments that did not have the kind of neat closure of mathematics or the order imposed by physical laws. My mentor in this journey was Evelyn Fox Keller, who strongly supported and guided me through this unconventional path.

I dreamt of being a unique kind of intellectual, a professional scientist and insider, actively working at cutting-edge research while simultaneously having the distance and perspective of an outsider, as a trained philosopher of science interrogating the practice of science.

A chance meeting with the eminent astrophysicist Martin Schwarzschild further altered my plans dramatically. He convinced me that I could think and write meaningfully about science without doing an entire PhD in the humanities. His encouragement and advice led me to apply to the University of Cambridge and to work with the cosmologist Professor Sir Martin Rees. That fateful meeting with Martin Schwarzschild put me on the path to my next set of mentors who deeply impacted me and my science at Cambridge: Martin Rees, Donald Lynden-Bell, Stephen Hawking, and Jim Pringle. This pattern of unexpected turns has been the hallmark of my life and scientific career thus far.

FIGURE 31.1. Priyamvada Natarajan paddling on the River Cam in Cambridge, UK, in 2003. Photo by Lalitha Natarajan.

When I first arrived at the porter's lodge of Trinity College on a damp and cold afternoon, little did I know that this institution would end up being the most nurturing academic environment that I would ever encounter in my entire career. Looking back, I finally understand the impact of that embrace for my intellectual growth and my creativity, that unique atmosphere of collegiality, collaboration, and respect. I immersed myself in research and had an exceedingly good run. Prior to finishing my PhD thesis, I was elected for my original research contributions to a prestigious junior research fellowship at Trinity College, Cambridge in 1997. I was the first woman in astrophysics to be elected to this fellowship. These were some of the most intellectually stimulating and productive years of my life. I crossed paths with thinkers across disciplines and met many incredible artists, writers, musicians, composers, and scientists at Trinity. I felt at home. It is here that I met the sculptor Antony Gormley, with whom I would end up collaborating decades

FIGURE 31.2. Priyamvada Natarajan in her office at the Department of Astronomy at Yale in 2019. Photo by Vincent Liota.

later on a Virtual Reality installation titled *Lunatick* that would be exhibited at the Venice Biennale in 2019. After handing in my PhD thesis and spending a few months at the Canadian Institute for Theoretical Astrophysics in Toronto on a fellowship, I was offered a junior faculty position at Yale University and have remained there since.

I have always been fascinated by real and metaphorical faraway places, as well as terrains that lie just beyond reach. Studying the visible and invisible Universe has offered me the ideal opportunity for just such adventures of exploration.

My key scientific contributions to date center around the development of novel theoretical ideas and new methodologies that permit direct comparison with observational data to answer fundamental questions in astrophysics. Exotica in the Universe—dark matter, dark energy, and black holes—have been topics that I have been exploring in my research. Given the timing of my career, coincident with the availability of new tools and data from the Hubble Space Telescope and the rise of powerful computing facilities, I have been very fortunate to have made many early key scientific contributions in several areas.

Dark matter dominates the Universe, yet is detected only indirectly, and its true nature remains elusive. The entire visible Universe is believed to be structured by the scaffolding provided by dark matter, and this concept remains one of the main pillars of modern cosmology. I proposed a brand-new framework, a conceptual model that enables mapping the detailed distribution of dark matter within galaxy clusters that are so massive that they bend light. I proposed modeling the overall mass distribution of clusters as superpositions of large-scale and small-scale lumps in order to take the granularity of dark matter into account. Analysis of the observed light bending—gravitational lensing—enables mapping of the lumpiness of dark matter that can then be directly compared with theoretical predictions from cosmological simulations.

Building upon my early work, I devised several new precision metrics for use in confronting the predictions of simulations of the cold dark matter model with the best available observational data. Applying these tests to high-resolution data from the Hubble Frontier Fields Project and state-of-the-art cosmological simulations, we not only created the highest spatial resolution dark matter map to date but also found hints of potential sources of tension between observations and theory.

In more recent work, we report on an intriguing new gap between observations and theory. Anchored to my earlier work using the conceptual framework for mapping the small-scale lumping of dark matter in galaxy clusters, we found a deep disagreement between theory and observations. Clusters in simulations contain dark matter clumps that are far less centrally condensed than in the real Universe. There are two potential resolutions for this major discrepancy, both of which have far-reaching implications. Systemic issues with simulation methods might be implicated, or alternatively this could signal a potentially deeper and more fundamental problem with the assumed properties of dark matter.

Using clusters of galaxies as nature's telescopes and as astrophysical laboratories, my work has also been instrumental in establishing gravitational lensing as a powerful cosmological probe. Gravitational lensing methodologies developed during the course of my career thus far to probe the properties of the Universe have served as important science drivers to several ongoing and past survey projects on current and planned space observatories.

The confluence of ideas and instruments coupled with the timing of my scientific career have fortunately led to many firsts in terms of scientific results.

Once again, fortuitous timing has permitted me also to make key scientific contributions to our current understanding of the formation, fueling, and feedback from black holes. Data revealing correlations between the masses of central supermassive black holes and their host galaxies were first reported in 1998 as I was wrapping up my PhD. For me the interesting question these correlations raised was whether they were causal. Motivated by this key question, my work has brought into sharper focus the role that supermassive black holes play in impacting their immediate vicinity and beyond, as most galaxies in the Universe appear to harbor a black hole in their centers.

I wrote a paper, now considered very impactful, that laid the methodological groundwork for incorporating black hole growth into the larger framework of the assembly of galaxies in the Universe. Over the years, the ideas that I proposed and elaborated on in my thesis work and since, that black holes and their host galaxies grow and evolve in tandem and that the black hole affects the larger-scale environment beyond just the inner parts of the host galaxy where it resides, have become accepted. What were once considered radical ideas now form the basis of our current understanding of the relationship between massive black holes at the centers of galaxies and their galaxy hosts. My subsequent refining and honing of this framework has permitted modeling populations of black holes and direct comparisons with multiwavelength observations.

Quasars, powered by growing black holes, are some of the brightest objects in the Universe. Motivated by the observational discovery of a population of incredibly bright quasars in the farthest reaches of the Universe, I proposed and developed a potential new channel for the formation of seed black holes—direct collapse black holes (DCBHs)—from the collapse of gas in the early Universe. DCBHs can account for the origin of the earliest quasars found so far. Once again, I worked on synthesizing and embedding DCBH formation into the larger framework of the formation of galaxies, providing a new causal basis—arising now out of the initial conditions—for the observed local correlation between the black hole mass and the properties of the galaxy host. This model will be tested with data from the upcoming James Webb Space Telescope (JWST). I have worked on many interesting problems in black hole physics that explore astrophysical processes on a range of spatial scales, involving spinning black holes and the role of gas in catalyzing the mergers of black holes. These calculations laid the groundwork for understanding the electromagnetic fireworks accompanying the mergers of black holes that should be detected via the emission of gravitational waves, detectable by the planned future Laser Interferometric Space Antenna (LISA) mission. My work on these fundamental physical processes is central to the emerging new field of multimessenger astrophysics.

Furthering my work on understanding the connection between black holes and their environments, on physical scales beyond the galactic nucleus, in papers deemed speculative at the time, I proposed that these quasars would drive gas outflows that would leave an unusual detectable imprint. Twenty years later, the first detection of these imprints using the Atacama Large Millimeter Array (ALMA) telescope was recently reported. I had also calculated the fate of this swept-up gas and shown that it would fragment and likely form abnormal dwarf galaxies—ones that were deficient in dark matter. I also proposed that supermassive black holes would stunt their own growth, which implies that their masses would be capped. This prediction of the absence of black holes with masses in excess of a few tens of billions of solar masses in the brightest galaxies in the local Universe has also been observationally verified.

My trajectory and life in science has been non-traditional in more ways than one. I have never been motivated by careerism and have followed my interests and passion in the topics and problems on which I have chosen to work. Countering the pressure to move toward joining and working in large scientific collaborations, my own intellectual style has involved working in a small group—with a few collaborators and students. This has afforded me freedom and permitted me to take intellectual risks while working on what were deemed somewhat speculative ideas at times. My choice to work in this mode was driven by temperament—a natural aversion to politics and hierarchy—but was also shaped by personal circumstances. I made a conscious decision to carve out an intellectual space where I would fit, belong, and be able to make my own mark. Also, the decision to work on two fundamental open questions—the nature of dark matter and the understanding of black holes—rather than on one narrowly focused research question aligns with my nature but runs counter to the academic style in the field. Convention dictates that expertise is garnered, authority and recognition conferred, only to those who establish their credentials as experts in very tightly defined and limited domains. These personal choices have not been without costs in terms of access to research funding and recognition.

As a woman, and a woman of color at that (sadly a rarity in astrophysics still), perhaps I ought to have paid more heed. My career path has been challenging, with many obstacles, some systemic and some rooted in personal prejudice. Fortunately, these have been superseded by the many more numerous moments of unparalleled pure joy derived from the thrill of discovery and the excitement of figuring things out. The feeling that I first experienced when I was a teenager who had coaxed a Commodore-64 to make a star map is what I always imagined a life in science would be all about—luckily I still feel that from time to time. The personal challenges that have come with an intersectional identity have catalyzed me in turn to become a very active mentor for younger scientists and to be deeply invested in issues of equity and inclusion in the field. On the flip side, my background and educational training in the history and philosophy of science have enabled me to have some intellectual impact beyond astrophysics with my writing. My writing outside of astrophysics has focused on critically examining the process of scientific discovery and demystifying the process and practice of science.

Most of all, I am thrilled to have the opportunity to do what I love for work and be part of a community of scientists, most of whom are fellow adventurers who share the very same insatiable curiosity to learn continuously. I have been very fortunate to have had the unwavering support and affection of my parents and family through the ups and downs of my life in science despite their being thousands of miles away. Meanwhile, my built family in science and beyond includes many of my close scientific collaborators; my students and the close friendships that have endured over the years are ever a source of inspiration and warmth. While I have yet to resolve the tug-of-war between feeling like an insider and an outsider simultaneously, with age I have come to realize that this tension has actually been productive and enriching as well as empowering. I remain as excited and open as I was as a child to new cosmic adventures of discovery.

Chapter 32

Dara J. Norman (PhD, 1999)

On Becoming an Astronomer and Advancing Science

Dara Norman, now an Astronomer and Deputy Director for the Community Science and Data Center and the National Science Foundation's National Optical-Infrared Research Laboratory (NOIRLab) in Tucson, conducts research in observational cosmology, in particular on observational biases associated with the study of quasars and on the influence of AGN on the evolution of galaxies. Her work now focuses on diversity and inclusion in astronomy and astrophysics. She is a Legacy Fellow of the AAS.

In graduate school, I participated in many outreach activities. One of them was Project Astro, in which an astronomer is paired with an elementary or middle school teacher to teach topics in astronomy. We first spent a workshop day with our teachers, getting to know each other and learning about possible activities to do with our class. In an icebreaker activity, each person shared what they wanted to be when they were ten. I had wanted to be an astronaut, but poor eyesight ruined my chances. It was clear that the teachers' youthful goals were all over the map! They had wanted to become doctors, firefighters, dancers, police officers, nurses, pilots, and actors . . . everything except teachers! The astronomers, however, all said they had wanted to grow up to be either an astronaut, a scientist, or an astronomer. I had never before realized that despite my unconventional background and the meandering route I had taken in my career, I was part of a privileged group of people who had known what they had wanted to do since age ten and had had the opportunity to stay on that path.

My mother, sister, and I moved to Chicago just after my parents split up. The city of Chicago is not optimal for stargazing, but when I was young, it did offer opportunities to develop a love of science and art through access to the many local museums. I remember being interested in space in third grade and had set my sights on becoming an astronaut, even though the boys in my fourth grade class insisted that girls could not be astronauts. Although the nighttime sky proved inaccessible from the city, I could still see it at Adler Planetarium. Even in middle school, my friends and I would take public transportation there or to the Museum of Science and Industry to see shows and play with the exhibits. For the price of a round-trip bus ticket, we were being tricked into learning about science! Sadly, the opportunities to naturally develop an interest in science through informal learning are dramatically less accessible now to young people growing up in socioeconomic circumstances similar to my own.

I was excited to leave the Midwest for college and head east to Boston. In 1984, my cohort had the largest percentage of women (25 percent) yet for an incoming first-year class at MIT. Because I had already taken some college-level physics classes, I signed up for the more advanced physics class. The class was smaller than the regular freshman intro class, with about fifty people. I was one of only two women in the class.

Despite some difficult adjustments, including being ignored by some men in the class and having others joke about having me, a Black woman, in the class, I did fine that first year. I was confident and excited for sophomore year when I started coursework for my major. I couldn't decide between physics and electrical engineering, so I thought I would keep my options open by taking introductory classes in both. I selected six classes and presented this schedule to my assigned mentor in the physics department for review. He looked at it and, without any comment, approved it. I had no idea that six classes were considered a heavy load for a semester. I was miserable, struggled to get everything done, and steadily fell further behind. My anxiety led to panic attacks. After this hellish semester, I ended up on academic probation. Limited to taking only four classes, I decided to take *only* the classes that genuinely interested me. One of them was an observational astronomy class; the instructor was Professor Jim Elliot.

Jim became my academic advisor, and it is not hyperbole to say that I would not have made it through MIT without him! In Jim, I finally had an advisor who truly cared about me. He took time to go over my classes and my university career plan. He gave me tips on how to reduce stress, and he was a friend who loaned me a bike when I moved off campus. He made me feel that I belonged both at MIT and in science. He invited me to stay at MIT as a graduate student working with him, but when I told him that I had had enough of school for a while, he kept me in the field of astronomy by helping me find a job at the Goddard Space Flight Center (GSFC) as a research assistant and public outreach officer.

At the "Jimboree" in 2010 (a celebration of Jim's life and work organized by his former students just a year before his death), it was interesting to learn that so many of us—especially women—had similar stories of being ready to leave MIT until we met Jim!

I worked at GSFC for two years on the team supporting the Goddard High Resolution Spectrograph (one of the five original instruments built for the Hubble Space Telescope [HST]) and had established myself as competent in working in the field. I had even co-authored a paper. HST had just launched, and the major problems with the mirror design had already been discovered. Plans were being made for a shuttle mission to fix the problem, but were it not to go well, it was unclear to me if I would still have a job. So I decided it was time to return to school.

I was accepted into two astronomy PhD programs, at the University of Colorado-Boulder and the University of Washington in Seattle. When I visited Boulder, it rained for two of the days I was there, something the locals insisted was not common in March. What gave me pause about UC Boulder was a comment from the professor who showed me around town. When I commented that Boulder was small, but Denver was nearby and I could find more big-city things to do (museums, ethnic food) there, the professor said, "Denver? Oh, we don't go there!" Hmmm . . . I thought, what does that mean? Who are "we" and what is wrong with going to Denver? Might there be too many people in Denver who look like me?

In Seattle, although I arrived in early April, the sky was completely clear, with the ghostly image of Mount Rainer visible off in the distance. The professor who was my contact made sure to introduce me to women and other minority graduate students in the department and to point out that, if I attended, I would have fellowship funding through the Office of Minority Education, given to encourage more underrepresented minorities (URM) in graduate school. I selected UW.

At UW, I found that the most important resource was the graduate student cohort! During my time, the grad students were all quite supportive of one another. One of them and his wife had had a baby, and for one semester, for about two hours twice a week, neither of them could cover childcare because of their classes. They offered to buy lunch for anyone who could watch their daughter during these hours. The great thing was that so many students stepped up to help out! We took turns through that semester keeping little Kayanne, who later returned to do her undergraduate work at UW!

Graduate school can be intense under any circumstances, but I was sometimes aware of situations that were unique to me (bad and good) as a Black woman in a field that is predominantly White and male, and failure was not an option. While I definitely had support from my fellow students, and a few of the faculty were mindful and aware of the issues of race that might weigh on me as the sole Black graduate student in the department, I had to navigate many micro-aggressions. My attempts to identify a research project with one professor by showing him the research I had done, similar to his area of expertise, during the summer at another institution were met with baffling redirects to find someone else with whom to work. Another professor, with whom I was doing a small initial project, would just not show up for scheduled meetings. I would find out, from office staff, that he was actually out of town.

Then there were some difficult undergraduates, one of whom made no secret of his contempt for having to be taught by a Black person, and a woman at that! He accused me of both not knowing what I was doing and explicitly downgrading him on test questions. Interestingly, because of the class grading strategy, each TA graded a single question for the whole class and I had not graded the questions he complained about. Unfortunately, for a Black woman in STEM these kinds of slights and dismissals in one's professional life are made easier to handle because they are an extension of the larger world in which we have to survive. My goal was to channel my anger and disappointment into a kind of revenge: "Dismiss me? I'll show you!"

My dissertation research concerned gravitational lensing of distant, background quasars by foreground masses, the mass doing the lensing being a group of galaxies in between us and the more distant quasars. Although astronomers often speak of quasars as if they are objects, in fact, a quasar is a phenomenon that occurs at the center of a galaxy because a galaxy has a massive, central black hole. Some of these black holes are sitting passively in their galaxies, but others are "active," which means that they are surrounded by large amounts of gas and dust spiraling into them. The gas and dust are heated as they swirl around the black hole and, as a result, shine very brightly before being swallowed by the black hole. Magnetic fields corral charged particles from the swirling material into jets that launch glowing ions away from the black hole before the black hole has a chance to capture them. Those galaxies with active black holes at their centers are known as active galactic nuclei, and the brightest of these are known as quasars.

But bright quasars may not be all we think they are! We know that the gravitational force exerted by an object is a result of its mass, and more massive objects have stronger gravitational pulls. We frequently see examples where the gravity of massive objects, like clusters and groups of galaxies, bends the light traveling from objects behind that mass. This bending of light rays works similarly to a lens in prescription eyeglasses; therefore, we refer to this phenomenon as gravitational lensing. Gravitational lensing can do two interesting things: it can magnify objects, making them appear brighter, and it can distort objects and the space around them, changing their perceived shapes or locations. Extreme examples of the phenomenon are the famous Einstein Cross,

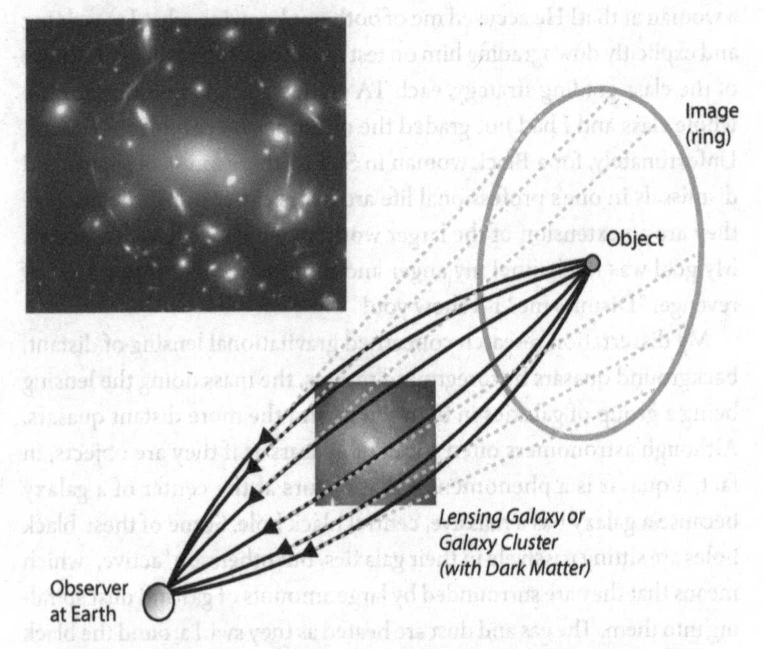

FIGURE 32.1. Example of gravitational lensing. The light from a distant bright object (solid curved lines) is bent by an intervening mass as it travels to an observer on Earth. The intervening mass can be a galaxy or a cluster of galaxies. The projection of the light rays back to the source (dashed lines) appears to come from an image that looks like a ring. The inset image is a Hubble Space Telescope image of Abell 2218, a massive cluster of galaxies where light from some of the object galaxies has been made to appear as elongated arcs (partial rings) because of gravitational lensing. *Credit*: STScI.

which shows how a gravitational lens splits the light from a single distant quasar into four separate images projected in the shape of a cross, and the Hubble image of galaxy cluster Abell 2218, which shows giant arcs that are elongated images of background galaxies.

The gravitational lensing phenomena that I explored as part of my thesis is called statistical gravitational lensing, and magnification bias is an example of it. Magnification bias is important, however, because evidence of its existence reminds us that, because of our vantage point here on Earth, when we look out into the heavens we have to be aware that we are looking through a plethora of foreground mass, and that mass can unknowingly distort some of our measurements, which in turn can distort our understanding of the Universe. Therefore, if we want to understand our data, we have to know about the magnification and distortions induced by gravitational lensing. My thesis was about measuring the existence of magnification bias by measuring correlations between bright quasars (perhaps made to appear brighter because of lensing) and foreground galaxies.

I found a significant statistical correlation of the presence of a high number of galaxies in fields with quasars compared to a much smaller number in fields without quasars. However, for any individual quasar it is hard to say how much it is magnified due to lensing. While this statistical study was interesting and gave me insights into the idea of understanding bias in science, I was frustrated in not being able to say anything about any individual quasar, its evolution, or the initial triggering that turns a galaxy from having a passive black hole to one having a full-on active event at its center. My frustration prompted me to switch my scientific interests away from gravitational lensing studies toward asking questions about the active galaxies themselves. I especially wanted to understand how the large-scale environment, beyond the galaxy's borders, might influence what happens at the galaxy's center, near its black hole.

After I had completed my PhD and spent five years doing research in Chile, my husband was offered a tenure-track staff position at the North base of the National Optical Astronomy Observatory (NOAO-N) in Tucson; I was offered a postdoctoral position, also at NOAO-N. While I was glad to have a job, it was clear to me that my years of trying to identify a niche in collaborations through outreach activities had led to an undervaluing of my work as a researcher. Typically, I was not included in collaboration activities; when I had suggestions about areas in which a collaboration should branch out, others were brought on board to lead those efforts. I was annoyed and angry and had to decide what I would do after this three-year job opportunity. In frustration, I threw myself into work on diversity and inclusion issues in astronomy and astrophysics through the Committee on the Status of Minorities in Astronomy and the National Society of Black Physicists (NSBP). With

FIGURE 32.2. "Dr. Dara" Norman on the lawn of the U.S. White House for the first WH star party, in 2009. I met both Sally Ride and Mae Jamison at the event. Later that evening, President Barack Obama and his family stopped by to look at Jupiter and its moons through the "time machine," a reference to the time delay in receiving light that has traveled across space.

the 2010 decadal survey in astronomy approaching, a more senior colleague suggested to me that someone should write a white paper on the lack of minorities in the field of astronomy. I suggested to two colleagues that we should co-lead the writing of that paper. However, as we got organized, both of them were asked to serve on decadal panels, and due to the rules neither could be an author on a white paper. Thus, I was left to lead the effort on my own.

At about the same time, the president of the Associated Universities for Research in Astronomy (AURA) required the directors of each of the AURA centers to appoint diversity advocates at their centers. David Silva, director of NOAO, asked for volunteers, and though I had no intention of applying, apparently others at NOAO suggested me. Silva came to me and asked if I would accept the role as diversity advocate.

I told him I would not take the job as a postdoc but would accept if offered a staff position. Subsequently, I was hired as an assistant scientist with 80 percent administrative duties and 20 percent research time. With 25 percent of my time targeted at diversity work, I now was able to timecard the diversity, equity, and inclusion activities in which I was already engaged. The more important part of the position was the opportunity to report directly to the director on issues of diversity. I was able to speak frankly with him on some policy matters and establish myself as the lead person in an area key to both the internal and external mission of the center.

In my role as diversity advocate at NOAO/AURA, I had the opportunity to work with a number of excellent social scientists. As a result, I became better versed in issues of diversity and inclusion in STEM. I began hearing more about issues that contribute to the paucity of URM scientists in the field. Ideas such as unconscious bias, unearned privilege, stereotype threat, and imposter syndrome were things I had experienced, but previously I lacked the language to name or express them. I began giving talks on these topics within NOAO, as well as to other groups. When I speak about unconscious bias, I point out that I am amazed that as scientists we understand the idea of bias in our data and methods and how bias can lead to wrong conclusions. We work tirelessly to identify such biases, and when they are found we collect more data and try new methods to mitigate or eliminate that bias. However, when confronted with bias in our profession and its workforce, despite clear examples of challenges, problems, and biases, many of us continue to deny the existence of the issue instead of looking at the data and then changing practices and modifying policies. As scientists, we should be and do better; we cannot continue to deny the existence of bias and hide our heads in the sand on these issues.

Despite the "liberal" natures of my colleagues, unintentional microaggressions pop up often. One person told me that postdoctoral researchers hired at the southern location of our organization, where I had started, are inferior to those hired in the north. Many of my suggestions have been dismissed with comments like, "Oh, I think *we* know what we are doing." I found out, when I asked a friend in accounting just to tell me if I was the lowest-paid staff member, that I was being underpaid compared to male staff members employed at the same level. Her answer: not only was I lowest, but I was barely making more than postdocs. After two years of going through proper channels, I made the decision to go through "improper" channels in order to receive the compensation I deserved. It became clear to me that I was underappreciated and undervalued within the NOAO organization.

In 2014, I visited a colleague at Howard University whom I had met through the NSBP. He invited me to meet with female professors who were "interested in talking with me." As I went to the meeting, my friend excused himself to teach a class. As the conversation continued, it became evident that I was being interviewed! Toward the end of the meeting, they asked me if I would be interested in spending a sabbatical year at Howard as an Advance-IT Fellow. My position did allow for sabbatical time, but with two small children and a husband, I struggled to figure out how to make a sabbatical far from them work. My sponsor, Professor Sonya Smith, allowed me to use money from the fellowship to support additional childcare for my children back in Tucson. This flexibility was crucial in enabling me to take this sabbatical.

My time away from NOAO in 2015 was unbelievably freeing! It gave me perspective on the micro-aggressions I had been experiencing, because in DC I was valued for the experience and expertise I brought, not only to Howard but to other nearby organizations that I had opportunities to work with while there. The leadership I had shown in organizations outside of NOAO was being recognized and rewarded! Shortly after my sabbatical, I was offered another position back in DC. I used that offer to leverage an expansion of my duties at NOAO and a move into a leadership role.

As I have proceeded through my career, I have found that the traditional path was not for me. As I tried to stay on it, I found myself marginalized and alienated. The research I did both as part of my thesis and after seemed too niche; the few people most interested in it would just spend their time telling me what I had done wrong. Ultimately, indulging in this research seemed to pale in comparison to that which I have experienced as a Black American woman just trying to *be* a scientist, an astronomer. I now get the most pleasure from my work enabling others to succeed in their science while remaining true to their identities, both within science and elsewhere. My current work on improving the culture of astronomy and astrophysics to better encourage, support, and welcome *all* researchers, students, and professors will, I hope, make a meaningful difference and is the best way for me to advance astronomical discovery.

Chapter 33

Sara Seager (PhD, 1999)

Adventures in the Search for Other Earths

Sara Seager, now the Class of 1941 Professor of Physics and Planetary Science at MIT, has laid the foundation for the field of exoplanet atmospheres, thereby helping transform the field of exoplanet discovery into a full-fledged part of planetary science research. Her work on space missions includes being principal investigator of the CubeSat ASTERIA, Deputy Science Director of NASA's TESS mission, and a lead on NASA's Starshade Rendezvous Mission concept. She is an Officer of the Order of Canada, a MacArthur Fellow, a Fellow of the National Academy of Sciences, the American Academy of Arts and Sciences, and the American Philosophical Society, and a Legacy Fellow of the AAS, which also awarded her the Helen B. Warner Prize in 2007. Her asteroid is 9279 Seager.

My mind was always in its own little universe, puzzling over and daydreaming about people, places, and things. As a small child I wondered, "Why does the Moon keep following me?" In the car no matter how far we drove or which way we turned, the Moon stayed fixed on me like a giant eye in the sky. One weekend evening my dad took me to a "star party." Star parties are not where Hollywood celebrities gather but rather are where amateur astronomers set up their telescopes and invite the public along. When I first saw the Moon through the telescope I couldn't believe my eyes—a whole new world! Later, on my first camping trip, I saw the dark night sky and a shocking vista of an endless number of stars. What is out there? Surely something giant and mysterious and bigger than us all.

I lived with my dad on weekends in the Toronto suburbs, hence the long car rides. Back in the inner city, life was not as dreamy. My stepfather was a monster—mentally torturing me and my siblings at every opportunity. I wondered how people could be so evil and why my mother watched without stopping him. Although I couldn't articulate it at the time, I learned to accept that life is unfair. Even though I was terrified of my stepfather, as I grew older I eventually saw him as pathetic, such that the only good thing to come out of the situation was my lasting disrespect and even deep disdain for the white male authority.

One day while walking to high school during my normal cut-through of the University of Toronto campus, I saw a sign for an open house weekend, specifically for the astronomy department. At the open house professors and graduate students told me about their work. "Wait, what?" I thought. "I can be an astronomer? For a job?!" I rushed home to excitedly tell my dad about my new career plan. He paused, then immediately launched into a long harsh lecture: "Sara, you need to get a job, support yourself, and NOT RELY ON ANY MAN!" My dad did not know if astronomers could find secure employment. He wanted me in a well-paying job and one that would enable me to have enough free time for a family.

I entered the University of Toronto to study pre-medicine to become the doctor my dad wished for. Soon enough, however, I found myself majoring in math and physics instead. I loved my summer job at the David Dunlap Observatory, located on the outskirts of Toronto, cataloging variable stars during the day and observing a program of variable stars with a 24-inch telescope at night. At the university, while working on a senior-year research project, I managed to get my rudimentary computer algorithm, which described the Sun's magnetic field and demonstrated the flipping of the Sun's eleven-year solar magnetic cycle, to work literally moments before my final presentation. The thrill of discovery—even for things already known to others—was incredibly satisfying. My path was now clear: astronomer.

In the fall of 1994, I headed to Harvard graduate school in Cambridge, Massachusetts. I supposed I would pursue "magneto-hydrodynamics," building on my undergraduate project and with the attitude that harder is better. But the Universe had other plans.

The discovery of the first exoplanet orbiting a Sun-like star was announced in October 1995, making headlines around the world. The new planet, 51 Peg b, was a Jupiter-mass planet orbiting seven times closer to its star than Mercury is to our Sun. It was nearly inconceivable to astronomers at that time that planets could form so close to the host star. Indeed, astronomers had been searching for Jupiter-mass planets with Jupiter-like orbits (over one dozen times farther from the host star than Mercury is from our Sun). Yet discoveries of more "hot Jupiters" were soon announced, their signals having been present in already-collected data.

Observational searches for exoplanets were a fringe novelty back then, and most astronomers thought it was a research topic to be avoided. The planets had not actually been "seen"; rather, their existence had been revealed by their indirect effects on their host stars. A star wobbles as the planet and star orbit their common center of mass; astronomers can measure the resulting, tiny back-and-forth motion of the star along our line of sight (called the radial velocity). The expected, natural skepticism of scientists focused on the arguments that a new kind of stellar pulsation probably caused the star's motion or that stellar variability mimicked the effects of a planet.

My research advisor, Dr. Dimitar Sasselov, was interested in the new exoplanets and suggested I work on the specific topic of their atmospheres. I loved the idea of exploring something brand-new, something that no one had worked on before. Our goal would be to study how the starlight heated the atmosphere of a hot Jupiter and what the resulting planet's spectrum might look like. Dr. Sasselov handed me an archaic computer code that was built for two stars orbiting each other and suggested I adapt this code to one star and one planet in orbit around each other.

I struggled in grad school, at first because I did not have the key skill I needed, how to write computer programs, and later because graduate students often receive little to no short-term positive feedback.

Graduate school projects are very long-term investments. Although I had made some friends, and my boyfriend Mike had moved from Canada to the United States to live with me, I often felt depressed. "It's normal," people told me.

One late winter afternoon, after many months of rewriting my computer code to work for planets instead of stars, I finally got my code to converge to a solution. I stared at my screen, my heart beating fast with excitement. I knew my results were correct because the planet looked like a fainter version of the star, as it should at visible wavelengths because the planet's light is mostly reflected starlight. I felt the triumphant thrill of discovery.

More and more exoplanet discoveries were being announced, including some orbiting far enough from their host stars that their signals could not be explained away by stellar pulsations. The detractors were now muted. Nevertheless, people, even including a professor at my own institution, continued to attack my work and plans. "We can barely detect giant exoplanets. Why are you bothering to study exoplanet atmospheres if we have no hopes of ever observing them?" "Why would you simulate clouds on exoplanets, a level of detail we'll surely never need?" My answer was that clouds dampen spectral features so we must understand their presence, but my reasoning fell on deaf ears. I had no reason to listen to or agree with "authority" so I pressed on, ignoring the naysayers and my own lethargy, successfully completing my PhD work and delivering my required PhD defense seminar to a packed lecture hall.

Mike and I, now married, moved to Princeton, New Jersey, in the fall of 1999, where I began my postdoc at the renowned Institute for Advanced Study. When I arrived all the other postdocs used their standard small-talk opener, "What's the next big thing?" A transit, I told them. If a planet-star orbit is fortuitously oriented edge-on from our viewpoint, the planet will go in front of the host star, or "transit." A transit will cause a small drop in the star's brightness equal to the planet-to-star area ratio. A transit detection would be a huge milestone because we could measure the planet's size to go along with its mass, thus measuring the planet's density. In turn, the planet's density can yield information on the planet's bulk composition. Moreover, if a planet transited at the moment predicted by the radial velocity measurements, there would be no doubt of its planetary status. The closer a planet is to its host star, the greater the probability for the planet to produce transits (assuming randomly distributed orientations of the host star's equatorial plane). Each hot Jupiter orbiting a Sun-sized host star has about a 10 percent probability to transit. At the time, about seven hot Jupiters were known (out of a total of about 30 known giant exoplanets).

I thus started working on what I called "transit transmission spectra." If a planet transits its host star, some of the starlight will be absorbed by the planet's atmosphere. In other words, the gases in the planet's atmosphere could be detected by their imprint on the light of the star. I used my exoplanet atmosphere and chemistry model to predict (among other things) that sodium gas should produce a strong signal at visible wavelengths. That same fall of 1999, right on time, the first detection of a planet via the transit technique, HD 209458b, was announced, and I rushed to complete my paper.

In my naive little postdoc bubble, I had no inkling of the cutthroat actions of other scientists. The peer review referee report on my new paper started well enough but unraveled to an ending with X-rated language: "You are like a %$#@-$%#@!@# bureaucrat at NASA." I burst into tears, then comforted myself with the thought that the reviewer is probably a pathetic old man. (Later, at a conference, I received an apology from the reviewer's colleague who had urged the older man not to submit such an offensive report.) Soon after, a friend, a graduate student, let me know that he was leading a proposal to the Hubble Space Telescope, using my prediction to observe sodium gas in the new transiting exoplanet's atmosphere. He explained that his team thought the proposal had a better chance at success with an older established (male) scientist rather than me as a member of the team. My blood boiled.

The proposal and the observation of sodium gas in HD 209458b's transmission spectrum were successful—ushering in the birth of the brand-new field of exoplanet atmospheres.

I was ready to move on from a postdoc to a faculty position. One university rejected me because at the time I had "never worked with real data." At another, a senior professor started out by saying my successful transiting planet atmosphere prediction was neat, but then dismissed it as a "one object, one method success." Either my work just was not good enough for me to get a faculty job or people thought there was no future in exoplanets, or both. A year later nothing had changed. I was once again dismissed from a faculty job interview with a statement echoed by most of the professors: "there will never be very many transiting planets."

As it happened, I knew how close we were to unleashing untold numbers of transiting planets. A Princeton University postdoc, Gabriela Mallén-Ornelas, and I were working to find transiting planets by a new method: a wide-field telescope survey simultaneously monitoring tens of thousands of stars for the tiny drop in brightness that might indicate the presence of a potential transiting planet. My postdoc mentor John Bahcall encouraged me to think big and take risks, supporting this ambitious project in what was a very competitive field with others racing for the same goal.

For a variety of reasons our project failed, but not before we found what could have been the first planet detected by the new method! Alas, the signal turned out to be from an eclipsing binary star. What should have been a deep transit signal mimicked the signal of a shallow, planet-like transit because of the presence of contaminating light from another star, either a background object or a third star that was part of the same eclipsing binary star system, whose light fell on the same pixel. We had discovered a "blend," what would turn out to be an insidious but very common type of false positive signal for a planet transit.

This blended transit looked awkward and led us to explore the equations that describe the light curve of a transiting planet. Working close to midnight one Saturday and furiously scribbling through algebra, we both burst out laughing at the same time. The equations told us that we could measure the density of the host star using the transit light curve alone. Although we never discovered a planet, we did establish one of the best methods to rule out false positives: if the host star density as measured from the transit light curve disagrees with the star density as measured from the conventionally determined mass and radius of the star, then the transit light curve indicates a false positive. The work remains one of my most highly cited papers today.

I received a job offer from the Carnegie Institute of Science in Washington, D.C., a small, flexible research organization that recognized the potential for exoplanet research. At Carnegie I found a remarkable group of scientists working at the frontiers of many topics in astrobiology. Outside of Carnegie I was focused on the work of the NASA Terrestrial Planet Finder (TPF) study team. TPF would be a sophisticated space telescope, with innovative new technology to block out the starlight so we could observe a planet directly—a nearly impossible task for a true Earth twin that is 10 billion times fainter than its host star.

I also wanted to start a family, now that I had a permanent job with a decent income. Less than a year after I'd arrived at Carnegie, on a gorgeous summer solstice day, I gave birth to baby boy Max. He was adorable with blue eyes and fair hair that appeared to have been cut just so. He was calm, good-natured, and a sound sleeper and brought so much joy. I could even get some work done while he napped. I was proud to be such a great mom. Two years later my son Alex was born, adventurous and feisty, demanding, and not a great sleeper. I had to concede nature over nurture; Max being an easygoing baby and sound sleeper had nothing to do with my parenting skills. Most mornings, just getting myself to work was a major accomplishment.

Meanwhile, other academic research institutions began acknowledging that the rapidly growing field of exoplanets was here to stay. The Massachusetts Institute of Technology (MIT) had been courting me for a faculty position, as someone who could jump-start exoplanet research university-wide.

On a cold December day in 2006, Mike and I and the kids moved back to the Boston area so I could start my new job at MIT. My beloved dad had died just days before, a huge blow to my psyche. An earlier professional-type death also weighed heavily on me: the TPF mission that was being funded at $50 million per year had been canceled. Now, lost without these two anchors in my personal and professional lives, I had to shift gears.

First, I hatched a backup plan to the TPF. If an Earth-sized planet is in an Earth-like orbit that transits a nearby, bright Sun-like star, we have no way of finding it. Furthermore, a single large telescope could never

FIGURE 33.1. Sara Seager in her office at MIT in 2013. Photo by Justin Knight.

observe stars spread out all around the sky for the two or three continuous years needed to find planets with year-long orbits. I dreamt of a constellation of tiny, Earth-orbiting telescopes, each monitoring its own bright, Sun-like star. With a team of students at MIT and the help of Draper Lab, I would soon develop the prototype of this telescope. The second plan was to embark on an exhaustive study of literally all possible biosignature gases, gases made by life that could accumulate in an exoplanet atmosphere and potentially be observed remotely with next-generation telescopes. With my students at MIT, I further explored the exoplanet research frontiers on more near-term topics. For example, Leslie Rogers and I worked to understand sub-Neptune-sized planet interiors, before any of the now numerous planets in that category had been discovered. Nikku Madhusudhan and I invented atmospheric retrieval for exoplanets, a method in which we use observations combined with computer simulations to constrain the range of atmospheric properties (such as temperature and the amount of different gases present), and later students helped develop this concept further.

A few years after I had moved to MIT my husband, Mike, had an ongoing stomachache that rapidly worsened. Following surgery, Mike was diagnosed with a rare, small intestine cancer. Our lives spiraled downward as chemotherapy had failed to prevent the cancer's spread. I took care of Mike the best I could so he could die peacefully at home—it broke my heart to say good-bye to the best friend I'd ever had. Just eighteen months after his initial diagnosis I was widowed at age forty, with our children now six and eight years old. In the many condolence letters I received from colleagues across the country, the best were the rare few that did not try to sugarcoat the situation. One simply sympathized, "Life is unfair. You've been dealt an extremely bad hand of cards." Another, "This is a terrible trial."

Some tough years followed. Because Mike had worked only part-time in order to support my ambitious career, I had never really experienced the plight of the working mom. Now I was drowning in a depressive grief with a steep learning curve for accomplishing what should have been simple household tasks. I burned through my savings just to pay for household help and babysitters so I could work. I stopped traveling to give talks because the cost of 24/7 childcare was ridiculous. I had to drop out of my leadership role on NASA's MIT-led Transiting Exoplanet Satellite Survey (TESS) Mission that had been selected for launch, and I cut back on most nonessential work activities.

To exacerbate the situation, Mike's death had catastrophically shifted my sense of purpose. Most everything seemed pointless. People appeared to me to be like guinea pigs spinning on exercise wheels of drudgery, with no escape possible. I had to admit that the field of exoplanet research was rapidly maturing and getting more and more crowded with duplicated research. It is my style to pioneer new things, not turn the crank on established topics. So I needed to discard some of my research portfolio and intentionally focus my direction. In order to get out of this midlife career crisis, I decided to work with renewed purpose on the most impactful topic I could imagine: to find another Earth orbiting a Sun-like star, one possibly with signs of life by way of atmospheric biosignature gases.

Months passed, and ever so gradually I began to feel better. Students and postdocs became my extended family, joining us for vacations after far-flung conferences to which I had to drag my kids along. My three babysitters and housekeeper became extended family as well, and I couldn't have made it through those difficult years without them. I look back and can fairly say I became so mentally tough and strong, having survived the terrible trial and made the most of a poor hand of cards. In the fall of 2013 I was awarded a MacArthur "genius" grant and addressed Congress, making the first official statement for our nation about the status of the search for other Earths. When I was trying to explain all of this to my kids one evening, my younger son, by then eight years old and trying out his new vocabulary on me, said, "Mom, you are . . . arrogant." I laughed for the first time in what felt like years.

Then an amazing thing happened. After I responded to a NASA call for participation on concept study teams for two separate "reduced scope TPF" concepts, formally called Probe-class missions, I was asked to lead the Starshade study. Since I had dropped all other obligations during my husband's illness and since his death, I had time to begin a new, big endeavor. Starshade is a giant, specially shaped screen that is tens of meters in diameter and must fly in formation with a space telescope located thousands of miles distant. Starshade will block the intense glare from the starlight that would otherwise prevent the telescope from observing the planet. In fact, a Sun-like star is 10 billion times brighter than an Earth-like planet, so the starshade will have to *block* 999,999,999 out of a billion photons and let pass only 1 in a billion to enter the telescope. My team was able to bring Starshade from a perceived, unworkable, "dead idea" to a mainstream, viable piece of technology now ongoing toward a real mission.

Just before my team's first face-to-face Starshade meeting, another beautiful thing happened. I traveled to Thunder Bay, Canada to give a keynote lecture to the Royal Astronomical Society of Canada, a national umbrella organization for the same amateur astronomy club that held the star party at which I first viewed the Moon, so many years earlier. At the reception that kicked off the weekend-long event, I saw a tall, dark, handsome man across the room and just thought, "Wow, who is that man? I have to meet him!" An instant friendship blossomed into a tremendous romance. Charles and I married in 2015, and he adopted my boys.

Today in 2021 we know of many thousands of exoplanets. A large fraction are transiting planets, thanks to NASA's now retired Kepler mission, numerous ground-based wide-field transit surveys, and the TESS mission (for which I was given my position back and worked as its Deputy Science Director). On the order of one hundred exoplanet atmospheres now have been observed via my method of transmission spectroscopy. (When I recently visited one of the esteemed universities that rejected one of my faculty applications, the same professor who stated there would never be many known transiting planets arrogantly claimed, "I always knew exoplanets would be big.") My prototype tiny telescope, eventually implemented and built by NASA's Jet Propulsion Laboratory and named ASTERIA, successfully completed a two-year mission in low earth orbit, and the related constellation study is ongoing. Starshade continues to be technically developed toward flight readiness. Today my work on biosignature gases remains one of my team's main threads of research.

For some reason my new work is still often rejected as unorthodox, while most of my offbeat work from one and two decades ago is now as mainstream as it gets. I now adopt the expression, "In exoplanets, the line between what is considered crazy and what is mainstream is constantly shifting." I hope this inspires scientists to spend some part of their time pursuing bold new ideas, especially those that are joyful and that may lead to the unparalleled thrill of discovery.

Chapter 34

Hiranya Peiris (PhD, 2003)

From Serendip to Serendipity

Hiranya Peiris, now Professor of Astrophysics, University College London, Professor in Cosmoparticle Physics, Stockholm University, and Director of the Oskar Klein Centre for Cosmoparticle Physics, Stockholm, combines observations, theory, and advanced statistical techniques to conduct research that sheds light on the physical origin of cosmological structure in the first moments of the Universe. She was awarded the Fowler Prize by the RAS in 2012, the Fred Hoyle Medal by the UK Institute of Physics in 2018, the Göran Gustafsson Prize in Physics by the Royal Swedish Academy of Sciences in 2020, and the Eddington Medal by the RAS and the Max Born Prize from the Deutsche Physikalische Gesellschaft in 2021 and is a Fellow of the APS. She shared the Gruber Cosmology Prize in 2012 and the Breakthrough Prize in Fundamental Physics in 2018 as a member of the WMAP Science Team and the Gruber Cosmology Prize again in 2018 as a member of the Planck Collaboration.

In June 1996, when I was just a second-year undergraduate, I stood at the gates of NASA's Jet Propulsion Laboratory (JPL) in California, ready to embark on the adventure of a lifetime. I remember the scent of the flowers, the dry warmth of the sunshine, and the deep blue sky. I remember seeing the people going about their everyday business, working on missions to the outer solar system and finding myself, to my great surprise, one among them. I was to spend the summer working on the Galileo mission to Jupiter, and that experience would change the course of my life.

JPL was very far from where my life began, as a shy, self-contained kid growing up in Sri Lanka. One of my oldest memories, from the early 1980s, was watching a rerun of Carl Sagan's *Cosmos* television series, and the dizzying span of space and time it portrayed caught my imagination. One of my aunts gave me a translated copy of *2001: A Space Odyssey* (by science fiction author Arthur C. Clarke, who had made his home in Sri Lanka), and I was instantly hooked. When I was not devouring any science fiction story I could get hold of, I could be found peering through a small telescope, sketching what I could see in the deep, dark skies of my homeland. With my country consumed by civil war, I lived a lot in my imagination, which put me out among the stars—it was a way to escape from the daily horrors in the news and in the streets. I eventually found some kindred spirits at the Young Astronomers' Association of Sri Lanka, which had Clarke as its patron. I made up my mind that I would grow up to be an astronaut.

My parents tolerated these fanciful hobbies and strongly encouraged my interest in science. They were both civil engineers—my mother was one of Sri Lanka's first women in that profession. I was very lucky to have her as a female role model growing up. One of my earliest and strongest memories is the sight of her, elegantly dressed in a saree, supervising a large team of workmen in the construction of a major bridge she had designed. At school, I disliked both exams and responsibility, but was good at both. To my chagrin I was constantly pushed into feeling I should "compete" to become top of the class in exams and also frequently assigned the role of class monitor, which was fundamentally a crowd-control role, and which I deeply hated. In middle school, things started looking up. Mrs. Mendis came with a formidable reputation, having taught my mother before me. She taught a very general subject called "Science," explaining physics, chemistry, and biology with skill and enthusiasm. She tried to answer my constant stream of "why" and "how" questions with great patience, even when it must have been rather annoying. I started looking forward to school, and I tagged along when my mother took a computer programming course after work, finding immense pleasure in coming up with coding projects on which I whiled away many hours.

By the end of the 1980s, reality was impossible to escape any longer. The civil conflict had dramatically escalated, and universities were closed. My parents set a very high priority on the education of my sister and me. In 1990 they left behind everything they knew and moved the family to Manchester, England to start again from scratch, with the hope of giving us a better life. Those were hard years. We were not well off. However, my sister and I flourished in our inner-city high school. In particular, my teacher, Dr. Egan, started my love affair with mathematical physics. For the first time ever, having previously attended a girls' school in Sri Lanka, I experienced being a minority in maths and physics classes; the experience was not always positive. The boys complained to Dr. Egan that I was racing through the problem sets much faster than they could. He told them that they could work at my pace if they wanted, and he gave me university-level textbooks. He encouraged me to apply to the University of Cambridge, which I would never have dreamed of doing without his suggestion. Following my high school exams, I became the first person from my school to be accepted by Cambridge.

The road to actually getting there was not easy. I could not afford to go right away, and I had to work for a while. It is very difficult to recall the times I sat alone in the house trying to study on my own, after all my friends had graduated from school and gone on with their lives. The UK economy was not doing well in the early 1990s, and few positions were available for a high school graduate. I accumulated a binder full of dozens upon dozens of rejection letters I had received. Eventually I was offered an apprenticeship scheme called "The Year in Industry"; I got to work for the company that ran the UK's nuclear reactors. I was assigned the task of modelling the turbo-generator system's response to vibrations and monitoring for cracks, which I could compare with my model. I taught myself a computational method called finite-element modelling to complete the task. The team (all male and much older than me) treated me as an equal and as a colleague. In hindsight, I feel it was an incredible stroke of luck to have landed in this working environment and have had this positive experience as the calibration for my future expectations. It must have been an extreme outlier for the time, considering the contrast I encountered when I visited nuclear plants to gather my data, where no women were to be seen anywhere, and the managers had topless pin-up calendars on their office walls. I got a taste for working both independently and as part of a team, and for taking on responsibility for important things, which I had not previously realized was enjoyable.

After two years I was finally able to go to Cambridge where I had resolved to study computer science, which was generally held to yield very good career prospects (something my recent experience had taught me to consider as important). I was still on this path when I went to JPL, but my summer there completely changed my direction. I worked for the Galileo Photopolarimeter-Radiometer (PPR) Team, with Terry Martin, Leslie Tamppari, and Glen Orton, and was tasked with making temperature maps of Jupiter and its moon Ganymede. Once again, I was given a computer and was expected to work out how to do this on my own. Over the weeks I worked out how to translate a time-stamped stream of temperature measurements from the spacecraft to its viewpoint at that time, and hence was able to create surface heat maps. I remember taking these maps to my mentors and the infectious frenzy of excitement that followed. Large images of Ganymede and Jupiter's Great Red Spot were taken off the shelves and unrolled onto desks and even the floor. We started correlating hot and cold features on my maps with visible features on the surface of Ganymede and in the atmosphere of Jupiter. I experienced firsthand the thrill of seeing something no human eye had seen before and working out something new for the first time. The maps I made were published in an article in the journal *Science,* and I realized that I could *do* science. Back in Cambridge, I immediately changed my study programme to physics and set my sights firmly on becoming an astrophysicist.

In the fall of 1998, I arrived at the Department of Astrophysical Sciences at Princeton to begin my PhD. My JPL experience had convinced me that I preferred the less hierarchical U.S. academic system for my further studies, and the money I had saved from the Caltech Summer

Undergraduate Fellowship that funded that JPL summer job paid for the fees that allowed me to apply to U.S.-based PhD programmes. I chose Princeton because, when I visited the institution after receiving an offer, a young professor, David Spergel, turned up personally at the train station late at night to pick me up and introduce me to a welcoming cohort of students. It felt like coming home to one's family, and I was deeply moved by the kindness shown to me. I made a decision to go there based on gut instinct there and then, which turned out to be one of the most important decisions in my life. Princeton offered me a scholarship for my studies. For the first few weeks after I started there, I kept detailed accounts, eating cheap kids' meals at Burger King (never before or since) and carefully watching my bank balance, until it gradually dawned on me that I did not have to worry about my finances. That discovery was a milestone for me—I feel grateful to society that I have been able, ever since then, to do work that brings me such joy while not having to be concerned about money.

In April 2002, I received a short but life-changing message from David Spergel. Following my qualifying exams, he had already agreed to be my thesis advisor—and now he offered me the opportunity to work on data from the NASA space mission now known as the Wilkinson Microwave Anisotropy Probe (WMAP). The mission mapped the cosmic microwave background—the light left over from the Big Bang. WMAP is now a cornerstone of modern cosmology, and the fact that it would become so was evident to everyone in the field at the time. Most cosmologists would have given their eyeteeth to have this opportunity—and here Spergel was, expressing a vote of confidence that I, a "mere" PhD student, could make a useful contribution to the very first analysis of this data.

I moved from my student office in the basement of Peyton Hall into an upstairs office with Licia Verde and Eiichiro Komatsu, both then postdocs working with Spergel on WMAP data. Once the data were in, we had to produce results to a tight deadline. Spergel and the three of us formed a tight-knit team, working more or less around the clock in shifts, solving problems in parallel. I was given the responsibility of creating a pipeline to compare the data with theoretical predictions from a range of cosmological models, resulting in precise measurements of the basic defining properties of the Universe. As the suggested route to doing this, for the first time I encountered Bayes' Theorem, a deceptively simple equation which in my mind embodies the scientific method—updating one's knowledge about a system using new information. The deep ramifications of Bayes' Theorem underlie a lot of my subsequent work. But in those early days of cosmological surveys, the best way to apply the Bayesian method was not clear, with the descriptions in the statistical literature obscured by unfamiliar jargon. This was my first exposure to something I have done repeatedly since in my career—having to translate the different languages used to describe the same ideas in different research fields, so that one can answer scientific questions that need input from both fields. Verde and I travelled to New York to talk with the eminent statistician Andrew Gelman and worked out how to accurately apply the Bayesian method in practice. She and I formed an efficient team to code up the required pipeline, reducing the potential for mistakes by reinventing "pair coding"—one person watching while the other coded, often working late into the evening. We soon had the supercomputers at NASA's Goddard Space Flight Center blazing away producing results, though on one infamous occasion we brought the computer cluster down with our code (prompting many half-serious jokes of expecting the CIA to come knocking at our door). Our efforts allowed the WMAP Team to work out the age of the Universe, its rate of expansion, and its basic composition. More broadly, I saw how people from different backgrounds all over the world could work together so effectively to unravel the mysteries of the Universe. The work was intense but fun. I was hooked—the experience converted me from a loner to a team player and collaborator.

I was also given the opportunity to lead one of the key cosmological results papers, on which I worked with Komatsu, using the WMAP data to try to figure out the physics of the Big Bang and understand the origin of all the structure we see in the Universe. This has since become a landmark paper, which brought the high-energy physics and cosmology communities closer together. This work triggered a passion in me for understanding why there is something rather than nothing in the

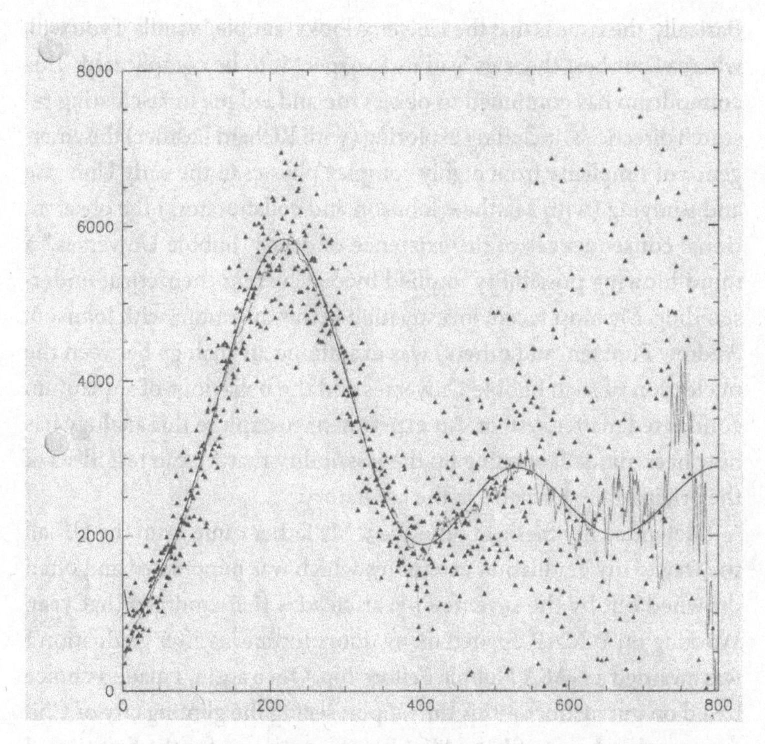

FIGURE 34.1. A page from Hiranya Peiris's lab notebook in the fall of 2002, showing one of the earliest analyses of the WMAP data. What we are seeing in the patterns imprinted in the ancient light of the cosmic microwave background is a cosmic tug-of-war, a battle between pressure and gravity that sets up sound waves in the early Universe. We see these sound waves frozen at the time the Universe became transparent, so this image essentially shows the frequency content of the cosmic sound waves, from low to high. The data are shown as dots, with their scatter representing noise. The solid line is the prediction from a cosmological model, given the geometry and age of the Universe and its basic constituents.

Universe and where everything in the Universe came from. Starting in 2009 I got a second crack at the problem using data from the European Space Agency's Planck mission, the successor to WMAP. The next generation Planck data were a leap forward from WMAP in sensitivity—when I first saw them I could not wipe the smile off my face for a week. However, while Planck eliminated many theories for the origin of cosmic structure, it did not find a "smoking gun" indicating the right theory.

Basically, the issue is that the Universe looks "simple," vanilla if you will, whereas our best theories lead us to expect it to be complicated. This conundrum has continued to obsess me and led me in fascinating research directions, including exploring (with Richard Easther) the emergence of simplicity from highly complex physics in the early Universe and studying (with Matthew Johnson and collaborators) the observational consequences of the existence of other "bubble Universes," a mind-blowing possibility implied by our current theoretical understanding. My most recent investigation in this direction (with Johnson, Andrew Pontzen, and others) was examining an analogy between the nucleation of such bubble Universes and the behaviour of a quantum condensed matter system. An experiment to explore this analogy has now been funded, opening up the possibility that we can test ideas of the origin of the Universe in the laboratory.

I defended my thesis in June 2003. My father came from the UK all the way to my graduation ceremony, which was punctuated and often drowned out by the seventeen-year cicadas that emerged that year. Working on WMAP opened many doors for me, as after graduation I was awarded a NASA Hubble Fellowship. Once again, I made a choice based on gut instinct—this time, upon seeing the glinting city of Chicago on the shores of Lake Michigan from the air for the first time, I decided that this was where I wanted to take my fellowship. But first I had to spend a summer in the UK while I waited for my U.S. visa to be approved. I decided to spend the summer visiting the Institute of Astronomy in Cambridge, where I lived in a tiny room in the gardener's cottage at the back of the Institute. There and then, another life-changing event occurred at an inopportune time—that summer I met Daniel Mortlock, who since then has shared the subsequent steps of my journey. Coming from opposite sides of the planet, we somehow instantly clicked. Off I went to Chicago, where I had a wonderful scientific and cultural experience. But my heart was back in Cambridge, and in 2007 I moved back there, to start an advanced fellowship at the Institute. By then Daniel, also an astronomer, was working at Imperial College in London, and I missed living in a big city. In 2009 I was offered a faculty position at University College London (UCL)—so we moved to

London. Solving the "two-body problem" for academic couples is difficult; doing so took us seven years. On the subject of partners, I firmly believe that an important reason for my being able to have a successful and fulfilling career in science is because my partner values my career as much as he does his own.

I currently split my time between UCL and the Oskar Klein Centre (OKC) for Cosmoparticle Physics in Stockholm, where I serve as the director. The OKC is an interdisciplinary institute with a broad scope, connecting physics, astronomy, mathematics, theory, and experimentation to answer deep questions about fundamental physics. In this role, my broad interests at the intersection of these fields help me facilitate new collaborations between researchers in different areas and incubate new research directions. In terms of my own research, I am still working hard to understand the mind-bending properties of the Universe that were revealed to be part of the Standard Model of Cosmology, including through my work on WMAP. We now know not only that our Universe is expanding, but also that it is expanding at an ever-increasing rate. Fuelled by a mysterious "dark energy," galaxies are accelerating away from each other at such enormous rates that not even gravity can hold the Universe together forever. I am now working to understand the evolution of our Universe by looking at comprehensive maps of the sky that will be produced by the Vera C. Rubin Observatory and its Legacy Survey of Space and Time (LSST). Over ten years starting in 2023, billions of objects in the sky will be imaged in six colours in an unprecedented large volume of our Universe. This survey, which covers over half the sky, will also record the time evolution of these sources: we will soon have the first motion picture of our Universe. Over the coming years, my collaborators and I will focus on turning raw data from LSST into an understanding of dark energy and the way in which the cosmos lit up the underlying "scaffolding" of dark matter filaments as it produced galaxies.

In addition to working on such long-term projects, every couple of years I investigate a new area, starting as a complete novice. For example, right now I am looking at "deep learning," an advanced technique of artificial intelligence. Often, I have started these new directions with

FIGURE 34.2. A portrait of Hiranya Peiris by photographer Max Alexander (2014); the setting is a modern recasting of Vermeer's painting *The Astronomer*.

a new graduate student so that we can both learn together; this has resulted in new ways of looking at hard problems, because we approached them with a fresh perspective compared to the standard lore in the field. I find this beginner's mindset very helpful in what I expect to be a lifelong journey—becoming a better scientist. In a similar spirit, I find that the most rewarding part of my job is helping young scientists flourish and achieve their potential.

I have always had an interest in science communication, but in 2014 my attitude to public engagement underwent a major change. I was invited along with space physicist Maggie Aderin-Pocock to discuss cosmology in primetime on the BBC's *Newsnight*, an appearance seen by 1.5 million people. The response was overwhelming and bifurcated—on one hand, a national newspaper alleged that we had been invited by the "politically correct" BBC on account of our race and ethnicity, while on the other hand I received hundreds of emails from people who said they were inspired by my presentation. This experience was an awakening—being a strongly individualistic person, I somehow had never realized that some people would only ever see me through the lens of irrelevant personal characteristics, no matter what my accomplishments were, and while for others it mattered that there was someone who *looked like them* on TV, talking about astrophysics. After that, in addition to focusing on outreach via mass media appearances, I have devoted significant time to community leadership roles that strengthen the connections between academia and society, most recently as Vice President of the Royal Astronomical Society and now as a member of the governing council that funds particle physics and astronomy in the UK.

I did not spend much time contemplating being a woman in science until I was a fairly senior faculty member and saw for myself how imbalances in the field impacted junior women. In hindsight I recognize that I have experienced stereotyping based on gender and ethnicity, but each time I had shrugged it off and powered through. Not everyone has such thick skin. Negative comments at conferences, being left out of meetings and collaborations, having your ideas dismissed, these little things add up over time—and it can get far worse. Having recognized the problem as a systemic one, I now spend a lot of time on efforts to fix the system. I am fortunate to be part of a huge groundswell in the field working towards change.

My story feels to me like a series of fantastic strokes of luck, helped along by mentors who saw in me qualities worth supporting. Growing up in Sri Lanka, I was taught that if someone does something good for you, you are bound to give something back. I would not be here except for the generosity of my mentors, and so I work to pay it forward by helping young scientists start off on their own adventures—feeling full of excitement and sensing limitless possibilities, just like I felt that sunny June morning long ago, standing at the gates of JPL.

Chapter 35

Poonam Chandra (PhD, 2005)

A Train to the Stars

Poonam Chandra, now Associate Professor of Astrophysics at the National Centre for Radio Astrophysics in Pune, India, conducts research on magnetic massive stars, the end stages of life of massive stars, and their exotic explosions that lead to supernovae and gamma-ray bursts. She was awarded the Kumari L. Meera Medal in Theoretical Physics for the best thesis in theoretical physics at the Indian Institute of Science in Bangalore, the Indian National Science Academy Medal for Young Scientists in 2006, and the International Union of Pure & Applied Physics Young Scientist award in 2010. She is also a recipient of the Swarna Jayanti Fellowship of the Department of Science and Technology in 2014 and is the first woman to receive this fellowship in astrophysics. She was awarded the Modali Award in 2021 by the Astronomical Society of India for outstanding astronomical work done in India over the previous decade. Her leadership role at the Great Metrewave Radio Telescope in Khodad, India, has made it the source of the longest wavelength measurements of many transient astrophysical events, including of the gravitational-wave event GW170817.

Writing my autobiography turned out to be a more difficult task for me than writing scientific papers. It brought back so many memories, memories buried deep inside, but beating inside me like a second heart. My journey to becoming a scientist is not woven around science or around dreams in which I could say that I loved watching the night sky and wanted to be a scientist since my childhood. Even if I had dared to dream big, being a scientist was not something my dreams could reach. My journey began with me accepting the reality that prevailed in the society around me at that time, preparing myself to fit into that reality, and gradually developing the courage to realize that I did not want to fit into that reality. Hence, in a sense, my journey started with what I did not want to be, rather than what I wanted to be.

This is only a part of my story.

I was born in a small town in northern India where the gender disparity is so deeply ingrained in day-to-day life that no one even thought about it as an issue. Like all the girls around me, I internalized the many examples and interactions in our society that the men were the superior gender, the bread earners, the decision makers. A girl's life was defined by two stages: before marriage and after. The former was to be spent in preparation for the latter. Arranged marriages and dowries were the norms, and marriage was the pinnacle every girl dreamt to achieve. From childhood, a girl was raised to be obedient and well-behaved so that her (future) in-laws would like her. Happiness was not something girls/women could dream of for themselves, but they could dream to marry into a family considerate enough to allow them a small patch of happiness which was earned by being sweet, soft-spoken, and obedient.

Seeking higher education would make a girl more marriageable, as long as she remained less qualified, professionally, than her prospective husband. If she was allowed to have a career, her salary should be less than that of her prospective husband. Education came with many rules, all directed towards ensuring she never challenged a man's superiority. This mindset was not limited to my small town; a female friend's PhD degree was delayed by her supervisor "so that she can get married while she still has just her master's, otherwise her parents will have trouble finding a suitable groom." It strikes me now that neither of us thought anything amiss in this delay, since it fit perfectly with the norms of the society in which we were living. The need for social acceptance was paramount and as deeply ingrained in my mind as my existence itself.

Not surprisingly, I grew up thinking that the meeker a girl/woman was, the more likeable she was. I took pride in being described as meek and sweet. This behaviour on my part was reinforced by the blessings of my elders: their appreciation took the form of compliments, like "what a docile girl; she will keep her in-laws happy." It took me many years to realize that no one ever blessed me with the wish that I, myself, should be happy. My dreams were preassigned to someone else; the purpose of my life was to make others happy. For many years, this was all I knew, and I was ok with these boundaries to my existence. It was only much later that, inside my own head, I started to secretly question these ideas.

Alongside doing a significant amount of housework starting at the age of ten or eleven, I was a bright student; I did very well in my studies and was always near the top of my class. However, this was never attributed to my intelligence but rather to my sincerity, my hard work, and, most importantly, my ability to memorize the study material. In contrast, though my older brother didn't do as well on his exams as I did on mine, his lower exam scores were dismissed as "excusable" laziness and never attributed to a lack of raw intelligence. This is so ingrained in me that even now, many times, I think that way.

I was always passionate about science, especially maths and physics. In those days, in eleventh grade one had to take physics and chemistry and then choose either biology or maths as one's third subject. This posed a problem for me in high school, since my all-girls' school did not offer maths at this level—it seemed the system believed that girls had neither the aptitude nor the need to study maths. Thankfully, since we lived close to the state capital, Lucknow, my parents allowed me to move to a school and to the girls' hostel there so that I could continue to study maths. I was very happy to be in the hostel, as I could spend as much time as I wanted doing science and maths. Maths was incredibly de-stressing for me. It felt more like a religion than a school subject! However, dreaming of being a scientist was still far from my imagination. I aimed for what my father wanted for me—if a career, it had to be an administrative job in the government sector. Like a good daughter, my every thought was geared towards achieving this goal. My dream was to fulfil my father's wishes, and my deep fascination with maths was but a small luxury that I assumed I would someday have to toss aside.

My school education was entirely in Hindi, and this posed a challenge when I joined Dayalbagh Educational Institute, in Agra, for my BSc studies, as science classes there were offered only in English. I did not have proper guidance on how to learn English, and there were umpteen fruitless days when I got up at 4 A.M. to learn English words from a dictionary, thinking that doing so would improve my English. However, I somehow picked up English, overcame the language barrier, and did well in my undergraduate courses. During my senior year, I was selected for a two-month summer program in advanced mathematics, to be held at the Indian Institute of Technology (IIT) in Mumbai. That program of summer study would have included an added opportunity of being considered for pre-selection for a master's degree program at IIT. I had not even dreamt of going to a renowned institution like IIT! But it was not to be. I came down with chicken pox just before my travel date, and I was unable to attend the summer program. The disappointment was crushing. At that point it seemed like an end to my life, my dreams. I felt broken inside. I was in pain, a pain I could not share with anyone, as no one understood my passion for maths and physics. I was so shattered that I missed some entrance exams for master's degree programs elsewhere. So there I was, with no prospect of enrolling for a master's, when most of my friends had already secured admission for graduate school programs at good universities. My failure to secure admission to a good university so that I could pursue a master's degree was an extreme low point. I was very dejected, my hopes were crushed, and I could only see the dark tunnel of my life ahead of me.

A reprieve came when Aligarh Muslim University (AMU) delayed their academic session. I matriculated there for a master's degree in physics, where I was one of only two women in my class of over sixty students. I was getting more and more passionate about physics, and I took Quantum Field Theory and Astrophysics as my specialization subjects in my final year. I was elated because I had some amazing teachers who considered me genuinely good at physics. This was when I started to realize that I did not want to pursue a career in administrative services and that my passion lay in physics. This was the first time in my life when I knew that I wanted to do PhD work in physics, even though this conflicted with my father's idea of my career. Despite my interest in doing a physics PhD, he secured my admission for a Bachelor of Legislative Law (LLB) programme so that I could prepare myself for the administrative services. I still remember when he put me on the bus to join this programme. I stuck my head out of the window and assembled all the courage I had to meekly and shyly, but unsuccessfully, tell him that I was no longer interested in administrative services and that I wanted to pursue a PhD in physics.

Instead of attending LLB classes, I secretly began to prepare for a research fellowship exam offered by the Council of Scientific & Industrial Research that would allow me to join a PhD programme. When I met my (female) undergraduate physics teacher, I was very excited to talk to her about my dreams. I believed that, as a female physics teacher, she would understand my passion and mentor me. Her first question took me by surprise: she wanted to know my age. Then, she pointed out that I would be twenty-seven by the time I completed my degree, which would impede my marriage prospects. So saying, she tried to dissuade me from pursuing my dream. Thinking about it now, I do not know which surprises me more, the fact that someone who could and should have been a role model and mentor let me down, or that what she said sounded very normal and correct at that time and yet I did not let her advice deter me. I think, generally, being the meek person I was, all these incidents should have been enough to stop me from dreaming about earning a PhD. But to my own surprise, that did not happen.

More valuable advice came from Professor Irfan, my particle physics teacher, who recognized my passion for physics and suggested I write to the Tata Institute of Fundamental Research (TIFR) to enquire about their PhD programme. Full of hesitation and self-doubt, I did so, and I was overjoyed to receive a reply, a physical letter in those days before email. Another positive step was that the Variable Energy Cyclotron Centre, in Kolkata, organized a summer school, and I was selected for that and was able to attend. This was the first time I was in an environment where the women were not expected to leave for marriages or switch to administrative jobs; everyone around me was there because they wanted to do higher studies in science.

Outwardly, I continued to be the timid girl who did not speak much. But something inside me had become too strong to hold back. The dream of a PhD was precious to me, and I wanted to nourish it, to fulfil it. I applied for a few entrance exams. Appearing for these entrance exams was not an easy task. For a woman travelling from a relatively small town in India, appearing for an entrance exam posed multiple barriers, some terrifying and some practical. In order to arrive on time, I had to take trains to Delhi, where the exams were held, that departed around 4–5 A.M. Travelling alone after dark as a single woman was unsafe. To try to ensure my personal safety, I would arrive at the train station the prior evening, before dark, and spend the night anxious and scared in the women's waiting room, waiting for the next-morning train. I cleared the written entrance exam for the Indian Institute of Science (IISc) in Bangalore and then was called for the in-person interview. But the ticket was quite expensive and I was short of money. AMU students were eligible to purchase train tickets at a 50 percent discount for travel to their homes. I lied and said that my aunt lived in Bangalore and that my family was currently in Bangalore with my aunt, so I needed the discount concession to visit Bangalore. The authorized person only asked me, "Is it so?" and I broke down. I confessed that I wanted to do physics and that IISc had asked me to come for the interview, and I apologized for lying. If this official had not been compassionate and granted me the ticket-price concession, despite the fact that I had lied about it, it would have been difficult to go.

Another problem was that I had no place to stay in Bangalore; a hotel was not an option, and no arrangements for accommodations were provided by IISc. I was desperate to find a contact in Bangalore who could host me. I decided to go anyway and try to figure something out; I thought I might even stay in the waiting room at the train station. I had one slim hope: the name of a friend of a friend who was a PhD student at IISc. In the era before cell phones, I embarked on the forty-two-hour train journey from Delhi to Bangalore with hope in my heart and a burning desire to break out of my circumstances. My resolve was challenged after reaching Bangalore, when I waited for four hours outside the accommodations of this unknown contact! Luckily for me, she was kind enough to allow me to stay with her so that I could attend my interview. All these efforts brought a sweet reward: I was selected for admission to IISc under the Joint Astronomy Programme. After one year of coursework in this programme, I would be eligible to join one of the top six or seven research institutes in India to pursue a PhD degree. I resigned from the LLB program, went home, and told my parents about my decision to start a PhD programme and the fellowship I had obtained to pay for my expenses. And this was how I came to TIFR to work on a PhD in astrophysics.

Now, when I look back, I remember the many times when I doubted my choices. I think about how reckless I was to take such risks and to put myself in such situations. Recently, someone offered me a perspective that I had not been able to see for myself: that my passion to pursue the PhD and my courage to dare to dream big dreams and to follow my heart were what allowed me to break through barriers. I had transformed from a girl who needed to be docile and willing to please to a woman who would do anything to follow her dreams!

My biggest dream now within reach, I should have been in seventh heaven. But there was a price to pay for the long-suppressed issues that I had not allowed myself to think about in my single-minded quest to pursue physics. Once I settled in TIFR and started to do what I always wanted to do, I was surprised to find my concentration lacking and my focus wavering. My life choices started to haunt me; all the issues I had put on the back burner surfaced with a vengeance and would not quiet down. On the advice of one of my friends, I sought counselling. This helped me begin my long journey to wellness. This was the first time in my life I realized how important it is to be compassionate with yourself and to take care of your own personal well-being. I make it a point to detail these struggles because, although outwardly I had begun to garner markers of success, these successes were giving me little happiness. I was awarded IISc's best thesis medal in theoretical physics and the Indian National Science Academy's young scientist award for my PhD

FIGURE 35.1. Poonam Chandra bungee jumping in the Swiss Alps in 2015, during a break while attending a conference in Bern.

work. But none of these, not even being named a prestigious Jansky Fellow and starting postdoctoral work at the National Radio Astronomy Observatory, cheered me up—in part due to my depression, but also because I was still unwilling to internalize that I genuinely deserved these awards and opportunities.

Many students describe feeling enthusiastic and energized when they complete their PhDs and begin postdocs. I should have felt this way, yet when I reached Charlottesville, Virginia, for my postdoctoral work, I was, instead, at perhaps the lowest phase of my life. I realized, then, that I had to focus on myself first, and I made it my first priority to find a counsellor and a psychiatrist. Thankfully I had understanding friends who helped me get through those tough times.

Looking back, I feel I could have done so much better during my years as a PhD candidate and during my first postdoc, a regret I will always have. But I also am happy that I pulled myself out of this mental health slump when I thought that just taking another step forward was not possible.

I joined the National Centre for Radio Astrophysics, which is part of TIFR, in 2012, and have been there ever since. My work focuses on understanding stellar evolution, the demise of massive stars as supernovae (some of the biggest explosions in the Universe), and gamma-ray bursts. It excites me to know that the stars in the sky and we humans on Earth are made up of the same stuff. The oxygen we are breathing, the iron flowing through our blood, the calcium in our bones, all were made inside big and massive stars. I love to work towards understanding and explaining these mysteries. I find that the most exciting aspect of my research field is that this Universe is our laboratory, and only nature gets to design and do the experiments. All the events we see "today" have already happened in the deep past, and the light from those astrophysical phenomena is only now reaching our observatories. We are detectives, like Sherlock Holmes, trying to collect the circumstantial evidence to re-create what must have happened! I try to collect evidence as best as I can and in as many possible ways as I can, including with the best radio and X-ray telescopes. I study exploding stars and their post-explosion shocks as this fast-moving ejecta interacts with earlier episodes of ejecta from the dying star that is in the form of slower-moving surrounding winds, which in turn have imprinted in them the footprints of those exploding stars. As the shock speeds are approximately 1,000 times faster than the stars' wind speeds, in some sense one can view the course of a star's life fast-forwarded by a factor of 1,000, giving us a glimpse of the star's life cycle, but seen backwards, starting from its moment of death! I call it "the amazing time machine." If we study these ejecta interacting with the wind for one year, we can study the winds from up to 1,000 years before the star exploded. This course of study has allowed us to understand the environments and histories of the stars in various galaxies, both near and far. Sometimes, this is the only way to learn about the far, far away environments of our Universe.

As I finish writing this autobiography, my thoughts come crashing down from stellar explosions into the inner spaces of my mind. Although I am now a faculty member with tenure, my self-doubts have not gone away. I believe it's hard for many women to develop self-confidence. For me, this is probably due to my upbringing. Furthermore, I internalize

FIGURE 35.2. Poonam Chandra at the Square Kilometre Array (SKA) Organisation headquarters at Jodrell Bank Observatory, near Manchester, UK, in 2019.

and magnify any external criticisms. When someone tells me I am not good enough at one thing, I take such comments seriously and believe that perhaps I am not good enough at anything. My outward professional facade hides an inner voice that sometimes screams out, "You are undeserving!" Due to my inner struggles to keep proving myself again and again, I don't even realize that I sometimes allow this voice to take over my sense of self-worth. To my own eyes, my achievements seem to be beyond my own capabilities and more than I deserve. I am probably not alone in thinking this way. Imposter syndrome looms over my psyche, as it perhaps does, also, for others.

My ability to doubt my place as a legitimately successful scientist was reinforced when I was selected for the prestigious Swarna Jayanti Fellowship from the Department of Science and Technology, given by the government of India to scientists under age forty. This is the highest honour possible for young scientists in India, and I was only the second woman to receive this fellowship in physics and the first woman to be awarded this fellowship for work in astrophysics. This selection process was competitive, with the last two stages being oral presentations made before two different committees. An acquaintance from another institute, whom I met during these presentations, told me that he deliberately performed badly so that I would be selected instead of him. I don't know which is worse, his crass comment or that, initially, I believed him. Only much later I realized how this couldn't possibly be true, and it probably spoke volumes about his abilities, not mine.

My journey was not easy, and I have no words of inspiration to offer, as unfortunately my journey is likely similar to that which many women have experienced and which many more young people may yet experience. I cannot even begin to think about what members of the LGBTQ community go through, who in many cases don't have the privilege of societal acceptance or supportive families.

Objectively speaking, I know I have an enviable job, and I have achieved more than I ever dared to dream of as a child. I am at a place that I had not even allowed myself to dream of for several decades. Yet when I look back, I can't stop myself from seeing the many failures in my life. My biggest fight is and has been with myself, to believe in myself. The struggle inside doesn't end, the "shock winds" of my experiences bring my past hurtling into my present, and I struggle to piece together what was, or should have been. The restlessness persists, as does the quest.

Acknowledgements: Writing down my autobiography was an extremely difficult task for me. I would like to thank Professor Shubha Tole, who meticulously went over the draft and made several suggestions, which improved my telling of my story tremendously.

Chapter 36

Xuefei Chen (PhD, 2005)

Staring at the Stars

Xuefei Chen, Professor of Astrophysics and Deputy Director of the Yunnan Observatories in China, specializes in the study of the evolution of stars in binary systems, including eclipsing millisecond pulsars, blue stragglers, and EL CVn-type binaries. She received the Natural Science Award of Yunnan Province (first prize) in 2011 and the State Natural Science Award of China (second prize) in 2013. In 2014 she was the first female astrophysicist to be named one of China's Top Ten Young Women Scientists. She also was selected for China's Ten Thousand Talents Program as a leading researcher in 2018 and for a National Science Fund for Distinguished Young Scholars grant in 2021. She is the PI of the Binary Project of the China Space Survey Telescope (CSST) and a vice president for both the Committee of Women Astronomers and the Committee of Stellar Physics of the Chinese Astronomical Society.

I was born and grew up in a small village by the Yangtze River. The village is quiet and beautiful. Many small rivers run through the village, and so it is like a tiny Venice. We could hear frogs and crickets singing on summer nights and see fireflies dancing with the Milky Way across the sky. Ms. Zhaojun Wang, one of the Four Beauties in Chinese history, lived in this region about two thousand years ago. The emperor of the Han dynasty presented her to the king of the Xiongnu (Hun) government, and the relations of the empire of the Han dynasty with the Xiongnu subsequently improved due to Zhaojun's marriage with the king.

My birth in 1976 violated the population control policy, as I already had an elder brother and two elder sisters. I was considered redundant and so my family received no food ration for me from the government for two years. Keeping secrets in my small village was impossible, and I was told about my redundancy when I was barely old enough to understand human words. I was hurt and felt lonely. This feeling is my earliest memory, and it accompanied me throughout my childhood.

I was not particularly eager to play with other children. What I did most was lie down on the grass at night, stare at the stars, and imagine stories about the heavens. I was so excited when I would see a shooting star fly across the sky, because we believed that all wishes blessed at that moment would come true. I wished I could leave the village to find my poems and pursue my dreams.

The first important choice for me occurred when I graduated from junior high school. In the Household Register System of China, I was from an agricultural family, which meant, based on the structure of Chinese society at that time, I was fated to live in poverty, work hard every day, and have no medical insurance. My parents wanted me to go to a technical secondary school rather than a senior high school. Doing so meant that I would become a non-agricultural-family person immediately and in the future have a less physically demanding and more rewarding job.

In the early 1990s in China, a college education was not typical and not as eagerly desired by Chinese parents as it is today. Yet even in small or middle-sized towns, due to future uncertainties, parents wanted their children to attend a technical secondary school rather than a senior high school if their children could pass the entrance examination. The decision was harder for a village girl since most girls who lived in the countryside had no chance or did not wish to continue their education after graduating from junior high school. Instead, they would go to work or help their parents earn money for their families and themselves, just as my elder sisters did. But that was not my dream. I chose to attend a senior high school, and three years later, in 1995, I passed the National

College Entrance Examinations and became the first-ever college girl from my village and the areas nearby.

I left my hometown in September for Kunming and entered Yunnan University as an undergraduate student majoring in physics. Kunming, the capital of Yunnan Province, is located at the northern edge of Dian Lake and is surrounded by temples and lake-and-limestone hill landscapes. The altitude is 1,900 meters above sea level, the latitude just north of the Tropic of Cancer. The city has a pleasant temperature all year round and is known as an Eternal Spring City. As a southwest province in China, Yunnan borders Vietnam, Laos, and Myanmar and is rich in natural resources, with both tropical rainforests and snow on the mountains all year round. It is rich in cultures, with the largest number of ethnic groups in China. The province is very much underdeveloped compared to those in the eastern and middle parts of China. However, the people there are kind and straightforward and enjoy comfortable lives. I felt relaxed and happy.

My second choice arose when I graduated from Yunnan University. Choosing to become a graduate student, and especially deciding to study astrophysics, was a big challenge for a woman. Astronomy was distant from our daily lives and was not well known to the public, even among college students. What led me to make the final decision to pursue astrophysics was the beautiful and quiet starry sky that remains one of my most profound memories. And so I came to Yunnan Observatory to be a graduate student and begin my life with stars and star couples.

Star couples, or binaries, consist of two stars orbiting each other due to their mutual gravitational attraction. The two stars interact and may exchange matter during their lifetimes. These interactions make the world of stars mysterious, exotic, and marvelous. My goal is to figure out the underlying physical processes that drive these interactions.

Yunnan Observatory (it expanded recently and became the Yunnan Observatories by establishing a few new observatories in Yunnan Province) was one of five observatories in China. It is situated on Phoenix Mountain in the eastern suburbs of Kunming. Both the Institute of Astronomy and the Institute of Physics of the Academia Sinica had been based here from 1938 to 1945, when their staffs escaped from Nanjing

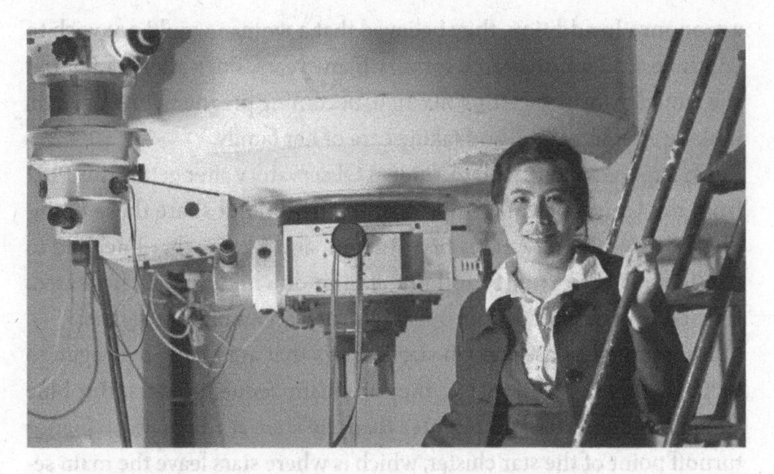

FIGURE 36.1. Xuefei Chen and the one-meter telescope at the headquarters of the Yunnan Observatories on Phoenix Mountain in Kunming, December 2014. The telescope was bought from East Germany and was the biggest telescope in China in the 1980s.

during the war when Japan invaded China. During the same period, three top universities, Peking University, Tsinghua University, and Nankai University, moved to Kunming and established the National Southwest Associated University. The University is known world-wide since it has trained many academic masters who have had a profound impact on Chinese society.

During my first few years at Yunnan Observatory, I became very nervous and could not sleep well, since I found that doing research in astronomy is not as poetic and romantic as my simple memories of the starry skies of my childhood. I did not know whether I was suited for this kind of work, and I did not know how far I could go along this path. I had always been told, since I was born, that for a woman the best choice was to find a stable and well-paid job. In this way, I could have a good life and enough time to take care of my family after getting married.

Not many women worked in astronomy in China at that time. Some teachers refused to accept female students because they did not think that women could achieve much in astronomy. One reason for this attitude was that they did not think women had the talent or aptitude for astronomy. In addition, they believed that a woman would not wish to continue academic research for very many years, since after graduation she would prioritize getting married, becoming pregnant, giving birth, looking after her child, and taking care of her family.

My supervisor returned to Yunnan Observatory after getting his PhD from the University of Cambridge, and he did not share these prejudices. I was lucky. I was his first student, and I knew that I needed to cherish this opportunity and not let him down. I insisted, hesitated, insisted, and hesitated again, until I learned about blue stragglers.

Blue stragglers are stars in star clusters that are fusing hydrogen to helium in their cores, that is, they are main-sequence stars. But blue stragglers are hotter and brighter than the stars at the main-sequence turnoff point of the star cluster, which is where stars leave the main sequence and become red giants after they exhaust their primary fuel source—the hydrogen in their cores. The existence of blue stragglers is not expected from the canonical theory of stellar evolution. Those stragglers have longer lifetimes than expected for stars of their brightnesses and temperatures. The reason for their long lifetimes is that they have swallowed matter from or merged with their companions, and as a result they have extra nuclear fuel to burn in their centers. Is this like *The Twilight Saga*? Yes, it is *The Twilight Saga* among stars—blue stragglers are often described as vampire stars by the media! I remembered many of the stories I had imagined while lying on the grass in my hometown. In my mind, I heard Stephen Hawking murmuring to me: "Remember to look up at the stars and not down at your feet. Try to make sense of what you see and wonder about what makes the Universe exist." My curiosity returned to me, and I did not hesitate anymore.

In 2005, I completed my PhD degree and gave birth to my son. From my point of view, this is the most critical period for a highly educated mother. What I did next would determine whether I could pursue my career in China. Due to the one child per family policy,[1] children have become precious for Chinese families. Preparing for pregnancy and taking care of children takes most of the time and energy of Chinese mothers because they have no chance to make mistakes in raising their children. Until my son was three years old, I devoted most of my time

FIGURE 36.2. Xuefei Chen playing with her son in the Green-Lake Park in Kunming, September 2010.

and energy to him, as most mothers in China do. So I made no progress on my research and had no publications in those three years. After my son started kindergarten, we just needed to look after him in the evenings. My husband was a teacher in a junior high school and he had time to take care of our son after work. More importantly, my husband did not refuse to do this. This situation is only possible in China in college-educated families.

Now, I could think more about blue stragglers, the mysterious vampire stars, and the beautiful stories I imagined in my childhood. I published a series of papers on blue stragglers. These works have been used as canonical examples to show that blue stragglers can originate from the evolution of two stars in binary systems.

Meanwhile, I helped mentor two PhD students working on the progenitors of Type Ia supernovae. Type Ia supernovae are thermonuclear explosions of carbon-oxygen-rich white dwarfs that are accreting matter from their companions. Because one star is gaining matter that is lost by the other, such systems are similar in some ways to blue stragglers.

Due to their extreme brightnesses and remarkable uniformity, Type Ia supernovae can be used to measure distances in the Universe. Indeed, they are excellent cosmological distance indicators and can be used to determine cosmological parameters. However, because the intrinsic properties of these supernovae are affected by their progenitors, understanding more about the progenitors could impact the precision of the distances we measure for them.

We devised a theory in which a white dwarf grows in mass by accreting helium from a companion star, which results in the white dwarf exploding as a Type Ia supernova. The resulting supernovae are from young stars. About half of Type Ia supernovae may arise from the presence of helium donor stars, and we predict that we should see more young Type Ia supernovae in the early Universe. This work was highlighted by the *Monthly Notices of the Royal Astronomical Society*, a leading journal in astrophysics, and was widely reported by the media, including by the BBC and *China Daily*. We also found that the intrinsic luminosities of Type Ia supernovae depend on how rich or poor they are in elements heavier than helium, which astronomers call metals. Since the abundance of metals was low when the Universe was young, correcting this bias should lead to more accurate distance measurements for distant galaxies.

I also helped mentor a PhD student who was working on the evolution of binary stars. Two long-standing, unsolved problems exist in the theory of binary star evolution. One is called dynamical mass transfer and the other common envelope evolution. The two unresolved issues result in some apparent conflicts between observations and theory, one of which is that theoretical models are unable to provide accurate predictions about the numbers and space densities of binary black holes and binary neutron stars. Mergers of black holes or neutron stars in binary systems produce gravitational-wave radiation, and some of these mergers have been detected by gravitational-wave observatories starting in 2015. Earlier work did not keep track of energy, momentum, and angular momentum during the transfer of mass from one star to the other. We were determined to improve the treatment of these dynamical problems, and therefore, in 2005, we started to develop a model in which mass transfer occurs but no heat is lost from the system (this is known as an adiabatic system). These calculations are hard and take a long time; hence, our publication rate on this project has been low, only about one paper every five years. However, the results are quite promising and may form the basis for a textbook. In recognition of the work I've done and am doing, I received the Natural Science Award of Yunnan Province (first prize) in 2011 and was named one of China's Top Ten Young Women Scientists in 2014.

At the same time, my students and I devised formation models for many kinds of binary objects. Some of the models are referred to as canonical. In particular, based on my understanding of binary evolution, I asked one of my students to plot a figure for binary white dwarfs. Such objects are divided into two groups on that figure. These groups clearly correspond to two different evolutionary paths for these binaries. This fantastic mapping makes me even more excited about and fond of astronomy.

Yes, binaries seem to be the magic wands behind many interesting astrophysical phenomena, and the modeling work I've done has contributed to a better understanding by astronomers of the critical roles played by binary evolution in astrophysics. But observations are needed to test whether my modeling work is correct. Initially, I collected already-published observations to help constrain my models, but the observational results were not as good as they needed to be. So I decided I must become an observer. This choice is challenging for me, because I was trained as a theoretical astronomer and was not familiar with observations. Such a change from theoretician to observer will require time and could be costly to my career. But I made my decision and have begun a new journey.

In 2015, the science committee of the Large Sky Area Multi-Object Fiber Spectroscopic Telescope (LAMOST) began to plan a survey that would be carried out during the second five-year period of operations (LAMOST-II). LAMOST is the first "big" telescope designed by Chinese astronomers. It is a 4-meter Schmidt telescope and is capable of collecting data simultaneously from 4,000 celestial objects. We proposed a binary project called LAMOST-MRS-B, but a fierce competition exists for projects that will be done with LAMOST-II. We will need to convince the scientific committee that binaries are an essential topic of study.

These experiences have broadened my horizons considerably, such that my activities as an astronomer are no longer confined to the world of binary stars. Soon after, I became Vice President of the Stellar Physics Committee of the Chinese Astronomical Society and organized the writing of a white paper on stellar astrophysics for the China Space Survey Telescope (CSST). The CSST will be launched in 2024 and will co-orbit with the Chinese Space Station. It will feature a 2-meter-diameter primary mirror and have a field of view 300 times larger than that of the Hubble Space Telescope. The telescope will take images and spectra of up to 40 percent of the sky over ten years. And I am now the Principal Investigator for the Binary Project of the CSST.

In 2018, I was selected for the Leading Talent of Technological Innovation of Ten Thousand Talents Program and in 2020 became deputy director of the Yunnan Observatories. Recently, I received a National Science Fund for Distinguished Young Scholars grant, which is highly respected in the Chinese academic community, because these grants are awarded based on a stringent evaluation system. It is the goal of every aspiring science researcher in China to receive such a grant, and receiving one means that my peers have recognized my academic accomplishments. My path to my present position is similar to that taken by most prominent Chinese astronomers, but it has been much harder for women to follow this path. Since the National Science Fund for Distinguished Young Scholars program was set up in 1994, more than fifty astronomers have been funded, but few have been women. I am only the fourth.

With the rapid development of China's economy, more and more women are choosing to continue their studies after graduation from college. However, it is still difficult for a woman in China to study astronomy. A few years ago, when I interviewed some college graduates, a young woman asked me, repeatedly: "Must I finish my PhD if I want to study astronomy?" "Yes," I said, "if you would like to do research in astronomy." She thought again and again and finally gave up on the idea of becoming an astronomer. She did not want to get a PhD degree because her relatives and friends told her that a female with a PhD is a monster who will have a hard time finding a boyfriend and getting married. It was a great pity. She had a solid foundation in physics and sound logical thinking skills, and she could have been an excellent scientist if she had chosen to pursue astronomy.

A quote attributed to Georg Wilhelm Friedrich Hegel that is popular in China today is this: "A Nation is hopeless unless it has people who look up at the stars with lively interest." Now is the golden time for the development of astronomy in China. At Yunnan Observatories, more than two hundred astronomers and more than one hundred MSc or PhD students are doing research on subjects ranging from stellar physics to solar physics, astrometry, planetary science, active galactic nuclei, and astronomy technology. More and more top Chinese universities have established astronomy departments. The number of people engaged in astronomy in China is ten times greater now than it was twenty years ago. Teachers don't refuse to take on female students anymore. As a consequence, more and more women are becoming astronomers. However, for both natural and objective reasons in Chinese society, the vast majority of women are unlikely to devote themselves as fully as men to work. For now, at least for astronomy, the top of the academic pyramid is nearly completely occupied by men. To change this, China urgently needs to adopt policies that encourage female astronomers to develop more fully.

In 2017, the Chinese Astronomical Society set up the Committee of Women Astronomers, and I was selected as vice president. The committee's purpose is to promote communication among women astronomers, create better platforms for them to show their abilities and to support their activities, and advocate for rule changes that would enable women to play increasingly important roles in astronomical research. I organized a national meeting for women astronomers at the end of 2019 in Xinjiang Uygur Autonomous Region. It was fruitful, and we therefore plan to hold such meetings annually.

In the summer of 2021, the China Ministry of Science and Technology and twelve other ministries jointly issued a circular on measures to support women scientists playing more significant roles in science and technology innovation. The Ministry proposed that priority should be given to female researchers for the election of academicians, the selection of awards, the support of talent programs, the approval of research funds, and the chairing of big science projects. The Ministry also proposed setting up projects and awards for women, relaxing age limits for female researchers in various projects and programs, and gradually increasing the fraction of female experts on various scientific committees. These proposals are exciting news for women and the academic community in China. But I know that we still have a long way to go.

Observations play a crucial role in astronomy, and every year remarkable discoveries are made. More and more big survey projects, such as the Sloan Digital Sky Survey, LAMOST, the Gaia mission, the Vera Rubin Observatory Legacy Survey of Space and Time, and the Chinese Space Station Telescope Project, have begun operations or will begin operations in the near future. These surveys will take images and spectra of billions of stars and other objects. In comparison to men, women are more careful and more patient. Their skills are well suited for making discoveries from the data that are becoming available from big surveys and contributing to the advancement of modern astronomy.

The stars are flickering, murmuring, and ejecting material into space, and they explode occasionally. They are waiting for us to discover their secrets. On July 17, 2021, I was invited to participate in the opening ceremony of the Shanghai Astronomy Museum, which has the largest planetarium in the world and can accommodate about six thousand people per day. When I came into the *Cosmos* theater, I lay down under the starry sky of my childhood and watched the seasons change and listened to stories about them. The feeling was so familiar and so fantastic. After that, I attended the Starry Concert, which attempts to connect our souls to the vast Universe. I hope more and more women will embrace the study of the boundless Universe. A nation is only a true civilization when both men and women are free to look up at and study the stars.

Notes

1. Changed to two children per family in 2016 and three children per family in 2021.

Chapter 37

Shazrene S. Mohamed (PhD, 2009)

The Sky Is for Everyone

Shazrene Mohamed, who holds a joint faculty appointment at the South African Astronomical Observatory and the University of Cape Town, conducts computational astrophysics research on winds and outflows from single and binary stars. She earned the Leo Goldberg prize at Harvard twice, for the best junior astronomy thesis and again for the best senior astronomy thesis, before becoming a Rhodes Scholar and then an Argelander Fellow in Bonn, Germany. She helped lead the effort to bring the IAU General Assembly meeting to Africa for the first time (in 2024) and now serves as the science co-chair for that meeting. She founded the group Astronomy in Colour as an African community actively working to transform astronomy to make it more inclusive and welcoming for all.

Growing up in Zimbabwe, I remember being captivated by the dark, star-filled night sky from a very young age. "Why do stars shine? What keeps them 'up there'? Where do they go during the daytime?" I had so many questions and searched for answers at every turn.[1] In grade 4, we studied the planets in our solar system, so I decided that my project that year was the perfect opportunity to learn more about space. Having waited a week for the only book on the topic to be returned to the library, I leafed through the pages. I do not remember anything else in that book except staring at the picture of an astronaut floating in space, with hints of our own cloudy, blue atmosphere in the background. I am not sure if I knew what an astronaut was. I was certainly familiar with the idea of alien visitors to our planet, as one of my favorite cartoons was *Dodo, the Kid from Outer Space*. But something about seeing that "real" image of an astronaut, "the reciprocal Dodo," made the concept come alive for me; I vividly remember having a huge *Eureka!* moment. It was obvious: the only way to get answers to all my questions would be to go to space to see for myself. And so began my journey . . . that faceless, bulky figure in a white suit would haunt my every moment—asleep and awake. I paid little attention to who the astronaut was and to this day still do not know for sure. What was important was that there was nothing to suggest that it could not be me in that space suit, my smiling face behind that visor.[2]

Sometimes what you do not know is just as important as what you do. My naive assumption that I would be able to go into space went unchallenged. No one told me it was improbable, and there was simply no evidence presented to suggest that. On the contrary, my parents encouraged my learning and told me that I could do anything if I worked hard enough. Despite not having education beyond high school, they had enough to know its power, particularly for raising children in the highly racialized society of Zimbabwe. They made huge sacrifices, were derided for driving and fixing old cars ("jalopies") while "wasting money on school fees—particularly a waste on girls." It was difficult for them to make ends meet, let alone give us all the material things that we encountered at school, but they instilled in us the value of education: that no one can take your education away from you, and that an A on a test, was an A on a test. It did not matter that it was earned in a school uniform five sizes too big (you had to grow into it), with torn socks and worn-out shoes . . . an A was an A!

I took my parents' advice to heart. I remember the headmaster on prize day asking me, "Do you need a wheelbarrow for all your prizes?" I was lucky to attend an all girls' high school, and although there were certainly issues with old-fashioned ideas of "what and how" young ladies should be, many of the gender stereotypes that drove young women from science early on never took hold in that environment. Someone at

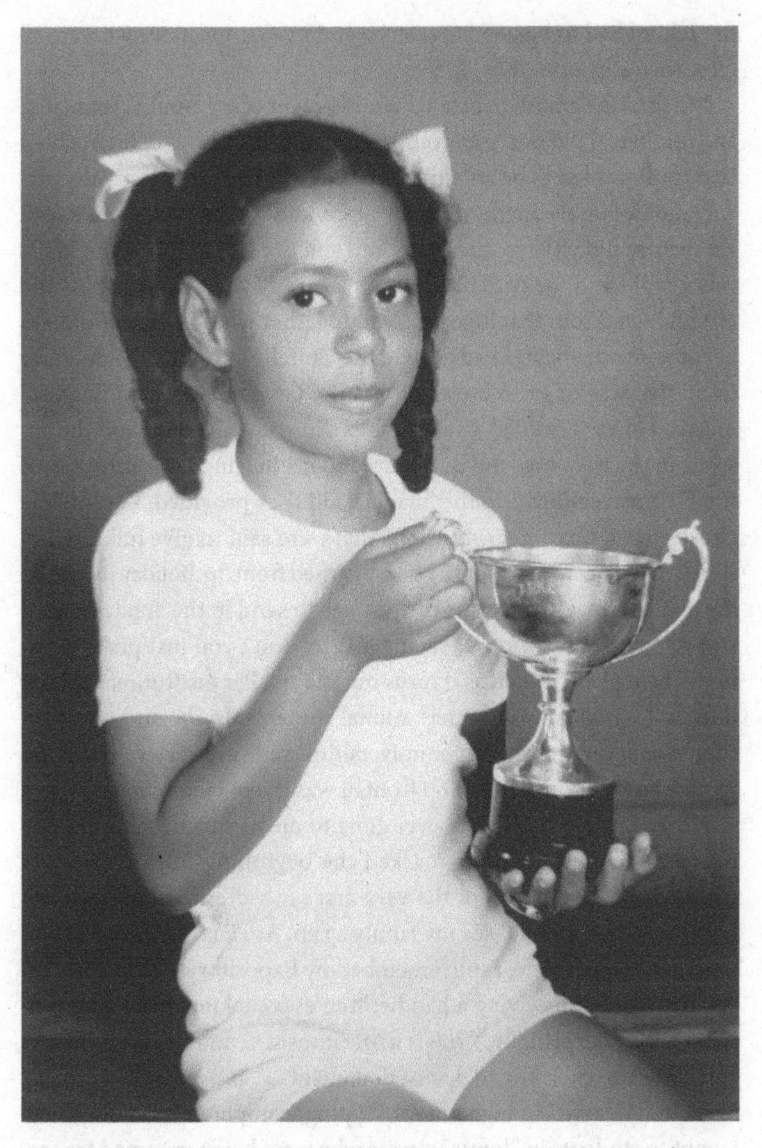

FIGURE 37.1. Shazrene Mohamed posing with a trophy in grade 3 (at age eight). "I asked my mother what the award was for and she said, 'I don't remember, you won so many every year!'"

the school had to be good at science and maths, and since we were all girls, then why not me?

Much to the amusement and befuddlement of my family, classmates, and teachers, I did not "grow out of it," stating at the end of high school that I still wanted to be an astronaut "when I grew up." Clearly my parents did not know anything about space, and it seemed no one else in Zimbabwe did either—there was and still is no space program. What was visible and identifiable in that picture of the astronaut was the NASA logo. From that logo, I pieced together what I needed to do to become an astronaut: I had to go to the United States, and the only way to do that was to go to university there.[3] Given our limited financial means, I knew I needed a scholarship. This time, I would find the answers in the Bulawayo public library. From a bulging book listing all of the U.S. universities and the financial aid they provided, I eventually chose eleven possible institutions that were said to give full scholarships. I used my whole month's salary earned from my holiday job working in a pharmacy to pay for the postage to send in the applications. I still remember my mother saying, "Why don't you just pick one or two?" I am glad I didn't, as it turns out the smaller institutions did not offer full scholarships to international students. In the end, I was accepted at several of the larger universities, and since Harvard did not require any fees or payment up front, it was the obvious choice.

No one in my family had ever gone to university, never mind one overseas. September 2000 marked the beginning of many firsts; I climbed aboard a plane for the very first time, drenched in tears, not knowing when I would see my family again. At Harvard I used a computer for the first time. I still remember my Expository Writing preceptor, bemused at receiving a handwritten essay, taking me to the computer lab and explaining, "*This* is a Macintosh . . . and *this* is a windows machine. You click *here* to open a document . . . and remember to click *here* often to save." In high school, "typing/computers" had only been taught to students in "lower" classes who were being groomed for secretarial and receptionist jobs. I spent the first couple of months in the computing lab, clawing my way up a very steep learning curve. Eventually, one of the students in my dorm gave me a monitor, and my host parents, the Childs, gave me a CPU and keyboard. Thus, my beloved Frankenstein's monster was born. Frankenstein's monster would be my trusted companion for several more years (they just don't make them like they used to), occasionally getting "upgrades" until I got my first laptop at the beginning of my senior year, a sleek, titanium-encased machine, which I bought with my award money after earning the Leo Goldberg award for the best junior thesis in astronomy. I would go on to write another Goldberg award-winning senior thesis on that very same machine.

While at Harvard I pursued my dream of becoming an astronaut with almost reckless abandon. I majored in astronomy and mathematics and earned a language citation in Russian by taking intensive classes every day for three years; just in case the Americans wouldn't take me to space, I hoped the Russians might. By now I had also learnt NASA astronauts had to be U.S. citizens and many also had military backgrounds. Joining the U.S. Army Reserve Officer Training Corps (ROTC) would give me both, so I travelled to MIT twice a week for pre-dawn training. I enjoyed the physical fitness, leadership, and teamwork aspects, but I was not cut out for the military. "Cadet, you need to fire your weapon!" was a familiar refrain. While thankful that I never had to go to war, the choice to leave the corps was made for me. After the September 11, 2001, terrorist attacks only U.S. citizens were allowed to stay in the program.

My approach to dealing with this setback was to keep busy. In addition to enrolling in the fast-paced physics and mathematics classes and intensive Russian courses, in my attempt to take advantage of every opportunity (my personal mantra), I pursued extracurriculars from playing the trumpet and gumboot dancing to JV soccer and Taekwondo, and also squeezed in three to four jobs (evening receptionist at the American Repertory Theater, dorm crew student janitor, house librarian, and tutor). Although they never asked, I sent most of my money home to help my family (though unbeknownst to me at the time, my mother saved a lot of it and purchased a small house in my name just a few blocks from our family home). I also took six classes each semester (the norm was four) in order to keep both my dream of becoming an astronaut and my family's dream of me becoming a "real" doctor alive.

FIGURE 37.2. Shazrene Mohamed in fall 2000, up bright and early for ROTC training at MIT.

In an effort not to disappoint them, I remained pre-med until my fourth year. It was only then that I felt confident enough to pursue my own dream.

That confidence was in part forged while working on my summer research project in Professor Patrick Thaddeus's spectroscopy lab. My time hunting for carbon-chain radicals in the lab was truly formative, and I discovered that what I really wanted to do, coming up with interesting questions and finding answers to them, was exactly what research was all about. I graduated two years later with three refereed publications to my name, but more than that, I left with a sense of purpose and identity. I was a scientist and a researcher, and I was completely hooked! Equally inspiring were the gigantic blow-up posters of Hubble Space Telescope images of planetary nebulae[4] that emblazoned the walls of the Center for Astrophysics (CfA). I had never seen anything like them; I was mesmerized and had so many questions: "How does a round star end up looking like a butterfly? Are those jets? Where do jets come from? Just how do stars die?" Through those halls, my dream underwent a metamorphosis and the astronaut in the white space suit grew smaller and smaller, until it was only a small white dot, a white dwarf surrounded by a glowing, intricate nebula, or perhaps a bright supernova explosion off the edge of a galaxy. My course seminars did little to reduce the number of questions; if anything they multiplied: "wait we don't know how white dwarfs grow to the critical Chandrasekhar mass (~1.4 times the mass of our Sun) and explode as supernovae? But they are used for cosmology . . . ?" These "new" mysteries filled my imagination and launched me on my journey to becoming an astrophysicist.

I ended up at the University of Oxford as a Rhodes Scholar, ready to search for the answers to these nagging questions. But life post-undergrad brought its challenges, particularly since we were very spoiled at Harvard. The transition to working toward earning a PhD and learning to be independent was difficult, and working on tough problems led to endless battles with imposter syndrome. Navigating sometimes culturally hostile environments made it more difficult, but I was grateful for mentors like Professor Jocelyn Bell-Burnell, friends, the African community, and for my husband, Ross-Sinclair Pinto, who followed me first to England and then to Germany. At the time, there were few women and even fewer people of colour in astronomy, and, unfortunately, we rarely discussed issues of gender or race, in part because I/we really didn't know how. That said, we were present and there for one another.

It would be the grit that my parents passed on to me, thoughts of their sacrifices and the desire to set an example for my younger sisters,[5] intertwined with the thrill of solving problems, that would help see me through the tough times. For my PhD, supervised by Professor Philipp Podsiadlowski, I was particularly interested in a puzzling system called Mira AB, a very old binary star system consisting of a pulsating, red giant star and a white dwarf companion. The stars orbit around each other with a period of hundreds of years. They are well separated, and yet all the observations seemed to suggest that the white dwarf companion strongly affects the outflow from the red giant. They appeared to behave like much tighter, closer binary stars. I had to develop a simulation code to model their interaction, and once again I faced a steep "computing" curve. As undergraduates, we were not required to do any programming; that was only for computer science majors! The code took years to develop, and I loved that it incorporated nearly all of the topics that I had so enjoyed as an undergraduate: the chemistry of the small molecules I had hunted in the lab, mathematics, and lots of physics. My first simulations and parameter tests were run "incognito" on the twenty or so desktops in the student lab,[6] but we eventually got time to run the models on supercomputers with thousands of cores. How far I had come from my dear old Frankenstein's monster!

Not only were we able to understand the strong interaction in Mira AB, we demonstrated that this new mode of interaction, which we called wind Roche-lobe overflow, could be quite common in binary star systems. In a Mira-type system, the red giant's outflow is funneled onto the companion through effects of gravity, and so the white dwarf can capture large amounts of material and grow more easily to the Chandrasekhar limit, at which time it explodes as a supernova. The material that is not captured by the white dwarf also forms a torus-like structure around the system that acts like a belt, restricting subsequent faster outflows in the orbital plane but allowing material to flow freely in the polar directions, creating the beautiful butterfly nebulae that had so captivated me on the walls of the CfA. Also imprinted on the outflowing gas was an Archimedes spiral structure, created by the orbital motions of the stars.

Armed with this powerful new code that had lots of realistic physics, I joined Professor Norbert Langer's group as an Argelander Fellow in Bonn. Interest in my models grew and led to invitations to present my research at major conferences on red giants; it was my goal to convince the community that these binary interactions were important. The breakthrough came with data from a telescope in Chile called ALMA: I collaborated with observers to study the detached shell surrounding the red giant star R Scl. We detected the shell . . . and . . . *surprise!* a spiral structure within it. I immediately knew what that meant: R Scl had a hidden companion! I felt like a football player who had scored the winning gooooooоaaaaaaal! If I could, I would have done somersaults and backflips for days! We used the models of the spiral to infer the changes that had occurred in the red giant's outflow, creating the detached shell. The latter had previously been only very poorly constrained, and so our results were published in *Nature*. Our work and ALMA's capabilities have catapulted binaries to the forefront of research on red giant stars, and the models provided the scientific rationale for further successful ALMA observations, in particular, the first sample of interacting binary stars. The work was enthralling, and I could easily have spent several more years in the group, but my husband and I decided to leave Germany, not least because as a foreigner he was unable to find work due to the lingering economic impact of the 2008 global financial crisis.

We entertained several options: stay in Europe, go to the United States, or go to South Africa. Yearning for the feeling of belonging and making a difference "back home," we decided to move to Cape Town (as close as we could get), where I took up a postdoctoral fellowship at the South African Astronomical Observatory (SAAO). What timing! Not only would I get to work directly with the first female director of SAAO in almost two hundred years, Professor Patricia Whitelock, but also the 10-meter-class Southern African Large Telescope (SALT) was ramping up operations, and it had just been announced that South Africa would host the lion's share of the Square Kilometre Array, which will be the world's largest radio telescope once completed. My research network mushroomed and further collaborations developed around my simulations, particularly of the runaway red supergiant star, Betelgeuse, leading to a second publication in *Nature*. I should note that when I started my PhD, studying "stars" was not in vogue; comments like "I thought we already solved stellar evolution" were not uncommon. However, the search for exoplanets (planets around other stars), the recent detection of gravitational waves from merging black holes (the dead remnants of massive stars), and the dawn of time-domain astronomy, which will revolutionize studies of stellar variability, have raised so many new questions and will shine light on so many old ones that have gone unanswered until now—these leave me convinced that there has never been a better time to be working in stellar astrophysics.

The confluence of time and place has also made this a golden age for other reasons, particularly for change. Questions of who is doing astronomy and why, as well as who we are as a discipline, are coming to the fore. Seizing this opportunity, I have consciously worked not only to contribute to the rapid rise of astronomy in Africa, serving as part of national committees and on international delegations, leading the successful bid together with colleagues in South Africa to bring the IAU General Assembly 2024 to Africa for the first time in the IAU's hundred-year history, and helping set up the first astronomy degree programme in my home city of Bulawayo, but also to transform its institutions by taking up a joint faculty position with the University of Cape Town. I founded and chair Astronomy in Colour, a group working to address the "double jeopardy" of race and gender in the field with the hope that soon I will no longer be "the only"—the only black woman to hold a faculty position in astronomy in South Africa, the only black woman on various committees, and the only black woman "in the room."

Through my teaching and supervision, I work with many first-generation students who remind me of my younger self, using stipends and part-time jobs to support family and trying to navigate the system with limited social and cultural capital and familiarity with academia. I know too well what their firsts mean for them and their families. This inspired me to lead an extensive process to redefine and restructure the postgraduate programs, including introducing thesis committees/assessments for better oversight and support, mentoring, student and supervisor training, fixing funding gaps, and tackling mental health issues. Although we have made some progress, much more needs to be done. The same students who have to overcome the greatest challenges to be "at the table" are also the ones most affected by the policies made there, particularly in education, science, and economics. A more inclusive cohort of researchers and policymakers will not only make the questions being asked more meaningful but also give us the best chance of answering them, too. A poignant and urgent example is the contribution of astronomy, and supercomputing in particular, to climate change, where we need the brightest minds to drive innovation and design greener machines and better algorithms in order to reduce our emissions and impact on the planet; at the same time, greater inclusivity in science and policymaking will undoubtedly result in better science.

Outreach is critical to ensuring that the best minds are aware of all these challenges and opportunities. I have been fortunate to spend a lot of my time (more than 50 percent in the early days) "taking astronomy to the people," discussing exciting new developments with the public, from kindergarteners to grandparents, and also sharing the story of my journey. I realize that in many ways my story is unusual; for now I am an anomaly, but I shouldn't be. From indigenous astronomy, we know that people have been trying to understand the night sky for thousands of years; astronomy is part of our cultural and scientific heritage. We all live under one sky; it is for everyone, and my goal is to make sure that everyone gets to learn about it, to look at it through telescopes, or to dream of blasting off into space. I see my role as a facilitator, giving encouragement, inspiration, and support to those who have already found their "floating astronaut in a bulky space suit," and helping guide those who haven't yet and are just starting on their journeys of exploration and discovery. I relish the thought of all that will be achieved by the next generation of astronomers.

Because I am mid-career, writing this chapter comes perhaps a little too early in my journey for me to fully reflect on my evolution from a researcher to a teacher, facilitator, and policymaker, and even more so, on my newest role as a mother. The latter definitely has a much steeper learning curve than computing (oh how I wish babies came with a manual!), but it is also overwhelmingly beautiful at the same time. Looking at my son, I know tremendous challenges lie ahead, but I also look back at from where I have come and I am filled with hope. I will nurture his curiosity and teach him to look up with wonder at the night sky . . . and to ask questions . . . questions that will lead him to answers and more questions, and eventually to discover a whole Universe full of possibilities.

Notes

1. "Let's ask Siri!" my four-year-old nephew's favourite retort, was not an option during those days.

2. Prompted by this opportunity to reflect, I have since done some digging and, given the dates involved, managed to convince myself that the picture was likely that shown here: https://en.wikipedia.org/wiki/Astronaut#/media/File:Bruce_McCandless_II_during_EVA_in_1984.jpg (or at least one that was very similar).

3. I have since realized that there are many paths to becoming an astronaut, and if I had to pursue it now, I would rather work to become a wealthy business person and simply pay for a space flight!

4. In their final phases of evolution, stars like our Sun become red giants. During this phase, they eject their outer envelopes and the hot cores that are exposed ionize the material thrown off, creating planetary nebulae. The cores become white dwarfs and, no longer doing nuclear fusion, eventually cool. The planetary nebula material also recombines and disperses into the interstellar medium and will become part of a new generation of stars and planets.

5. Nazia is now an English teacher in Poland and the country's +35 squash champion; Naseemah went to Harvard and Oxford, was also a Rhodes Scholar, and now has her PhD in education policy; and Aisha, the youngest, is in her final years of study at a United World College in Germany.

6. The jobs were "niced" so that they had a low priority and were able to run unimpeded at full tilt at night and so that the undergrads were never disturbed during their classes in the day.

Chapter 38

Yilen Gómez Maqueo Chew (PhD, 2010)

Flipping Tables from the Sonoran Desert to the Stars

Yilen Gómez Maqueo Chew is a researcher at Universidad Nacional Autónoma de México and coordinator of the SAINT-EX Project, the first telescope in Mexico dedicated to the search for exoplanets, which was installed in 2018 at the Observatorio Astronómico Nacional de la Sierra de San Pedro Mártir, in Baja California. In her research, she studies extrasolar planetary systems and low-mass stars, using eclipsing binary systems to determine the physical properties of low-mass stars and brown dwarfs in order to understand the physical processes that govern these objects throughout their lives. As a member of the WASP team, she is co-discoverer of many exoplanets, and in her work on eclipsing binary stars, she has measured stellar masses to within 1 percent.

Becoming an astronomer and remaining in academia has been an adventure that has led me through great heights and deep lows. I was not intentional about becoming a scientist until I was finishing my undergraduate degree. While I did not exactly know what being a scientist was, what a scientist did, and even less how to become one, I attended private schools, was fluent in a foreign language (English) since I was a kid, and had sufficient financial and unconditional moral support from my parents and the collective care from my chosen family. My being here, as a tenured professor/researcher in the lead astronomy research institution in Mexico, happened through a series of events, opportunities, and choices, some of which are a product of circumstance, many a product of my privileges, and some a result of discrimination—and of course, it took a lot of hard work. Here is the story of my journey of discovery of myself, the world, and the Universe.

My favorite thing growing up was looking at the night sky in the Sonoran Desert. During daytime road trips, I would fall asleep almost as soon as I sat down. At twilight or at night, I would stare at the sky, the sunset, the Moon, and the stars. I am not even sure what I thought about as I did so. I was never intentional about basking in the moonlight or starlight. It just happened, and I enjoyed it immensely. Although I did not know being an astronomer was a profession, I have always enjoyed observing the world around me to figure out its patterns and meaning and the challenge of understanding new things. Science and math classes were my favorites since the beginning. My very first bibliographic project in first grade using my school's library (the best one in town and perhaps the only one with books in English) was about the planets in the solar system. In hindsight, I know that I have been interested in astronomy for a long time, but at the time, it seemed an unintentional and random choice of topic.

In my early teens, one of my favorite things in school was creating a herbarium of local plants of the desert around Hermosillo in Hans Bodenhamer's geography class. While exploring outside, every time I saw one of those plants I would call them out to whoever was with me. I still remember some of those plants and how to identify them. After those fantastic science classes, I wanted to be an astronaut. I wrote to NASA and asked them for information on how to become an astronaut. My math teacher, Mike Robinette, encouraged me to do so, posted the letter for me, and served as the contact person. NASA wrote back with informational pamphlets from which I learned that to be an astronaut, I needed to be either a pilot or some sort of scientist. It was one of the first times I thought of being a scientist as a profession.

I started studying physics engineering in college, in part because I had excelled in math and physics in high school but also because I was following my older sister Aline's example. I moved to Mexico's northeast and learned to live far away from my parents and hometown. I started to learn how to take care of myself on my own, what was important to me, what was widespread, and what was exclusive to my household. I did well as an undergraduate. I was not an exceptional student, but I worked and played hard. As one of the three physics students eligible for the initial class in a double-degree program, I went off to live in France. Living in a foreign country gave me an even broader perspective about myself, my country, and my culture. I became very aware of what it was like to be very different and not to understand every word that was being said nor the underpinning rules of social engagement. I learned to navigate with code-switching in order to be friendly, be approachable, and blend in as much as possible. I battled with stereotypes, and because Mexico is so far away from France, my experiences of discrimination were mostly of being exotic and tokenism. For my engineering specialization, I chose energy and environmental engineering because, at the time, I thought that I could contribute to optimizing my country's use of its vast natural resources and consequently help save the Earth. It was also the least engineering and most science-focused option available at my school, making it the first time I intentionally chose to orient my studies toward science.

As part of my French degree program, I had to do an engineering internship (known as stage *élève ingénieur*), consisting of working three months in a relevant engineering field. Environmental engineering was not as popular in the early 2000s, and the energy jobs in France were in nuclear energy, for which as a foreigner I was not eligible. As time was running out, I learned about the existence of unpaid internships at the Observatoire de Paris, so I applied and got one. This internship was my first encounter with astronomy as a scientific field and a profession. I spent three months calculating accurate positions of some of the moons of Jupiter and Saturn, based on new observations, in order to better constrain their orbits, working with Nicole Baron and Jean-Eudes Arlot. From the *Cupole Arago* in the middle of Paris, I looked at Saturn and its rings for the first time through the eyepiece of its 38-centimeter refractive telescope. It was magical and the most beautiful thing I had ever seen with my own eyes. I also participated in my first observing run at the Observatoire d'Haute Provence to observe the mutual events of the Galilean moons. We had a three-night observing run with no data because of clouds; instead, we consumed a lot of bread, cheese, and red wine. At the end of that internship, I was certain that what I wanted for myself and my life was to become an astronomer.

The summer of 2003, one year before graduating, I applied to and was admitted to the Verano del Observatorio in Ensenada, Baja California. I wanted to learn astronomy more formally. By then, I knew that doing astronomy research was what I liked, but I had never attended an astronomy class or lecture or read a book on astronomy. I visited the Observatorio Astronómico Nacional de la Sierra de San Pedro Mártir (OAN-SPM) for the first time and observed for the first time. That experience convinced me. I decided then to get a PhD doing astronomy research. Once again, I followed in Aline's footsteps; I took all the necessary exams and submitted applications to graduate programs in the United States. I was admitted to one graduate program, the PhD program in physics at Vanderbilt University. After graduating simultaneously from my undergraduate and engineering master's programs, having completed a final degree project on the visual impact of a wind farm in Santa Catarina, Nuevo León, I moved to Nashville, Tennessee.

I spent six years in graduate school studying the fundamental properties of young stars, precisely measuring their masses, radii, and relative temperatures and using these direct measurements to understand how stars form and how they evolve with time. Graduate school was one of the most challenging times of my life. Now I know that it is because academia can be a toxic place in which all people are not valued equally and equitably. Not for the first time, as a Mexican-born woman with Chinese heritage I was marked as an "other" by peers, colleagues, professors, and especially the culture of academia. I mostly did my research by myself and did not discuss my work with others, except my advisors. I learned from papers and books in unstructured ways.

I had to create the structure for my learning and understanding and find a way to survive in such a toxic environment. While in graduate school at Vanderbilt, I started going to therapy. I discovered feminism with my fellow graduate students, Martha Holmes and Gilma Adunas, my chosen sister in life. With a group of students, we spearheaded the creation of the Physics Department Climate Committee, raising awareness about diversity, equity, and inclusion, and establishing a protocol to denounce harassment. In my fourth and fifth years of PhD studies, to further separate myself from an unhealthy research group and, in general, from the southern United States, I moved to Philadelphia to complete my thesis research. Philly life was important and meaningful because I was able to recover from my depression and build my confidence and myself up again. I was at an undergraduate-only institution, so I had to be proactive about finding people with whom I could meaningfully connect. I started dancing socially for exercise, fun, and friends. In terms of making progress with my research, I started sharing time and space to work with friends in other science fields, particularly with Talia Young, a biologist who at the time was teaching, and Yonah EtShalom, who was then in nursing school. With them, in a healthy environment with people who cared about me and my well-being and were excited about science, I relearned that I could indeed talk about science, do research, and make progress on my thesis project and that I could enjoy all of that. Another important aspect of my time with these friends was that we could talk about discrimination and privileges in our fields and discuss our different perspectives and experiences, so that we might better learn how to navigate them.

For my last year in graduate school, I was back at Vanderbilt writing my thesis. Being back felt like a setback, as little had changed locally except for my personal growth. Bob O'Dell, a senior member of the research faculty who had always been supportive, reminded me that I had the ability to finish my PhD, that doing so would not invalidate my negative experiences, but that I had to work within the system. His advice was direct and real and helped me internalize that working toward finishing my degree, even if doing so was soul-crushing, did not mean that a career in astronomy needed to be like this forever and did not mean that I had to forget and forgive any and all the wrongs. Writing my thesis was still hard, of course, and led to my friend and fellow graduate student Brittany Kamai and me co-creating Hana Club, a collective space to co-work while doing the emotional work that otherwise might have blocked our continued academic progress. I was able to counteract my environment's toxicity with enough support from my community to finish my thesis. Hana Club was crucial and has been a launching mechanism for me ever since.

As I neared completion of my PhD, Leslie Hebb, at the time a post-doc in my research group, said that it would be easier to continue with research right after the PhD than to come back afterward and that the decision was reversible. I took this advice to heart and, much thanks to her, got funding to support my work as a postdoc in Belfast, in Northern Ireland, with Don Pollacco and his group. I also started doing research on exoplanets, which are the same kind of physical system as the stellar binaries I had worked on during my PhD years. Instead of one star orbiting another star, for exoplanets the simplest system is composed of one planet orbiting a star. The field of exoplanet study was growing rapidly, and later that year the 400th known transiting planet was found. By late 2021, the existence of more than 3,400 transiting exoplanets had been confirmed. Don's group was at the time composed mostly of women. For the first time, and even though the study of exoplanets was a new field for me, I was included in science conversations and discussions at all levels, mainly because another postdoc and friend, Francesca Faedi, was intentional about doing so and explaining how the field worked. I learned about working as part of a large team, and I started writing science proposals both for funding and for acquiring new observations. Living in Belfast showed me that city life is for me; a population of a few thousand people is too small to generate the activities that nourish me. Nevertheless, living in Belfast was a very challenging experience for me. Again, I had to learn how to make friends and build support networks in a different context. The underlying social conflict in Northern Ireland made it almost impossible to break into long-standing friendship groups with locals. And most people in Northern Ireland are white, making non-white people like me stand out everywhere. My support network consisted of other people from abroad also living in Belfast, like Francesca and Ana Claudia Ruiz.

In 2012, I moved with Don's group to the University of Warwick in Coventry, England. After my time in Belfast, I decided I wanted to continue living in a big city, so I commuted for more than one hour each way from Birmingham, the second-largest city in the United Kingdom, to the small city of Coventry. As a postdoc at Warwick, I was coming up with new ideas and doing good science. Collaborating with Francesca enabled me to talk about scientific ideas without fear, whether about basic principles or about asking science questions and learning how to answer them. Don's guidance and support were vital to rebuilding my scientific confidence and learning about the academic management part of science.

In late 2013, I applied for a tenure-track position at the Instituto de Astronomía at the Universidad Nacional Autónoma de México (UNAM); not only is UNAM the premier astronomy institution in Mexico, but UNAM is meaningful to my family and me. I was delighted when I got the job since I could live closer to family, have more job security, and contribute to UNAM. However, moving back to Mexico after spending most of my adult life abroad was a culture shock. On the one hand, machismo in Mexico, our implementation of patriarchy, is rampant and normalized and creates a country in which more than ten women are killed each day for being women. And on the other hand, the color of your skin determines the probability of getting a job, and for the first time, the national census of 2020 will enable naming, counting, and mapping the Black community in Mexico. This is the moment and social context in which we live.

The academic system is not set up to teach us to be leaders, mentors, advisors, or professors. As a new member of the faculty at UNAM, I had to learn how to teach as well as how to do research outside a large group and start a new research group in observational studies of exoplanets and low-mass stars at the same time, while still publishing papers in order to be considered productive. Academia has shown to me how extractive it is; I think it would be better if we considered all the contributions a person makes and not limit ourselves to using only publication

FIGURE 38.1. Left: The SAINT-EX telescope at sunset in Baja California, Mexico, during the summer of 2019 (*credit*: E. Cadena). Right: Yilen Gómez Maqueo Chew at SAINT-EX during first light on December 15, 2018, after the telescope was released to the astronomers.

numbers to assess productivity and determine advancement. I have built relationships with mentors who have been in the system for a long time, like Rafael Costero and Leticia Carigi, whose judgment I respect and trust. They keep it real for me and care about me as a multifaceted person. I also have the support of other feminist researchers/professors with whom I can scheme, plan, and take action for a better future, particularly Antigona Segura, Laura Oropeza, and Laura Serkovic. I have reconnected with my extended family, and my cousin Ethna has been my day-to-day rock in navigating this gigantic city. My community has been key to feeling supported and having my hand held when things get tough, as they are bound to do again.

During my time at the Instituto de Astronomía, I have continued to study low-mass stars in eclipsing binaries (both at the beginning of the star's life and during their adulthood) and to discover and characterize transiting exoplanets. I have been working toward establishing this relatively new line of research in Mexico. As part of this work, I am the project coordinator of SAINT-EX, a one-meter telescope located at OAN-SPM with a state-of-the-art camera that allows the very precise measurements of stellar light that are then used to search for new exoplanets. SAINT-EX is the first and only facility in Mexico dedicated to

FIGURE 38.2. Cerro Tetakawi in San Carlos, Sonora at sunset on December 15, 2020, with a waxing crescent Moon, Jupiter and Saturn. This picture, taken by Yilen Gómez Maqueo Chew during her self-quarantine at the beach, "fills me with joy at being able to do astronomy while enjoying the sea, the night sky, and the desert."

exoplanet science. In this role, I have learned more about bringing together people with very distinct expertise to get things done, how to develop instrumentation, and how my university works. I have depended on effective communication, which is a non-trivial skill that is often overlooked.

As I have moved up the academic ladder, I have tried to incorporate the lessons I have learned about systems of oppression and power dynamics into my daily interactions and how I live my life. We can now talk to some extent at the global scale about systems of oppression in astronomy and, more generally in STEM. I have hope for a more equal and equitable future, but I know the road ahead of us is long. In summer 2020, after more murders of Black people in the United States and elsewhere, after decades of unconscious bias training and diversity statements, it is clear to me that what we are doing now to make astronomy more diverse is not enough.

We need to do more, and we need to act radically differently. I was part of a group of leaders that co-created ShutDownSTEM, a day to pause and start taking action against racism in STEM and academia. My contribution was to ensure that ShutDownSTEM was not centered in the United States but instead centered on Black lives everywhere. Throughout my science journey, I have aimed to revolutionize how we are doing science, to make it more equitable, diverse, and inclusive. I intend to continue to embrace this goal in my teaching, mentoring, and research.

I wrote this story in the middle of the 2020–21 Covid-19 pandemic, a decade after defending my PhD thesis, while hanging out with my nieces Isabel and Victoria, living in Mexico City, being an astronomer, and institutionalizing transformations in STEM. I am now much more intentional and mindful in the ways I interact with others, in the work that I do, in the science I pursue, and in how I live my life. Hana Club is still going strong and has expanded as we bring more people into the fold. I will continue alongside my community on the infinite adventure of understanding myself, the world, and the Universe.

Chapter 39

Postlude

For much of the twentieth century, the considerable intellectual skills of women astronomers far too often were not enough to enable them to pry open doors of opportunity that were otherwise closed to them. Many women never went to college, let alone pursued advanced degrees. And marriage could either help or, more often, hinder participation in scientific research.

Over the last six decades, many of the formal barriers that kept women out of graduate schools, locked out of fellowships, and shut out of telescope control rooms, and some of the roadblocks that previously had maintained the world of professional astronomy as a place almost exclusively for men and kept others from participating fully, have been partially eroded. When the powers that be decided that affirmative action for women was in order, the first changes made were to offer committee memberships to women. Then women became eligible for election to society offices. Eligibility for grants, talks at conferences, and minor prizes came next. Opportunities to compete for tenure-track positions at high-profile universities and institutions came later. True leadership positions (e.g., department chairs, deans, observatory directorships) and consideration for the most prestigious prizes arrived last. In these ways, the era in which the women whose stories grace this volume came of age is a very different one from the times of Margaret Huggins, Williamina Fleming, Annie Jump Cannon, Cecilia Payne-Gaposchkin, and the many other women who aspired to but were denied opportunities for equal careers in astronomy in earlier decades.

The careers of many of the women whose stories are in this volume were shaped by the working landscape of astronomy when each of them entered the profession, but they also have played significant roles in reshaping that landscape into what it is today. They and other women have gained greater access to professorial appointments, pushed aside many of the barriers for the awarding of research grants and scientific prizes, and achieved positions of power in the administration of the modern enterprise of astronomy and astrophysics. As a result, women choosing careers in astronomy today have professional opportunities that previous generations of women could dream about but to which they could not realistically aspire.

Nevertheless, many of our writers suffered indignities, endured insults, and received slights to which their male colleagues were less prone. Others, more, in fact, than we expected, found encouragement and strong support from male—and eventually female—collaborators and leaders. For young women today who wish to enter the field of astronomy, the doors are more open than they have ever been, first having been pried open a crack and finally having been flung wide open by their tough and determined predecessors.

Though much more needs to change before equal opportunities are open to all, much progress has been made. One can hope that after more than half a century of difficult struggles and changes, soon the moment may arrive when women from across the international community of scientists who peer up at the sky and seek to unravel the astrophysical mysteries that lie above will finally hold up half the sky.

Further Reading and Additional Resources

Below are sources of information for a selection of women astronomers who either were born before 1960 and are, we are most happy to say, still with us or died in 1960 or later. Those interested in the stories of these scientists can start with these sources to learn about them. They are listed in chronological order by year of birth.

Margaret Harwood (1885–1979) Director, Maria Mitchell Observatory.

Clark, Thomas. "Littleton Astronomer Margaret Harwood Remembered for Achievements." https://www.wickedlocal.com/article/20120410/NEWS/304109549

Ida Barney (1886–1982) Made measurements for the *Yale Zone Catalogues.*

Bracher, K. "Barney, Ida M." In T. Hockey et al., eds., *Biographical Encyclopedia of Astronomers.* New York: Springer, 2014. https://doi.org/10.1007/978-1-4419-9917-7_9242

Hoffleit, Dorrit. "Ace Astrometrist." *Status: The Committee on the Status of Women in Astronomy* (June 1990).

Louise Freeland Jenkins (1888–1970) An author of the 2nd and 3rd editions of the *Yale Catalogue of Bright Stars.*

Altschuler, D. R., and F. J. Ballesteros. "Louise Freeland Jenkins (1888–1970)." In *The Women of the Moon.* Oxford: Oxford University Press, 2019.

Hoffleit, D. "Jenkins, Louise Freeland." In T. Hockey et al., eds., *Biographical Encyclopedia of Astronomers.* New York: Springer, 2014. https://doi.org/10.1007/978-1-4419-9917-7_720

Julie Marie Vinter Hansen (1890–1960) Editor, IAU Circulars at Central Bureau of Astronomical Telegrams.

Hansen, Julie Vinter. "The International Astronomical News Service." *Vistas in Astronomy* 1 (1955): 16–21.

Rasmusen, H. Q. "Obituary." *Quarterly Journal of the Royal Astronomical Society* 2 (1961): 38–39. http://adsabs.harvard.edu/full/1961QJRAS...2...38

Martha Betz Shapley (1890–1981) Eclipsing binary stars.

Bok, Bart J. *Harlow Shapley, 1885–1972: A Biographical Memoir.* Washington, DC: National Academy of Sciences, 1978.

Welther, B. L. "Martha Betz Shapley: First Lady of Harvard College Observatory." *JAAVSO* 15 (1986): 2.

Maud Worcester Makemson (1891–1977) An author of archaeoastronomy and astrodynamics.

Bell, T. E. "Makemson, Maud Worcester." In T. Hockey et al., eds., *Biographical Encyclopedia of Astronomers.* New York: Springer, 2014. https://doi.org/10.1007/978-1-4419-9917-7_891

"Maud W. Makemson." Vassar Historian, in *Vassar Encyclopedia.* http://vcencyclopedia.vassar.edu/faculty/prominent-faculty/maud-w-makemson.html

Maude Verona Bennot (1892–1982) Planetarium director.

Marché, J. D. "Bennot, Maude Verona." In T. Hockey et al., eds., *Biographical Encyclopedia of Astronomers.* New York: Springer, 2014. https://doi-org.proxy.library.vanderbilt.edu/10.1007/978-1-4419-9917-7_135

Emma T. R. Williams Vyssotsky (1894–1975) Measured motions of stars in Milky Way.

"Notable Scientists: Historical Astronomers." September 16, 2014. https://geekgirlcon.com/tag/astronomers-female/

Priscilla Fairfield Bok (1896–1975) Studied star clusters and the structure of the Milky Way.

Altschuler, D. R., and F. J. Ballesteros. "Priscilla Fairfield Bok (1896–1975)." In *The Women of the Moon.* Oxford: Oxford University Press, 2019.

Frances Woodworth Wright (1897–1989) Celestial Navigation lecturer, Harvard University.

"Frances Wright, 92, Harvard Astronomer." https://www.nytimes.com/1989/08/01/obituaries/frances-wright-92-harvard-astronomer.html

Charlotte Emma Moore-Sitterly (1898–1990) Author of Moore's *Multiplet Tables;* atomic spectroscopist.

"Interview of Charlotte Moore Sitterly." David DeVorkin, June 15, 1978. Niels Bohr Library & Archives, American Institute of Physics, College Park, MD. https://www.aip.org/history-programs/niels-bohr-library/oral-histories/4784

Roman, N. G. "Obituary." *Bulletin of the American Astronomical Society* 23, no. 4 (1991): 1492–94. http://adsabs.harvard.edu/full/1991BAAS . . . 23.1492R

Shore, S. N. "Moore-Sitterly, Charlotte Emma." In T. Hockey et al., eds., *Biographical Encyclopedia of Astronomers.* New York: Springer, 2014. https://doi.org/10.1007/978-1-4419-9917-7_976

Cecilia Payne-Gaposchkin (1900–1979) Showed that stars are made mostly of hydrogen and helium.

Moore, Donovan. *What Stars Are Made Of: The Life of Cecilia Payne-Gaposchkin.* Cambridge, MA: Harvard University Press, 2020.

Payne-Gaposchkin, Cecilia. "The Dyer's Hand: An Autobiography." In Katherine Haramundanis, ed., *Cecilia Payne-Gaposchkin: An Autobiography and Other Recollections*, 2nd ed. Cambridge: Cambridge University Press, 1996.

Henrietta Hill Swope (1902–80) Identified Cepheid variable stars.

Haramundanis, K. "Swope, Henrietta Hill." In T. Hockey et al., eds., *Biographical Encyclopedia of Astronomers.* New York: Springer, 2014. https://doi.org/10.1007/978-1-4419-9917-7_1355

"Interview of Henrietta Swope." David DeVorkin, August 3, 1977. Niels Bohr Library & Archives, American Institute of Physics, College Park, MD. www.aip.org/history-programs/niels-bohr-library/oral-histories/4909

Margaret Walton Mayall (1902–95) Director, AAVSO.

"Interview of Margaret Mayall." Owen Gingerich, August 11, 1986. Niels Bohr Library & Archives, American Institute of Physics, College Park, MD. www.aip.org/history-programs/niels-bohr-library/oral-histories/28323-1

Saladyga, M. "Mayall, Margaret Walton." In T. Hockey et al., eds., *Biographical Encyclopedia of Astronomers.* New York: Springer, 2014. https://doi.org/10.1007/978-1-4419-9917-7_919

Frida Elisabeth Palmér (1905–66) Measured proper motions of semi-regular variable stars.

Hoffleit, Dorrit. *Women in the History of Variable Star Astronomy.* Cambridge, MA: American Association of Variable Star Observers, 1933.

Nyholm, Anders. "Frida Elisabeth Palmér: Astronomer, Pioneering Female Ph.D. in Astronomy, Lecturer." *Svenskt kvinnobiografiskt lexikon*, 2020. www.skbl.se/sv/artikel/FridaElisabethPalmer

Helen Sawyer Hogg (1905–93) Studied globular clusters and variable stars.

"Interview of Helen Hogg." David DeVorkin, August 17, 1979. Niels Bohr Library & Archives, American Institute of Physics, College Park, MD. https://www.aip.org/history-programs/niels-bohr-library/oral-histories/4679

Jarrell, R. A. "Sawyer Hogg, Helen Battles." In T. Hockey et al., eds., *Biographical Encyclopedia of Astronomers.* New York: Springer, 2014. https://doi.org/10.1007/978-1-4419-9917-7_1223

Wilhelmina Iwanowska (1905–99) Director, Astronomical Institute of Toruń, Poland; stellar spectroscopist; IAU Vice President.

"Honorary Member: Dr. W. Iwanowska." Royal Astronomical Society of Canada. https://www.rasc.ca/honorary-member-dr-w-iwanowska

Trimble, V. "Iwanowska, Wilhelmina." In T. Hockey et al., eds. *Biographical Encyclopedia of Astronomers.* New York: Springer, 2014. https://doi.org/10.1007/978-1-4419-9917-7_9299

Helen Dodson-Prince (1905–2002) Measured solar flares and solar activity.

Lindner, R. P. "Obituary." *Bulletin of the American Astronomical Society* 41 (2009): 575. http://adsabs.harvard.edu/full/2009BAAS . . . 41..575L

"Memoir: Helen Walter Dodson-Prince." University of Michigan Faculty History Project. http://faculty-history.dc.umich.edu/faculty/helen-walter-dodson-prince/memoir

E. Dorrit Hoffleit (1907–2007) Editor, *Yale Catalogue of Bright Stars; General Catalogue of Trigonometric Stellar Parallaxes.*

Hoffleit, Dorrit. *Misfortunes as Blessings in Disguise: The Story of My Life.* Cambridge, MA: AAVSO, 2002.

Horch, E. "Hoffleit, Ellen Dorrit." In T. Hockey et al., eds., *Biographical Encyclopedia of Astronomers.* New York: Springer, 2014. https://doi.org/10.1007/978-1-4419-9917-7_636

"Interview of Dorrit Hoffleit," David DeVorkin, August 4, 1979. Niels Bohr Library & Archives, American Institute of Physics, College Park, MD. www.aip.org/history-programs/niels-bohr-library/oral-histories/4677

Trimble, V. "Obituary: E. Dorrit Hoffleit, 1907–2007." *Bulletin of the American Astronomical Society* 39, no. 4 (2007): 1067–69. https://ui.adsabs.harvard.edu/abs/2007BAAS . . . 39.1067T/abstract

Paris Marie Pişmiş de Recillias (1911–99) Studied galaxies and stellar clusters.

Manoukian, Jennifer. "An Armenian Supernova in Mexico: Astronomer Marie Paris Pishmish." *Ianyan,* March 24, 2003. http://www.ianyanmag.com/an-armenian-supernova-in-mexico-astronomer-marie-paris-pishmish/

Morrell, N. I. "Pişmiş, Paris Marie." In T. Hockey et al., eds., *Biographical Encyclopedia of Astronomers.* https://doi.org/10.1007/978-1-4419-9917-7_1094

Ruby Payne-Scott (1912–81) Pioneer in radar and solar radio astronomy.

Bhathal, R. "Payne-Scott, Ruby." In T. Hockey et al., eds., *Biographical Encyclopedia of Astronomers.* New York: Springer, 2014. https://doi.org/10.1007/978-1-4419-9917-7_9332

Goss, M., and R. McGee. *Under the Radar: The First Woman in Radio Astronomy.* New York: Springer, 2010.

Halleck, Rebecca. "Obituary." *New York Times,* August 29, 2018. https://www.nytimes.com/2018/08/29/obituaries/ruby-payne-scott-overlooked.html

Dorothy Davis Locanthi (1913–99) Studied red giant and supergiant stars.

Muir-Harmony, T. "Locanthi, Dorothy N. (née Davis)." In T. Hockey et al., eds., *Biographical Encyclopedia of Astronomers.* New York: Springer, 2014. https://doi.org/10.1007/978-1-4419-9917-7_337

Barbara Cherry Schwarzschild (1914–2008) Comparing high- and low-velocity stars.

Mestel, L. "Martin Schwarzschild." *Biographical Memoirs of the Fellows of the Royal Society* 45 (1999): 469–84.

Trimble, V. "Martin Schwarzschild (1812–1997)." *PASP* 109 (1997): 1289–97.

Katherine C. Gordon Kron (1917–2011) Light curves of binary stars; editor of *PASP*.

"Katherine Gordon Kron Papers, 1935–2011." Lowell Observatory. http://www.azarchivesonline.org/xtf/view?docId=ead/lowell/Kron_Katherine.xml&doc.view=print;chunk.id=0

"Oral History Interview with Katherine Kron." Antoinette Beaser, 2009. Lowell Observatory Archives. https://collectionslowellobservatory.omeka.net/items/show/518

Edith Alice Müller (1918–1995) Solar physicist; general secretary of the IAU.

Maeder, A. "Obituary." *Quarterly Journal of the Royal Astronomical Society* 37 (1996): 267–68. http://adsabs.harvard.edu/full/1996QJRAS..37..267M

Hockey, T. "Müller, Edith Alice." In T. Hockey et al., eds., *Biographical Encyclopedia of Astronomers*. New York: Springer, 2014. https://doi.org/10.1007/978-1-4419-9917-7_989

Alla Genrikhovna Massevich (1918–2008) Studied stellar structure and evolution.

Gurshtein, A. A. "Masevich, Alla Genrikhovna." In T. Hockey et al., eds., *Biographical Encyclopedia of Astronomers*. New York: Springer, 2014. https://doi.org/10.1007/978-1-4419-9917-7_9357

E. Margaret Burbidge (1919–2020) Developed theory for formation of elements in stellar nuclear fusion reactions and measured spectra of AGNs.

Clark, Stuart. "Obituary." *Guardian*, April 22, 2020. https://amp.theguardian.com/science/2020/apr/22/margaret-burbidge-obituary

"Interview of E. Margaret Burbidge." David DeVorkin, July 13, 1978. Niels Bohr Library & Archives, American Institute of Physics, College Park, MD. https://www.aip.org/history-programs/niels-bohr-library/oral-histories/25487

Trimble, V. "Obituary: Eleanor Margaret Burbidge (1919–2020)." *The Observatory* 140 (2020): 213–24.

Anne Barbara Underhill (1920–2003) Studied hot, young stars, esp. Wolf-Rayet stars.

Houziaux, L. "Underhill, Anne Barbara." In T. Hockey et al., eds., *Biographical Encyclopedia of Astronomers*. New York: Springer, 2014. https://doi.org/10.1007/978-1-4419-9917-7_9378

Roman, Nancy Grace. "Obituary: Anne Barbara Underhill, 1920–2003." *Bulletin of the American Astronomical Society* 35 (2003): 1476–77. https://ui.adsabs.harvard.edu/abs/2003BAAS...35.1476R/abstract

Giusa Cayrel de Strobel (1920–2012) Studied galactic evolution.

"Observatoire de Paris Obituary." http://giusacayrel.obspm.fr/

Martha Elizabeth Stahr (Patty) Carpenter (1920–2013) Radio astronomer.

Carpenter, Martha. *Bibliography of Extraterrestrial Radio Noise, Supplement for 1950*. Ithaca: Cornell University Press, 1952.

Larsen, Kristine, and Thomas Corbin. "Martha Stahr Carpenter (1920–2013)." *Bulletin of the American Astronomical Society* 45 (2013): 1. https://baas.aas.org/pub/martha-stahr-carpenter-1920-2013/release/1

Antoinette Piétra de Vaucouleurs (1921–87) Univ. Texas *Reference Catalogue of Bright Galaxies.*

Bash, F., et al. "Antoinette de Vaucouleurs." *Physics Today* (July 1988).

Capaccioli, M., and H. G. Corwin Jr., eds. *Gérard and Antoinette de Vaucouleurs—A Life for Astronomy*. Teaneck, NJ: World Scientific Publishing, 1989.

"In Memoriam Antoinette de Vaucouleurs, 1921–1987." University of Texas, 2008. https://www.as.utexas.edu/lectures/adv.html

Margherita Hack (1922–2013) Director, Trieste Astronomical Observatory; stellar spectroscopist.

Bergani, Stefano. "Obituary." *Astronomy & Geophysics* 54, no. 5 (2013): 5.38. https://academic.oup.com/astrogeo/article/54/5/5.38/250153

Erika Böhm-Vitense (1923–2017) Studied atmospheres and rotation of stars; mixing length theory.

Lutz, Julie, and George Wallerstein. "Obituary." *Bulletin of the American Astronomical Society* 49, no. 1 (2017). https://baas.aas.org/pub/erika-bohm-vitense-1923-2017/release/1

Nancy Grace Roman (1925–2018) Discovered high-velocity, metal-poor stars; "mother of the Hubble Space Telescope."

"Interview of Nancy G. Roman." David DeVorkin, August 19, 1980. Niels Bohr Library & Archives, American Institute of Physics, College Park, MD. https://www.aip.org/history-programs/niels-bohr-library/oral-histories/4846

"Nancy Roman: Astronomer/'Mother of Hubble.'" https://solarsystem.nasa.gov/people/225/nancy-roman-1925-2018/

Vera Rubin (1928–2016) Measured galaxy rotation curves; identified presence of dark matter in galaxies.

"Interview of Vera Rubin." Alan Lightman, April 3, 1989. Niels Bohr Library & Archives, American Institute of Physics, College Park, MD. https://www.aip.org/history-programs/niels-bohr-library/oral-histories/33963

Mitton, Jaqueline, and Simon Mitton. *Vera Rubin: A Life*. Cambridge, MA: Belknap Press, 2021.

Overbye, Dennis. "Obituary." *New York Times*, December 27, 2016. https://www.nytimes.com/2016/12/27/science/vera-rubin-astronomist-who-made-the-case-for-dark-matter-dies-at-88.html

Margaret G. Kivelson (b. 1928) Principal Investigator for the magnetometer on the Galileo Orbiter mission to Jupiter.

"Interview with Margaret Kivelson." Fran Bagenal, in *Status* (June 2005).

Kivelson, M. G. "The Rest of the Solar System." *Annual Review of Earth and Planetary Sciences* 36 (2008): 1–32.

Carolyn S. Shoemaker (1929–2021) Discoverer of 32 comets and more than 800 asteroids.

Chapman, Mary G. "Carolyn Shoemaker." *Astropedia,* May 17, 2002. https://astrogeology.usgs.gov/people/carolyn-shoemaker

Levy, David H. "Comet Hunters, Night Watchmen of the Heavens." *Smithsonian* 23, no. 3 (June 1992).

Beverly T. Lynds (b. 1929) Namesake of Lynds Dark Nebulae.

Lynds, Beverly. *Dark Nebulae, Globules, and Protostars*. Tucson: University of Arizona Press, 1971.

Lynds, Beverly. "The Littlest Astronomer." *Science* 141, no. 3581 (August 16, 1963): 594. https://science.sciencemag.org/content/141/3581/594.1

Suzy Collin-Zahn (b. 1938; also Suzy Collin-Souffrin) Astrophysicist and research director, Paris-Meudon Observatory.

"Interview with Suzy Collin-Zahn." Emeric Planet, in *Afis Science* (January 2020). https://www.afis.org/Entretien-avec-Suzy-Collin-Zahn

Andrea Kundsin Dupree (b. 1939) Associate director, Harvard-Smithsonian Center for Astrophysics; studies cool stars.

"Interview of Andrea Dupree." David DeVorkin, October 29, 2007. Niels Bohr Library & Archives, American Institute of Physics, College Park, MD. https://www.aip.org/history-programs/niels-bohr-library/oral-histories/33706

"Interview with Andrea Dupree." In H. Zuckerman, J. R. Cole, and J. T. Buer, eds., *The Outer Circle: Women in the Scientific Community*. New York: W. W. Norton, 1991.

Storey-Fisher, Kate. "Meet the AAS Keynote Speakers: Dr. Andrea Dupree." *Astrobites,* January 3, 2020. https://astrobites.org/2020/01/03/meet-the-aas-keynote-speakers-andrea-dupree/

Beatrice Muriel Hill Tinsley (1941–81) Pioneer in studies of evolution of galaxies and cosmology.

Hill, Edward O. E. *My Daughter Beatrice: A Personal Memoir of Dr. Beatrice Tinsley*. American Physical Society, 1986.

"Interview of Beatrice Tinsley." David DeVorkin, June 14, 1977. Niels Bohr Library & Archives, American Institute of Physics, College Park, MD. www.aip.org/history-programs/niels-bohr-library/oral-histories/4914

Overbye, Dennis. "Overlooked No More: Beatrice Tinsley, Astronomer Who Saw the Course of the Universe." *New York Times,* July 18, 2018. https://www.nytimes.com/2018/07/18/obituaries/overlooked-beatrice-tinsley-astronomer.html

Ene Ergma (b. 1944) Studies pulsars and black holes; former president of the Estonian parliament.

Hindre, Madis, and Susann Kivi. "Former Riigikogu President Ene Ergma Leaves IRL." March 6, 2016. https://news.err.ee/118275/former-riigikogu-president-ene-ergma-leaves-irl

"Once a Physicist: Ene Ergma." *Physics World* 24 (2011): 44.

Sandra Moore Faber (b. 1944) Henry Norris Russell Lectureship, AAS, 2011; National Medal of Honor, 2013; Gruber Cosmology prize, 2017; galaxy evolution.

"Interview of Sandra Faber." Alan Lightman, October 15, 1988. Niels Bohr Library & Archives, American Institute of Physics, College Park, MD. https://www.aip.org/history-programs/niels-bohr-library/oral-histories/33932

Jill Tarter (b. 1944) Director, Center for SETI Research.

Powell, Corey S. "First Person: Jill Tarter: The Origins and Evolution of SETI." *American Scientist* 106, no. 5 (September–October 2018): 310.

Scoles, Sarah. *Making Contact: Jill Tarter and the Search for Extraterrestrial Intelligence.* New York: Pegasus Books, 2017.

Margaret J. Geller (b. 1947) Pioneer in mapping the Universe.

AIP Institute of Physics: Margaret J. Geller: https://history.aip.org/phn/11510016.html

Margaret Geller homepage at CfA: https://lweb.cfa.harvard.edu/~mjg/

Paula Szkody (b. 1948) Observer of cataclysmic variables; editor-in-chief, *Publications of the Astronomical Society of the Pacific.*

"Interviews: Paula Szkody." Mike Simonsen, March 27, 2009, in AAVSO Cataclysmic Variables. https://sites.google.com/site/aavsocvsection/interviews/paulaszkody

Marcia J. Rieke (b. 1951; also Marcia J. Lebofsky) Principal Investigator, Near-Infrared Camera for James Webb Space Telescope.

"JWST Meet the Team: Marcia J. Rieke." https://jwst.nasa.gov/content/meetTheTeam/people/riekeMarcia.html

Carolyn Porco (b. 1953) Leader, imaging science team, Cassini-Huygens mission to Saturn.

Carolyn Porco's TED talks: https://www.ted.com/speakers/carolyn_porco

Carolyn Porco: http://carolynporco.com/about/biography/

In this section, we list a short selection of books and articles for further reading about women astronomers from the nineteenth and early twentieth centuries, and their contributions to the field of astronomy.

Bergland, Renee. *Maria Mitchell and the Sexing of Science: An Astronomer among the American Romantics.* Boston: Beacon Press, 2008.

Haley, Paul A. "Williamina Fleming and the Harvard College Observatory." *Antiquarian Astronomer: The Journal of the Society for the History of Astronomy* (June 2017): 2–32.

Hoffleit, Dorrit. "Maria Mitchell's Famous Students." *CSWP Gazette* 3, no. 4 (December 1983).

Hoffleit, Dorrit. "The Maria Mitchell Observatory—For Astronomical Research and Public Enlightenment." *JAAVSO* 30 (2001): 62.

Hoffleit, Dorrit. *Women in the History of Variable Star Astronomy*. Cambridge, MA: American Association of Variable Star Observers, 1993.

Johnson, George. *Miss Leavitt's Stars: The Untold Story of the Woman Who Discovered How to Measure the Universe*. New York: W. W. Norton, 2005.

Sobel, Dava. *The Glass Universe: How the Ladies of the Harvard Observatory Took the Measure of the Stars*. New York: Viking, 2016.

Zrull, Lindsay Smith. "Women in Glass: Women at the Harvard Observatory during the Era of Astronomical Glass Plate Photography, 1875–1975." *Journal for the History of Astronomy* (May 10, 2021): 115–46.

Below is a selection of astronomy books for children that relate the stories of women astronomers.

Barrett, Hayley. *What Miss Mitchell Saw*. San Diego: Beach Lane Books, 2019. (level: preschool–grade 3)

Bortz, Alfred B. *Beyond Jupiter: The Story of Planetary Astronomer Heidi Hammel*. London: Franklin Watts Publishing, 2005. (level: grades 6–9)

Burleigh, Robert. *Look Up! Henrietta Leavitt, Pioneering Woman Astronomer*. New York: Simon and Schuster, 2013. (level: preschool–grade 3)

Gerber, Carole, and Christina Wald. *Annie Jump Cannon, Astronomer*. New York: Pelican, 2011. (level: 8–10 years)

Ghez, Andrea. *You Can Be a Woman Astronomer*. Marina del Rey, CA: Cascade Pass, 2014. (level: grades 3–5)

Lasky, Kathryn, and Juliana Swaney. *She Caught the Light: Williamina Stevens Fleming: Astronomer*. New York: Harper-Collins, 2021. (level: preschool–grade 3)

Additional Resources: Chapter Authors (listed alphabetically)

Neta A. Bahcall

Bahcall, N. "Clusters of Galaxies." *ARA&A* 15..505B (1977).

Bahcall, N., and R. Soneira. "The Spatial Correlation Function of Rich Clusters of Galaxies." *ApJ* 270..20B (1983).

Bahcall, N., and X. Fan. "The Most Massive Distant Clusters: Determining Ω and σ_8." *ApJ* 504..1B (1998).

Beatriz Barbuy

Barbuy, B. "Oxygen in 20 Halo Giants." *A&A* 191, 121 (1988).
Cayrel, R., et al. "Measurement of Stellar Age from Uranium Decay." *Nature* 409, 691 (2001).
Barbuy, B., et al. "First Stars. XV. Third Peak R-process Element and Actinide Abundances in the Uranium-Rich Star CS31082–001." *A&A* 534, A60 (2011).

Jocelyn Bell Burnell

Hewish, A., S. J. Bell, J.D.H. Pilkington, P. F. Scott, and R. A. Collins. "Observation of a Rapidly Pulsating Radio Source." *Nature* 217, 709 (1968).
Burnell, J. B. "Presidential Address 2004: A Celebration of Women in Astronomy." *Astronomy & Geophysics* 45f..10B (2004).
Burnell, J. B. "Astronomy: Pulsars 40 Years On." *Science* 318, 5850, 579 (2007).

Ann Merchant Boesgaard

Boesgaard, A., and G. Steigman. "Big Bang Nucleosynthesis: Theories and Observations." *ARA&A* 23..319B (1985).
Boesgaard, A., and M. J. Tripicco. "Lithium in the Hyades Cluster." *ApJ* 302, L49 (1986).
Boesgaard, A., et al. "Boron Abundances across the 'Li-Be Dip' in the Hyades Cluster." *ApJ* 830, 49B (2016).

Catherine Cesarsky

Lagage, P., and C. Cesarsky. "The Maximum Energy of Cosmic Rays Accelerated by Supernova Shocks." *A&A* 125, 249 (1983).
Cesarsky, C., et al. "ISOCAM in Flight." *A&A* 315L..32C (1996).
Elbaz, D., et al. "The Bulk of the Cosmic Infrared Background Resolved by ISOCAM." *A&A* 384..848E (2002).

Poonam Chandra

Chandra, P., B. Cenko, and D. Frail. "A Comprehensive Study of GRB 070125, A Most Energetic Gamma-Ray Burst." *ApJ* 683..924C (2008).
Chandra, P., and D. Frail. "A Radio-Selected Sample of Gamma-Ray Burst Afterglows." *ApJ* 746..156C (2012).
Chandra, P., et al. "X-Ray and Radio Emission from Type IIn Supernova SN 2010jl." *ApJ* 810..32C (2015).

Xuefei Chen

Chen, X., and Z. Han. "Mass Transfer from a Giant Star to a Main-Sequence Companion and Its Contribution to Long-Orbital-Period Blue Stragglers." *MNRAS* 387, 1416 (2008).
Wang, B., et al. "Evolving to Type Ia Supernovae with Short Time Delays." *ApJ* 701, 1540 (2009).
Chen, H., et al. "Formation of Black Widows and Redbacks—Two Distinct Populations of Eclipsing Binary Millisecond Pulsars." *ApJ* 775, 27 (2013).

Cathie Clarke

Clarke, C., A. Gendrin, and M. Sotomayor. "The Dispersal of Circumstellar Discs: The Role of the Ultraviolet Switch." *MNRAS* 328..485C (2001).

Reipurth, B., and C. Clarke. "The Formation of Brown Dwarfs as Ejected Stellar Embryos." *AJ* 122..432R (2001).

Alexander, R., C. Clarke, and J. Pringle, "Photoevaporation of Protoplanetary Discs—II. Evolutionary Models and Observable Properties." *MNRAS* 369..229A (2006).

Judith (Judy) Gamora Cohen

Cohen, J. "The Lithium Isotope Ratio and F and G Field Stars." *ApJ* 171, 71 (1972).

Cohen, J., and K. Freeman. "The Tidal Radii of Globular Clusters in M31." *ApJ* 101, 483 (1991).

Cohen, J., et al. "Caltech Faint Galaxy Redshift Survey X: A Redshift Survey in the Region of the Hubble Deep Field North." *ApJ* 538, 29 (2000).

France Anne Córdova

Córdova, F., K. Mason, and J. Nelson. "X-ray Observations of Selected Cataclysmic Variable Stars Using the Einstein Observatory." *ApJ* 245..609C (1981).

Córdova, F., and K. Mason. "X-ray Observations of a Large Sample of Cataclysmic Variable Stars Using the Einstein Observatory." *MNRAS* 206..879C (1984).

Córdova, F. "Leadership to Change a Culture of Sexual Harassment." *Science* 367..1430C (2020).

Anne Pyne Cowley

Cohen, M., et al. "The Peculiar Object HD 44179 ('The Red Rectangle')." *ApJ* 196..179C (1975).

Cowley, A., et al. "Discovery of a Massive Unseen Star in LMC X-3." *ApJ* 272..118C (1983).

Phillips, M., et al. "The Type IA Supernova 1986G in NGC 5128: Optical Photometry and Spectra." *PASP* 99..592P (1987).

Bożena Czerny

Muchotrzeb, B., (B. Czerny), and B. Paczyński. "Transonic Accretion Flow in a Thin Disk around a Black Hole." *Acta Astronomica* 32, 1 (1982).

Abramowicz, M. A., et al. "Slim Accretion Disks." *ApJ* 332, 646 (1988).

Czerny, B., and K. Hryniewicz. "The Origin of the Broad Line Region in Active Galactic Nuclei." *A&A* 525, L8 (2011).

Wendy L. Freedman

Freedman, W. L., et al. "The Hubble Space Telescope Extragalactic Distance Scale Key Project. I. The Discovery of Cepheids and a New Distance to M81." *ApJ* 427, 628F (1994).

Mould, J., et al. "The Hubble Space Telescope Key Project on the Extragalactic Distance Scale. XXVIII. Combining the Constraints on the Hubble Constant." *ApJ* 529, 786 (2000).

Freedman, W. L., et al. "Final Results from the Hubble Space Telescope Key Project to Measure the Hubble Constant." *ApJ* 533, 47F (2001).

Yilen Gómez Maqueo Chew

Brothwell, R., et al. "A Window on Exoplanet Dynamical Histories: Rossiter-McLaughlin Observations of WASP-13b and WASP-32b." *MNRAS* 440.3392B (2014).

Gómez Maqueo Chew, Y., et al. "Fundamental Properties of the Pre-Main Sequence Eclipsing Stars of MML 53 and the Mass of the Tertiary." *A&A* 623A..23 (2019).

Niraula, P., et al. "π Earth: A 3.14 Day Earth-Sized Planet from K2's Kitchen Served Warm by the SPECULOOS Team." *Astronomical Journal* 160..172N (2020).

Gabriela (Gaby) González

Abbott, B., et al. "Observation of Gravitational Waves from a Binary Black Hole Merger." *Phys Rev Letters* 116f1102A (2016).

Abbott, B., et al. "Multi-messenger Observations of a Binary Neutron Star Merger." *ApJ* 848L..12A (2017).

Abbott, B., et al. "Gravitational Waves and Gamma-Rays from a Binary Neutron Star Merger: GW170817 and GRB 170817A." *ApJ* 848L..13A (2017).

Saeko S. Hayashi

Kaifu, N., et al. "The First Light of the Subaru Telescope: A New Infrared Image of the Orion Nebula." *PASJ* 52..1K (2000).

Fukagawa, M., et al. "Spiral Structure in the Circumstellar Disk around AB Aurigae." *ApJ* 605L..53F (2004).

Lucas, P. W., et al. "High-Resolution Imaging Polarimetry of HL Tau and Magnetic Field Structure." *MNRAS* 352.1347L (2004).

Martha P. Haynes

Haynes, M., and R. Giovanelli. "Neutral Hydrogen in Isolated Galaxies. IV. Results for the Arecibo Sample." *ApJ* 89..758H (1984).

Giovanelli, R., and M. Haynes. "Gas Deficiency in Cluster Galaxies: A Comparison of Nine Clusters." *ApJ* 292..404G (1985).

Roberts, M., and M. Haynes. "Physical Parameters along the Hubble Sequence." *ARA&A* 32..115R (1994).

Roberta M. Humphreys

Humphreys, R. "Studies of Luminous Stars in Nearby Galaxies. I. Supergiants and O Stars in the Milky Way." *ApJS* 38..309H (1978).

Humphreys, R., and D. McElroy. "The Initial Mass Function for Massive Stars in the Galaxy and the Magellanic Clouds." *ApJ* 284..565H (1984).

Humphreys, R., and K. Davidson. "The Luminous Blue Variables: Astrophysical Geysers." *PASP* 106..1025H (1994).

Vicky Kalogera

Abbott, B., et al. "GW 151226: Observation of Gravitational Waves from a 22-Solar-Mass Binary Black Hole Coalescence." *Phys Rev Letters* 116 × 1103A (2016).

Abbott, B., et al. "GW170104: Observation of a 50-Solar-Mass Binary Black Hole Coalescence at Redshift 0.2." *Phys Rev Letters* 118v1101A (2017).

Kalogera, V. "Too Good to Be True." *Nature Astronomy* 1E..112K (2017).

Gillian (Jill) Knapp

Knapp, G. "Observations of H I in Dense Interstellar Dust Clouds I. A Survey of 88 Clouds." *ApJ* 79, 527 (1974).

Knapp, G., T. Kuiper, and R. Brown. "Observations of Heavy-Element Recombination Lines in the Rho Ophiuchi Dark Cloud at 13 cm Wavelength." *ApJ* 206, 109 (1976).

Knapp, G. "Mass Loss from Evolved Stars VI. Mass Loss Mechanisms and Luminosity Evolution." *ApJ* 311, 731 (1986).

Shazrene S. Mohamed

Mackey, J., et al. "Double Bow Shocks around Young, Runaway Red Supergiants: Application to Betelgeuse." *ApJ* 751L..10M (2012).

Mackey, J., et al. "Interacting Supernovae from Photoionization-Confined Shells around Red Supergiant Stars." *Nature* 512..282M (2014).

Kasliwal, M., et al. "SPIRITS: Uncovering Unusual Infrared Transients with Spitzer." *ApJ* 839..88K (2017).

Carole Mundell

Mundell, C., et al. "The Nuclear Regions of Seyfert Galaxy NGC 4151: Parsec-Scale H I Absorption and a Remarkable Radio Jet." *ApJ* 583, 192 (2003).

Mundell, C., et al. "Early Optical Polarization of a Gamma-Ray Burst Afterglow." *Science* 315, 1822 (2007).

Mundell, C., et al. "Highly Polarized Light from Stable Ordered Magnetic Felds in GRB 120308A." *Nature* 504, 119 (2013).

Priyamvada Natarajan

Haehnelt, M., P. Natarajan, and M. Rees. "High-Redshift Galaxies, Their Active Nuclei and Central Black Holes." *MNRAS* 300..817H (1998).

Schneider, R., et al. "First Stars, Very Massive Black Holes, and Metals." *ApJ* 571..30S (2002).

Lodato, G., and P. Natarajan. "Supermassive Black Hole Formation during the Assembly of Pre-Galactic Discs." *MNRAS* 371..1813L (2006).

Dara J. Norman

Bruhweiler, F., et al. "The Galactic Halo and Local Intergalactic Medium toward PKS 2155–304." *ApJ* 409..199B (1993).

Norman, D., and C. Impey. "Quasar-Galaxy Correlations: A Search for Amplification Bias." *ApJ* 118..613N (1999).

Norman, D., R. De Propis, and N. P. Ross. "The Two-Point Correlation of 2QZ Quasars and 2SLAQ LRGS: From a Quasar Fueling Perspective." *ApJ* 695..1327N (2009).

Hiranya Peiris

Spergel, D., et al. "First-Year Wilkinson Microwave Anisotropy (WMAP) Observations: Determination of Cosmological Parameters." *ApJS* 148..175S (2003).

Spergel, D., et al. "Three-Year Wilkinson Microwave Anisotropy (WMAP) Observations: Implications for Cosmology." *ApJS* 170..377S (2007).

Aghanim, N., et al. "Planck 2018 Results. VI. Cosmological Parameters." *A&A* 641A..6P (2020).

Judith Lynn Pipher

Fazio, G., et al. "The Infrared Array Camera (IRAC) for the Spitzer Space Telescope." *ApJS* 154..10F (2004).

Allen, L., et al. "Infrared Array Camera (IRAC) Colors of Young Stellar Objects." *ApJS* 154..363A (2004).

Gutermuth, R., et al. "A Spitzer Survey of Young Stellar Clusters within One Kiloparsec of the Sun: Cluster Core Extraction and Basic Structural Analysis." *ApJS* 184..18G (2009).

Dina Prialnik

Prialnik, D. "The Evolution of a Nova Model through a Complete Cycle." *ApJ* 310, 222 (1986).

Prialnik, D., and A. Bar-Nun. "On the Evolution and Activity of Cometary Nuclei." *ApJ* 313, 893 (1987).

Prialnik, D. *An Introduction to the Theory of Stellar Structure and Evolution*. 2nd ed. Cambridge: Cambridge University Press, 2010.

Anneila I. Sargent

Sargent, A., and S. Beckwith. "Kinematics of the Circumstellar Gas of HL Tauri & R Monocerotis." *ApJ* 323, 294 (1987).

Beckwith, S., et al. "A Survey for Circumstellar Disks around Young Stellar Objects." *AJ* 99, 924 (1990).

Isella, A., J. Carpenter, and A. Sargent. "Structure & Evolution of Pre-Main Sequence Circumstellar Disks." *ApJ* 701, 260 (2009).

Sara Seager

Seager, S., and D. D. Sasselov. "Theoretical Transmission Spectra during Extrasolar Giant Planet Transits." *ApJ* 537, 916 (2000).

Seager, S., and G. Mallén-Ornelas. "A Unique Solution of Planet and Star Parameters from an Extrasolar Planet Transit Light Curve." *ApJ* 585, 1038 (2003).

Seager, S. *The Smallest Lights in the Universe*. New York: Crown, 2020.

Gražina Tautvaišienė

Pagel, B., and G. Tautvaišienė. "Chemical Evolution of Primary Elements in the Galactic Disc: An Analytical Model." *MNRAS* 276, 505 (1995).

Pagel, B., and G. Tautvaišienė. "Chemical Evolution of Primary Elements in the Solar Neighbourhood—II. Elements Affected by the S-Process." *MNRAS* 288, 108 (1997).

Pagel, B., and G. Tautvaišienė. "Chemical Evolution of the Magellanic Clouds: Analytical Models." *MNRAS* 299, 535 (1998).

Silvia Torres-Peimbert

Peimbert, M., and S. Torres-Peimbert. "Chemical Composition of H II Regions in the Large Magellanic Cloud and Its Cosmological Implications." *ApJ* 193..327P (1974).

Peimbert, M., and S. Torres-Peimbert. "Chemical Composition of the Orion Nebula." *MNRAS* 179..217P (1977).

Lequeux, J., et al. "Chemical Composition and Evolution of Irregular and Blue Compact Galaxies." *A&A* 80..155L (1979).

Virginia Trimble

Trimble, V. "Motions and Structure of the Filamentary Envelope of the Crab Nebula." *ApJ* 73..535T (1968).

Trimble, V. "The Origin and Abundances of the Chemical Elements." *Reviews of Modern Physics* 47..877T (1975).

Trimble, V. "Existence and Nature of Dark Matter in the Universe." *ARA&A* 25..425T (1987).

Meg Urry

Urry, C. M., and P. Padovani. "Unified Schemes for Radio-Loud Active Galactic Nuclei." *PASP* 107, 803 (1995).

Woo, J.-H., and C. M. Urry. "Active Galactic Nucleus Black Hole Masses and Bolometric Luminosities." *ApJ*, 579, 530 (2002).

Treister, E., C. M. Urry, E. Chatzichristou, F. Bauer, D. M. Alexander, A. Koekemoer, J. Van Duyne, W. N. Brandt, J. Bergeron, D. Stern, L. A. Moustakas, R.-R. Chary, C. Conselice, S. Cristiani, and N. Grogin. "Obscured Active Galactic Nuclei and the X-Ray, Optical, and Far-Infrared Number Counts of Active Galactic Nuclei in the GOODS Fields." *ApJ* 616, 123 (2004).

Ewine F. van Dishoeck

van Dishoeck, E., and J. Black. "The Photodissociation and Chemistry of Interstellar CO." *ApJ* 334..771V (1988).

Herbst, E., and E. van Dishoeck. "Complex Organic Interstellar Molecules." *ARA&A* 47..427H (2009).

van der Marel, N., et al. "A Major Asymmetric Dust Trap in a Transition Disk." *Science* 340..1199V (2013).

Patricia Ann Whitelock

Whitelock, P., et al. "AGB Variables and the Mira Period-Luminosity Relation." *MNRAS* 386..313W (2008).

Whitelock, P., et al. "Asymptotic Giant Branch Stars in the Fornax Dwarf Spheroidal Galaxy." *MNRAS* 394..795W (2009).

Whitelock, P., et al. "The Local Group Galaxy NGC 6822 and Its Asymptotic Giant Branch Stars." *MNRAS* 428..2216W (2013).

Sidney Wolff

Fraknoi, A., D. Morison, and S. Wolff. *Astronomy: A Free Open-Source Textbook*. OpenStax.

Wolff, S., and G. Preston. "Late B-Type Stars: Rotation and the Incidence of HgMn Stars." *ApJS* 37..371W (1978).

Wolff, S., et al. "The Origin of Stellar Angular Momentum." *ApJ* 252..322W (1982).

Rosemary (Rosie) F. G. Wyse

Gilmore, G., R. Wyse, and K. Kuijken. "Kinematics, Chemistry, and Structure of the Galaxy." *ARA&A* 27..555G (1989).

Belokurov, V., et al. "The Field of Streams: Sagittarius and Its Siblings." *ApJ* 642L..137B (2006).

Gilmore, G., et al. "The Observed Properties of Dark Matter on Small Spatial Scales." *ApJ* 663..948G (2007).

Index of Subjects

Index of People